EXOTHEOLOGY

EXOTHEOLOGY
Theological Explorations of Intelligent Extraterrestrial Life

Joel L. Parkyn

☙PICKWICK *Publications* • Eugene, Oregon

EXOTHEOLOGY

Theological Explorations of Intelligent Extraterrestrial Life

Copyright © 2021 Joel L. Parkyn. All rights reserved. Except for brief quotations in critical publications or reviews, no part of this book may be reproduced in any manner without prior written permission from the publisher. Write: Permissions, Wipf and Stock Publishers, 199 W. 8th Ave., Suite 3, Eugene, OR 97401.

Pickwick Publications
An Imprint of Wipf and Stock Publishers
199 W. 8th Ave., Suite 3
Eugene, OR 97401

www.wipfandstock.com

PAPERBACK ISBN: 978-1-7252-9148-5
HARDCOVER ISBN: 978-1-7252-9147-8
EBOOK ISBN: 978-1-7252-9149-2

Cataloguing-in-Publication data:

Names: Parkyn, Joel L., author.

Title: Exotheology : theological explorations of intelligent extraterrestrial life / by Joel L. Parkyn.

Description: Eugene, OR: Pickwick Publications, 2021 | Includes bibliographical references and index.

Identifiers: ISBN 978-1-7252-9148-5 (paperback) | ISBN 978-1-7252-9147-8 (hardcover) | ISBN 978-1-7252-9149-2 (ebook)

Subjects: LCSH: Life on other planets—Religious aspects | Space theology | Exobiology—Religious aspects | Religion and science | Outer space—Exploration—Moral and ethical aspects

Classification: BL254 P37 2021 (print) | BL254 (ebook)

Image of *HGC 12343* used with permission from the Vatican Telescope, Tuscon, Arizona, October 1996; from *Pulchritudo Naturae in Luce Ianthina*. Boyle, Burg, De Jong, Ponder, and Windhorst

08/24/21

Dedicated to

Judah and Leah

and

Steven Spielberg

"A single, free intelligent creature graced by God is more valuable than the entire material universe."

—St. Thomas Aquinas, *Summa Theologica* I–II, Q. 113, Art. 9.2.

Contents

List of Illustrations | ix
Acknowledgements | xi
Introduction | xiii

Chapter 1 A Roman Catholic and Thomistic Approach | 1
Section A A Thomistic Exotheology 8

Chapter 2 Cosmology, Astrobiology, and Modern Astronomical Technologies | 13
Section A The New Cosmology 13
Section B Science and Extraterrestrial Intelligence 28
Section C Ancient and Early Medieval Philosophy on the Plurality of Worlds 34

Chapter 3 Historical Literature on the Theology of Extraterrestrial Life | 52
Section A History of Theological Developments on Extraterrestrial Life: Antiquity to Nineteenth Century 52
Section B The Plurality of Worlds and Christian Theological Formulations 80
Section C Theological Thought on Extraterrestrials in the Twentieth and Twenty-First Centuries 82

CONTENTS

Chapter 4	Extraterrestrial 'Anthropology,' Xenobiology, Morphology, and Theological Systems	126
Section A	Extraterrestrial Exoanthropology	126
Section B	Extraterrestrial Religious and Theological Systems: Exotheology	154
Chapter 5	Exotheology and Traditional Christological Formulations	188
Section A	Review of the Four Major Historical Positions	188
Section B	The Varied Hypothesis	195
Section C	Christology and Intelligent Extraterrestrials	208
Section D	The Divine Pedagogy	229

Conclusion | 270

Appendix A: Historical Cosmology and Theological Timeline | 274
Bibliography | 287
Index | 317

Illustrations

Hubble Ultra Deep Field Image (HUDF) | 26

Peter Apian's Cosmographia. Renaissance woodcut illustrating the Ptolemaic system, 1524 | 38

Frontispiece of Fontenelle's Entretiens | 42

Flammarion Engraving | 46

Imago Mundi | 52

Ancient Hebraic Cosmology | 54

1579 Drawing of The Great Chain of Being by Didacus Valades, Rhetorica Christiana | 63

Exotheological Metanarrative Diagram | 265

Simulated large scale structure of the universe | 273

Acknowledgements

I express my deepest gratitude to Dr. Christopher Southgate of Exeter University, United Kingdom for his indispensable support on this project over the long term of its development, and Dr. Yayha Michot and Dr. Heidi Hadsell in cooperation with Exeter University for their acceptance and support of this important subject for Christian theology. I also thank Rody Bazzano of Saint Thomas Seminary Library, Bloomfield, Connecticut, for the extensive use of its resources.

Introduction

This book will discuss the theological implications of the existence of intelligent extraterrestrial life, representing one of the long unresolved issues of Christianity. The presence of life, particularly intelligent life originating outside Earth has never constituted a specific area of theological formulation. Historically, scholarship on the subject has amounted to sporadic and fragmented speculation, not a systematized theology, and presents a quandary that has been cautiously avoided by most theologians. Given our current astronomical and cosmological understandings, it is now apparent that humanity and its history represent an infinitesimally small part of the universe. For millennia Earth religions have made claims that transcend our physical world to encompass the entire universe. However, future contact with an intelligent extraterrestrial civilization could call into question certain cherished religious doctrines, and result in a reorientation of our limited terrestrial belief systems. Logic would dictate that humans on Earth are either entirely alone in a vast universe containing billions of galaxies or the universe contains other forms of life, intelligent or otherwise. Either possibility is highly significant for scientists and theologians; however, for two millennia theology has fully explored the former possibility to the near exclusion of the latter, resulting in an anthropocentric myopia which is at odds with what recent science indicates. Given that our world's religions were instituted amidst an anthropocentric cosmology, even now the magisterium of the Catholic Church, as well as other major Christian denominations and world religions, has no official teachings regarding Christian doctrines and extraterrestrial life. The New Testament, while speaking of the relationships between God and humanity in a cosmic context, is Christocentric and geocentric on the subject of extraterrestrial life. For example, Colossians 1:15–16 speaks of Jesus as the "first born of all creation . . . all things have been created through him and for him." This text will receive particular focus in chapter 5, in a discussion of hermeneutical issues raised by the possibility of extraterrestrial intelligence. Also, John's prologue echoes Paul, who proclaims the preexistence of Christ over the entirety of creation, and Revelation 5:11 records the existence of an unknown multitude of angelic beings who act as ministering spirits to humankind. In none of these cases is there reference to other biological intelligent beings existing outside the heavenly and earthly realms occupied by angels and humans. Despite Christianity's eventual acceptance of the scientific heliocentrism of Copernicus, it has remained theologically geocentric, lagging behind twenty-first century scientific advances in astronomy, astrobiology, and knowledge gained through space exploration. Further, theology since the nineteenth century has given little thought of the implications of intelligent extraterrestrials in the cosmos and their relation to Christianity, and for the past century has been preoccupied with the ideas and changes posed by Darwinism, Freudian psychology, and new methodologies of biblical criticism; the massive changes as a result of multiculturalism, ecumenism, and religious pluralism; the sociological and psychological impacts of industrialization and modernism; exponential advances in science and technology, the sexual revolution, secularism, the influence of the internet and mass media; and the general decline in religious belief and practice in Europe and North America. This theological narrowness of view ought to have been acknowledged and corrected early in the twentieth century with the advent of Einsteinian cosmology, the general acceptance of Lemaitre's Big Bang, and Hubble's findings of interstellar light absorption, distance indicators, and calculations of the age of the universe. Consequently, at present most Christians lack a rudimentary understanding of how to synthesize their religious beliefs with the concept of extraterrestrial intelligence. A responsible and coherent theology must incorporate all reality, and by confining its theological systems to terrestrial concerns and ignoring the actual context of humanity, Christianity risks a theology and world view that is divorced from reality. If Christianity is to speak of reality and maintain its credibility in an environment where religion has experienced severe disassociation in our age of technological achievements, ever-advancing scientific knowledge, and vast cultural change, it can do so only with a theological accommodation of the possibility of a plurality of inhabited worlds. The question of the existence of intelligent life within our modern understanding of the known universe is one of the most important subjects in which science and religious thought interpenetrate, representing the last possible consequence of an "extended Copernican principle" that first deprived human beings of the geometric center of the universe, their uniqueness of creation outside the biological history on Earth, and the possibility of the centrality of their consciousness within the cosmos.

Several centuries were required for the Copernican revolution to be accommodated in theology, and we remain in the transitionary period in our psychological and theological adjustment to the Darwinian revolution. The theological reckoning with the new universe is the next phase in achieving a true understanding of our place within the established corpus of human knowledge. Christian theology must therefore come to terms with this fundamental challenge as its reluctance to incorporate the implications of a universe which may include a variety of intelligent life forms more advanced scientifically, culturally, and/or spiritually will render it incomplete and short of realizing its next and natural stage of development.

Modern scientific discoveries have resulted in the realization that we no longer inhabit the Ptolemaic, Copernican, or Galactocentric universes (that is, centered upon the Milky Way). We now inherit the scientific discoveries of Galileo, Newton, and Einstein, in addition to telescopic advances which reveal a universe, according to recent estimates, as ≈ 13.8 billion light years in extent, full of billions of evolving galaxies existing in an Einsteinian space-time. Earth, rather than being the center of God's attention as in the geocentric models of medieval theology, appears to be one planet among billions of others, orbiting an average size star, within a typical solar system orbiting a vast but average spiral galaxy. Given the age of the universe and the genus Homo sapiens arriving only a short time ago according to cosmic timescales, it is possible that other intelligent races exist within the vast cosmos. Humanity therefore occupies an undetermined place within the ranks of a possible great continuum of species. Clearly, this view has profound implications regarding the works of the Creator from the perspective of our terrestrial religious traditions. The understanding of our true place within this greater context, realized by scientific achievements beginning with the first telescopes, has resulted in a great expansion in technological knowledge. In particular, the last fifty years represent a historically unprecedented increase in knowledge regarding the extent of and laws governing the physical universe. Most remarkably, the possibility of the existence of a plurality of worlds beyond our solar system, debated since ancient Greek philosophy until the modern age, has in the last two decades been confirmed by the extraordinary discoveries of exoplanets. The new *Kepler* space telescope as of September 2020 has indicated 2,662 extrasolar planets exist within a small field of view, within a limited scope of six hundred to three thousand light years outside our solar system.[1] This lends credibility to the decades-old working hypotheses of many of astrophysicists and scientists in the growing fields of astrobiology and bioastronomy of a universe where

---

1. "NASA's Kepler Marks 1,000th Exoplanet Discovery," NASA.

planetary systems are common, life exists where conditions permit, and this life can be intelligent.[2] The *Kepler* findings and subsequent planned telescopic missions are bringing about an astronomical revolution in extrasolar planet discoveries, and future evidence of life forms on remote exoplanets could lead to a corollary revolution in social and political thought about technologically advanced intelligent life in the galaxy and its impact on humanity. Clearly these discoveries demonstrate the validity of astrophysical theory that solar system formation as a consequence of star evolution is common and hence the likelihood of multitudes of rocky planets in our galaxy alone—many within the habitable zones of their parent stars. Given the scope of these breakthroughs, astronomers, social scientists, historians, and government sponsored studies have begun to speculate regarding the political, social and cultural, and (with very scant attention) the theological implications of a discovery of advanced extraterrestrial life. Additionally, Christian leaders have historically and remain to the present day taking little, if any part in developing a comprehensive theological response to the potential confirmation of the existence of extraterrestrial intelligence. Only recently, Vatican astronomers Fr. Jose Gabriel Funes, SJ, and Br. Guy Consolmagno made pronouncements regarding the compatibility of Catholic faith and extraterrestrial life,[3] and Pope John Paul II was known to have made a private affirmation in regard to extraterrestrials.[4] An intelligence-containing extraterrestrial reality could result in a reformulation of certain theological principles and may wholly redefine many of our conceptions of God and creation. Much depends on the mode of contact, information obtained, and reactions by individuals and religious institutions. Indeed, a positive outcome of contact with intelligent extraterrestrial life could be an expansion of our known laws governing the universe, a redefinition of the cosmic context of the human species, new understanding of the origin and diffusion of life, the nature of creation, the prehistory of Earth, the solar system, and our galaxy.

Apart from the impact on the sciences, evidence of a second Genesis could drastically call into question certain Christian foundational

---

2. Grinspoon, *Lonely Planets*; Dick and Strick, *The Living Universe*; Race, "Societal and Ethical Concerns"; Drake, "Extraterrestrial Intelligence," 474–75; Shklovskii and Sagan, *Intelligent Life in the Universe*; Davies, "Biological Determinism, Information Theory, and the Origin of Life," 15.

3. "The Vatican's Astronomer."

4. Pope John Paul II, when questioned by a child during his visit to the parish of St. Innocenzo I. Papa e S. Guido Vescovo, in the north of Rome on November 28, 1999. Question: "Holy Father, are there any aliens?" He answered, "Always remember, they are children of God as we are." Twitchell, *Global Implications of the UFO Reality*, 56.

theological teachings regarding creation, the Incarnation, and the Redemption. Such an event would firmly establish that humans are neither the biological nor theological center of the universe, and, present as it would new and unprecedented information of religious or theological significance, which would inevitably result in a profound reformulation or recontextualizing of theology, requiring it to be expanded to accommodate a new *Exotheology*—representing the next phase of Christian theological research. The ramifications for Christian theology are myriad. In this case, the theological adjustment will not have as its foundation the place of the physical world, as in Copernicanism, nor the biological world, as in Darwinism, but to the expanse of an immensely vast universe, where time, space, and intelligences possibly surpass our species' physical capabilities and/or spiritual perfections. Consequently, more difficult in this adjustment is Christianity whose messiah, prophet, and holy men and women of old are centrally linked to person, locality, and time. Traditional medieval and Renaissance arguments against a plurality of inhabited worlds have questioned the unity of the human family in conjunction with the extraterrestrial moral position with respect to original sin. Existential questions are also raised about extraterrestrial moral beliefs, human and extraterrestrial soteriology, the nature of the soul, of good and evil, and of the respective theological places of ourselves and other intelligent species within the known universe. As such, Christianity appears to be the most vulnerable since it is most species-specific due to the unique nature of the Incarnation. Therefore, belief in the God-human, the Christocentrism of all creation, as well as the Trinitarian image of God, present major areas of theological research in considering an extraterrestrial reality.

Central to this study is an exploration of the possibilities of an extraterrestrial economy of salvation. The Hebrew and Christian traditions affirm the existence of angels, whose existence precedes the creation of Earth, is subject to an economy of salvation separate and differentiated from that of humanity. This clearly allows for the possibility of other, unknown economies.[5] This perspective reflects the logical conviction that the Creator has his own ways to make himself recognized everywhere and is capable of making himself present within all his creatures. Also of fundamental importance is the possible theological 'anthropology' of intelligent extraterrestrial life, how their 'religion' compares to human religious perspectives, and how their existence can or cannot be correlated with salvation

---

5. Rev 12:7–13. This economy of salvation is understood as the creation of the angels, their test to gain entry into heaven, followed by their radical and irrevocable decision resulting in a final and unchangeable state in heaven with God or hell with Satan. *Catechism of the Catholic Church*, 392.

through Christ due to the universality of God's creation. These questions can only be answered by examining the role of original sin or its absence in an extraterrestrial civilization and if extraterrestrials sin—whether they have consciences similar to ours, struggle with ethical questions, and fear death, as well as the role or absence of the role of Christ's redemptive death for extraterrestrials and the meaning of the transcendence of divine plans beyond our geocentric and anthropocentric concerns.

Therefore, the central questions of this discussion contend with the inevitable consequences of a final abandonment of theological anthropocentrism, representing the end of theology's extended geocentric and homocentric adolescence and its coming to terms with a greater reality. These can be summarized as follows: what aspects of our theology would require reformulation and how would they be conceived to accommodate an extraterrestrial reality? What can be assumed of the spiritual and material natures of intelligent extraterrestrial life, their religious perspectives, and their capabilities given our current knowledge of astronomy, astrobiology, and history of our human religious traditions? What is the significance of the singular, terrestrial, historical Jesus and resulting Christian faith within an incomprehensibly vast physical universe, not to mention of the possible discovery or disclosure of an alternate, competing economy of salvation from a superior and possibly extremely ancient race? Can we consider whether the Incarnation was unique to the Earth, does it extend to other planets, or was there or will there be incarnations on other planets? How can the existence of extraterrestrial intelligence be reconciled with the notion of the universality of God's redemption of *all creation* within the cosmos, as mediated through the Old and New Testaments of the ancient near East? How do we consider the assertions of the universality of Christ's kingship of all creation as described by St. Paul and St. John the Evangelist within the context of intelligent extraterrestrials? What will be the reactions of Christians both as faith groups and as institutions to evidence of extraterrestrial existence? What role do the absolutist claims of Earth religions have within the context of a galaxy and universe that may be populated with other intelligent life forms? It must be emphasized that such consequences depend heavily upon the nature of extraterrestrial contact, whether via radio transmission, alien artifacts discovered on nearby planets, direct physical contact, or other.

An important contribution to contemporary theology would therefore be to lay the foundation for a contingency plan in theologically forecasting the consequences of an eventual contact or disclosure scenario. Failure to do so may have serious and unforeseen consequences for organized religion, most especially Christianity. Although recent sociological studies, such as the Peters ETI Crisis Survey[6] and Alexander Report,[7] have indicated a high level of confidence on the part of most believers and leaders in the resiliency of particular Christian denominations in the event of extraterrestrial contact, this confidence is not consonant with historical evidence of advanced cultures encountering the lesser advanced. Such a prediction may be grossly premature given the unforeseeable content or mode of a future contact situation. One need only think of the Native American or indigenous peoples of Mesoamerica, whose experience gave witness to the complete transformation, utter destruction, or marginalization of ancient and cherished religious beliefs as a result of imperialism and colonialism. In accommodating the actual fact of extraterrestrial intelligent life present within or without human society, rather than its mere hypothetical possibility, it is possible, and perhaps likely, that certain subsequent generations after contact may either abandon or make key modifications to certain terrestrial religions, as evidenced in historical analogs. This book therefore is a preemptive effort to prepare Christian theology to contend with a possible future discovery/contact event which could have wide and long-lasting consequences.

In regards to research methodology, this book is an interdisciplinary examination of the subject to include themes of philosophy, anthropology, cosmology, astrobiology, psychology, and sociology, among others. Chapter 1 it sets out and justifies its theological grounding in the doctrines and traditions of the Roman Catholic Church, especially the thought of Thomas Aquinas. It also includes a historical inventory of the plurality of worlds idea from ancient philosophical cosmology and theological speculation, to pertinent up-to-date astronomical data, space exploration, and recent astrobiological data. The first section of chapter 2 will survey scientific advances in cosmology, astrobiology, and modern astronomical technologies, providing the modern scientific context and the expanded physical, philosophical, and theological setting, and argue the continuing relevance

---

6. Conducted by Ted Peters, professor of systematic theology at Pacific Lutheran Theological Seminary and the Center for Theology and the Natural Sciences at the Graduate Theological Union, Berkeley, California, to test the belief that upon confirmation of contact between Earth and an extraterrestrial civilization of intelligent beings, the long established religious traditions of Earth would confront a crisis of belief and perhaps even collapse. The survey polled over 1,300 religious individuals, including clergy, concluding no crisis would result from such contact among individual believers as well as religious institutions.

7. Victoria Alexander surveyed U.S. Protestant, Catholic, and Jewish clergy in her 1994 study regarding their religious response to the confirmation of the existence of extraterrestrial life. Her conclusions contrasted with conventional wisdom that religion would collapse. The Alexander study differs from the Peters study in that the former is associated with Unidentified Flying Objects, whereas the latter covered a wider range of religious traditions and was prompted by discoveries in astrobiology.

for the contemporary theological discussion. The second section of chapter 2 will discuss key scientific organizations contributing to contemporary discussions in consideration of extraterrestrial life, relevant for theology and religion. The last section of chapter 2 will provide a literature review on the history of inquiries on extraterrestrial life, both scientific and philosophical, providing the necessary background for the historical theological and Christological discussions in chapter 3.

Chapter 3 will include pertinent literature from antiquity to the present. Historical discussions on theological and Christological issues arising with the concept of a plurality of worlds and related considerations of extraterrestrial life will be examined. Discussion of specific Christological formulations of key historical figures, as well as biblical, philosophical, ecclesiological, theological, and other influences for the rationale of a Christology in consideration of intelligent extraterrestrials will be examined, serving as a basis for review and evaluation of later theologians within the twentieth and twenty-first centuries. The literature review will provide compelling evidence that the central and fundamental concern for Christian theology from Origen to modernity in consideration of intelligent extraterrestrials has been the Christological doctrine of the Incarnation, and intimately related teachings on Original Sin, the Redemption, and Creation. In particular, with regard to the historical debate on the plurality of worlds and extraterrestrial intelligence, *the enduring and continued central and fundamental concern for Christian theology given the possibility of the existence of intelligent extraterrestrial beings was and is Christological.*

Therefore, in simple terms, the fundamental historical question for this new branch of theology, *Exotheology*, can be summed up as: if God created non-human intelligences, what is their relation to the Creator, to our terrestrial and human economy of salvation, and how would such creatures be saved? How does the Second Person of the Trinity figure within the larger, modern cosmological context that may contain other, non-human intelligences? Historically most thinkers have been required to either repudiate the idea of plural worlds and extraterrestrial inhabitants given the unique and unrepeatable nature of the Incarnation in human form on Earth, reject Christianity (as Thomas Paine argued) in favor of a naturalistic God, or accept one of four "solutions" offered by theologians.

This book will introduce four new soteriological categories of approaches to these questions. These are explored in detail and may be summarized as follows:[8] An *exclusive* type, one Incarnation of Christ on Earth for the redemption of humans alone, regardless of the absence or

---

8. Taxonomy is mine.

reality of extraterrestrial existence; an *inclusive* type, one Incarnation of Christ on Earth providing redemption for humans *and* for all intelligent extraterrestrials if they exist; a *multiple* type, where the divinity or Logos is incarnate in like manner on other planets and within alien civilizations according to their specific forms and nature, as had been accomplished on Earth for salvation; and a *varied* type, where the Creator manifests to his creation in a variety of ways according to his own designs for all intelligent creatures. For this thesis, specific attention and argumentation in support will be given to the *varied* type incarnational position, which historically has had only a handful of proponents and is the least theologically developed; this book will develop this argument and argue for its superiority over the other soteriological models.

The first section of chapter 4 will provide original material in examining the subject of possible extraterrestrials themselves, important for the theological discussion. Consideration will be given to xenobiological structures and social compositions. This data will be considered on the basis of known types of planetary habitats, evolutionary theory, competition models, and behavioral analogs according to taxonomical family, among others. The last section of the chapter will introduce, given the data in the previous two sections, possible extraterrestrial religions and theological systems and their relation (or lack of relation) to our terrestrial Christological doctrines and Trinitarian theology.

Chapter 5 will present new theological formulations of possible intelligent extraterrestrial life. The first section will provide extensive theological analysis of the four major earlier historical positions (*inclusive, exclusive, multiple, varied*), and argue the thesis that the most positive and fitting solution in reconciling the central teaching of the Incarnation with extraterrestrialism is the *varied type* of incarnational theology. The second section will discuss at length the *varied* soteriological position, as well as introduce a biblical hermeneutic of the *varied* view. The third section will present a cohesive, systematic Christology and incarnational theology to accommodate intelligent extraterrestrials. The discussion will center on the theology of Incarnation, and will necessarily include the relationship to Creation, Original Sin, and the Redemption. This section will propose a soteriology of extraterrestrials that incorporates modern scientific knowledge in collaboration with natural theology and Christological doctrine, and will be proposed as a necessary and natural expansion of established orthodox teaching. The section also endeavors to forecast the possible consequences of contact with an intelligent extraterrestrial civilization and its theological import for Christian theology. Since an event of such magnitude could result in a paradigm shift in our theological world view, and introduce new realities which must be accommodated within our terrestrial understandings, Christian theology may require a reorientation/reinterpretation to allow it to remain viable within this new context. Lastly, given the possibility of new information imparted by civilizations more technologically (and/or spiritually) advanced, an examination of the nature and place of humanity within the universe, and the role, history, and possible future and destiny of humanity within this expanded theological setting will be introduced. The final section of this chapter argues for the plausibility of the *varied* view as a new soteriological formulation for exotheology and its import for traditional Christian doctrinal positions.

Chapter 1
A Roman Catholic and Thomistic Approach

This book is framed within the Roman Catholic theological tradition: the *Catechism of the Catholic Church*, official church teachings, writings of the church fathers, and particularly the works of Thomas Aquinas's Christology and Soteriology.[1] There are several fundamental bases for an academic rationale for utilizing the Catholic intellectual tradition in examining the question of extraterrestrial life and theology.

First, the Church's claim as an apostolic faith is one of its four essential features, in its profession as one, holy, catholic, and apostolic, each which are inseparably linked with each other;[2] and received from the one divine source in Christ through the Holy Spirit and given to the apostles.[3] Accordingly the bishops have by divine institution take the place of the apostles as pastors of the Church by apostolic succession;[4] all members of the Church share in this apostolic mission through the communion of faith and life with its origin.[5] Further, the Magisterium, the authority of the College of Cardinals and the Pope, with the assistance of the Holy Spirit, claims for itself as guardian and the interpretive and teaching authority of divine revelation; derived from the commission given to the apostles in Luke 10:16.[6] It is the

---

1. References on conciliar creeds include: Tanner, *Decrees of the Ecumenical Councils*, 5, 24, 40–60; Davis, *The First Seven Ecumenical Councils*.

2. The Church claims the apostolic nature of the Church is founded on the apostles in three distinct ways: it is built on those witnesses chosen and sent on mission by Christ himself, the handing on of teaching with the assistance of the Holy Spirit, and the Church continues to be taught, sanctified, and guided by the apostles by their successors, the college of bishops. *Catechism of the Catholic Church*, 811; Matt 28:16–20; Acts 1:8; 1 Cor 9:1; 15:7–8; Gal 1:1.

3. *Ad Gentes*, 5.

4. *Lumen Gentium*, 20 § 2.

5. *Apostolicam actuositatem*, 2.

6. The Magisterium exists in service to the Word of God; what it proposes for belief as divine revelation is drawn from the single *depositum fidei*. Therefore, sacred scripture,

role of the Magisterium to interpret in a decisive way both scripture and tradition, utilizing the consensus of documents and teaching of the Church Fathers, councils, encyclicals, and pastoral documents.[7]

Second, the Roman Church has provided the longest continuous theological and scientific historical inquiry in the debate on the plurality of worlds and extraterrestrial intelligence. Its patristic authors and theologians were the first to offer formulations in response to certain theological, philosophical, and scientific arguments for or against plural worlds and extraterrestrials while maintaining the teaching on the uniqueness of humanity within an expanding knowledge of creation. Also, the Church has an extensive legacy of scientific research, led by its Pontifical Academy of Sciences, distinctive as the sole supranational academy of science in the world; and was founded in Rome in 1603 as the world's first exclusively scientific academy.[8] Its heritage of joint scientific and theological inquiry provides long-standing support for a theology engaged with scientific endeavors. The Vatican operates two important observatories, in Castel Gandolfo in Rome and Mount Graham, Arizona, as part of its research in space studies and whose scientist/theologians, including Vatican astronomers have expressed with certain degrees of confidence, the coherence of a putative existence of extraterrestrial intelligence with Catholic faith.[9] For this thesis in a work of Catholic theology engaged with science, data provided by astronomy and astrobiology and its consideration of xenologies will be used to inform theology as to the possible xenobiology, morphology, and environments of extraterrestrials, and to extrapolate potential theological anthropologies of divine action with creatures, considered alongside the record of divine action with humans.

Third, Catholic teaching holds two distinct modes of revelation from one common source; sacred tradition and sacred scripture.[10] The Church's

---

sacred tradition, and the teaching authority of the Magisterium work together under the action of the Holy Spirit contribute effectively to the salvation of souls. *Catechism of the Catholic Church*, 85; *Dei Verbum* 10 § 2, 3.

7. *Catechism of the Catholic Church*, 888–92.

8. First named the *Linceorum Academia*, of which Galileo Galilei was an appointed member in 1610, and reestablished in 1847 by Pius IX as the *Pontificia Accademia dei Nuovi Lincei*, and given its current name in 1936 by Pope Pius XI.

9. Most prominent among these discussed are Thomas O'Meara, Joseph Pohle, Pierre Teilhard de Chardin, Theodore Zubek, Karl Rahner, Yves Congar, John Haught, Guy Consolmagno, José Gabriel Funes, Chris Corbally, Marie George, and Ilia Delio.

10. The Council of Trent affirmed that the *depositum fidei* passed on orally and in writing by the apostles, preserved in the scriptures and unwritten traditions, and continued in apostolic tradition is both a product of the early Church and each a component is a foundational means of revelation. *Dei Verbum* 9, 10 § 1; cf. 1 Tim 6:20; 2 Tim

adoption of the modern methods of biblical criticism by The Pontifical Biblical Commission, founded in 1901, ensured the proper defense and interpretation of scripture according to scientific methods promulgated by Pope Leo XIII's directive in his encyclical *Providentissimus Deus* in 1893, and later reaffirmed in Pope Pius XII's *Divino afflante Spiritu* in 1943. Vatican II's *Dei Verbum*, on the *Dogmatic Constitution on Divine Revelation* in 1965, reaffirmed this approach and encouraged biblical criticism while asserting divine authorship, with interpretive emphasis on the content and unity of the whole of scripture. Accordingly, its interpretive framework is not bound by a literalist approach, rendering it more capable of evolving to integrate new information from other disciplines while remaining in continuity with long-standing teaching and tradition. In contrast, fundamentalist faiths in certain mainline churches have typically considered the notion of extraterrestrial intelligence contentious given its scriptural absence, in approaches based on *sola scriptura*. However considerable attention has been given by certain modern Protestant theologians.[11] This approach supports the efforts of exotheology in its engagement with the interdisciplinary field of astrobiology in rendering a more comprehensive and coherent theological soteriology engaged with modern science which demonstrates an ever-expansive universe. Therefore, as the extensive Catholic theological record on the subject of extraterrestrial intelligence demonstrates a lack of doctrinal, scriptural or traditional prohibitions against the possibility of intelligent life outside Earth, a theological reframing and expansion to accommodate outside intelligences will be argued in this thesis as a natural (and necessary) growth in the evolution of theological understanding of human life and its civilization within a vast and diverse universe. This groundwork allows for new development in scriptural interpretation given the recent scientific data; provided in this thesis is a biblical hermeneutic of the *varied* view in an effort to advance further a modern cosmology in conversation with theology.

Fourth, the Roman Catholic Church embraces a tradition of the development of doctrine, important for theological research in engagement with science.[12] Catholic theologian Vincent de Lérins's *Commonitorium* (c. 434) was the earliest effort to formulate two fundamental rules to guide the

---

1:12–14 (Vulg).

11. Prominent among these are David Wilkinson and Ted Peters. Others include Wolfhart Pannenberg, Jügen Moltmann, Paul Tillich, Alfred Whitehead, Lewis Ford, and John Jefferson Davis.

12. Important works on the development of doctrine within the Catholic tradition are Congar, *Tradition and Traditions*; Noonan, *A Church that Can and Cannot Change*; Theil, *Senses of Tradition*; Tilley, *Inventing Catholic Tradition*.

assessment of the proper growth of doctrinal development within orthodox teaching. His first rule argued that new data which bears on doctrine should be evaluated and validated by the standards of antiquity, ubiquity, and universality, and his second that such development in Church teaching must be understood through what precedes it:[13]

> [Progress] must be an advance in the proper sense of the word and not an alteration in faith. For progress means that each thing is enlarged within itself, while alteration implies that one thing is transformed into something else. It is necessary; therefore, that understanding, knowledge, and wisdom should grow and advance vigorously . . . in the whole Church and this gradually in the course of ages and centuries. But this progress must be made according to its own type, that is, in accord with the same doctrine, in the same meaning, and in the same judgment.[14]

Lerins's thought was later more fully expressed in John Henry Newman's, *Essay on the Development of Christian Doctrine* in 1845. Newman's hermeneutic of doctrinal continuity argued that ideas become more true as they develop in time; Catholic teaching develops in a progression in accord with new data in order to maintain continuity with the old. Newman states, "It changes with them [external circumstances] in order to remain the same."[15] He argued that the evolution of ideas is a natural process; therefore, the development of doctrine in Catholicism is necessary. This idea is encapsulated in his famous quote, "In a higher world it is otherwise, but here below to live is to change, and to be perfect is to have changed often."[16] Newman outlined seven 'notes' or tests of genuine doctrinal development, two of which are pertinent for this discussion. The first is his 'Preservation of Type':

> All great ideas are found, as time goes on, to involve much which was not seen at first to belong to them, and have developments, that is, enlargements, applications, uses and fortunes, very various, one security against error and perversion in the process is the maintenance of the original type which the idea presented to the world at its origin.[17]

His second on the 'Continuity of Principles' states:

---

13. See McGuckin, *The Westminster Handbook of Patristic Theology*, 348–49.
14. De Lérins, *Commonitorium*, 23.1–12.
15. Newman, *An Essay on the Development of Christian Doctrine*, 1.1.10.
16. Newman, *An Essay on the Development of Christian Doctrine*, 1.1.7.
17. Newman, *An Essay on the Development of Christian Doctrine*, 6. Introduction.

Doctrines grow and are enlarged, principles permanent; doctrines are intellectual, and principles are more immediately ethical and practical. Systems live in principles and represent doctrines.[18]

Accordingly, doctrines evolve but their foundational principles remain unaltered by new information. Illustrations of this development of doctrine in Catholicism in its theological formulations can be determined by examining its theological and philosophical record in how it contended with new, transformative modes of thought. Examples of these historical paradigms are Hellenism, Aristotelianism (preeminent in Aquinas's synthesis with Catholic theology), Copernicanism, Darwinism, the Enlightenment, Modernism, and Ecumenism. Each of these world views, to a greater or lesser extent, portended trouble for orthodox teaching. Each however in time was accommodated in varying degrees by Catholic theology, to the effect that theology was expanded and reoriented; and each new view significantly influenced Christianity in its dialog with these new understandings. Darwinian theory, in particular, is illustrative of this process. The teaching that God created the universe *ex nihilo*, a position founded upon patristic readings of the first verses of Genesis; and the special creation of humans from matter on Earth was held by the Church for centuries.[19] With the development of doctrine as a result of information brought forth by proponents of biological evolution, the Roman Church in time endorsed *theistic evolution*, where contingent natural processes can be understood as part of God's plan in the development of the human body;[20] while maintaining that the human soul remains a special, supernaturally created reality. This adjustment is directly relevant to this thesis in considering the evolution of planets, their biospheres, potential habitats and an evolutionary process for extraterrestrial beings possessing immortal souls and who inherit a creator-creature relationship. In this modification in doctrine God remains principally the creator of the body and soul in accordance with established teaching; although the process by which the body is produced is developmental rather than instantaneous. This thought on the development of doctrine is further implemented in chapter 5 in examining the historical evolution of Christian

---

18. Newman, *An Essay on the Development of Christian Doctrine*, 5.2.1.

19. *Dei Filius*, can.2–4: Denzinger, *Enchiridon*, 3022–24; Lateran Council IV (1215): Denzinger, *Enchiridon*, 800; cf. Denzinger, *Enchiridon*, 3025.

20. See Rahner, "Christology Within an Evolutionary View of the World," 184. The scientific method in certain terms is a product of Judeo-Christian revelation; belief in a Logos as the source of rationality and order and by which creation manifested, the universality and stability of natural laws, and the principle of creation which affirms the reality of physical time and space as derived from a first cause.

theology's contending with possible extraterrestrial intelligence in the *exclusive*, *inclusive*, and *multiple* soteriological formulations, culminating with the development of the *varied* view.

For Newman the appropriation of new data within the tradition, if done correctly, allows for expansion and congruity with existing knowledge, or dialectic between continuity and change in Catholic doctrine. As doctrine develops through gaining new insights into what had been revealed through new information, faith continues to gradually understand its full significance over the course of history.[21] Roman Catholicism, therefore, given the above discussion is uniquely positioned to explore the theological possibility of extraterrestrial intelligences in accord with its own Christological and soteriological teachings as argued herein. The new astronomy has brought profound attention to the universal claims of a terrestrial religion within an increasingly vast cosmos, presenting a new, major shift in world view which bears directly on these claims. This thesis will argue that the *varied* hypothesis continues this legacy of the development of doctrine, by offering an evolution in Christological doctrine while maintaining the foundational principles of Christianity as a bona fide supernatural religion within a context of potential outside intelligences.

Therefore, exotheological inquiry as expressed according to this thesis represents the latest development of doctrine in the accommodation of theology to a new contextual paradigm in the discoveries of the space sciences; particularly those encompassed by astrobiology, following earlier historical theological engagements. This evolution in this thesis is illustrated by arguments on extraterrestrial xenology, theological anthropology, and new soteriological formulations with regard to extraterrestrials examined in chapters 4 and 5. Specifically, information discovered since the beginning of the space age, and more recently the discoveries of exoplanets have motivated a handful of theologians to consider the theological implications of the new data to Christianity. The new astronomy tends to support the arguments of natural theology that creation is not geocentric nor anthropocentric as taught for centuries by the Church, but rather God as creator is active and present in all places and the sphere of humanity may be a single example of many possible places of divine activity; allowing for the possibility of environments for other created intelligences. Further, the Catholic teaching of an active, omnipotent, and omnipresent Trinitarian God I argue provides for a universe where creation of intelligence and salvation are integral parts of divine action on a cosmic scale.[22]

21. *Catechism of the Catholic Church*, 66.
22. The world created is the best possible for the creator's purpose of perfecting

Central for this thesis in considering intelligent extraterrestrial life, and which figures prominently in the hypothesis of the *varied* view is an acknowledgement of the 'omni-properties' of God, supported by Catholic teaching,[23] principal among them the absolute freedom of the creator to create. As in the case of human beings, I argue for a diversity of intelligent beings, all originating from the same creator and each ordered to his glory.[24] These 'omni-properties' described in the foregoing offers such a cosmic theological perspective, argued in the *varied* view. In such a putative biologically diverse universe, God works through the natural order in the creation of matter and in the emergence and support of all living organisms, in their reproduction and differentiation; in which all creatures possess their own particular goodness and perfection; and in the very nature of creation material beings are endowed with their own stability, truth, excellence, order, and laws.[25]

The model of the relationship of science and theology used according to this thesis recognizes the need for a profound and convincing synthesis between faith and reason in accordance with Catholic teaching.[26] Chapter 2 illustrates the importance and relevance of the new scientific data for theology. Chapter 3 provides the historical narrative of a Christian theology in dialog with early, philosophical and theological questions and later, with scientific information which impacted long-held doctrines, most directly those of Christology and soteriology. Exotheology combines divine revelation, reason, and the historical precedents in philosophy, theology, and the sciences. By extrapolation, this thesis proposes the formulation of certain Christological and soteriological possibilities with regard to extraterrestrial life, intelligent or otherwise while maintaining the supernatural legitimacy of

---

human beings in a world "*in staue viae*"; it is relatively but not absolutely perfect nor eternal. *Catechism of the Catholic Church*, 310; Aquinas, *Summa Theologica*, I, Q. 25, Art. 6; *Summa Contra Gentiles* 3, 71.

23. The Fourth Lateran Council defined that God is the sole principle of all things visible and invisible, the creator of all, as personal a priori First Cause possessing infinite power and creative productivity, indivisible, spiritual, personal, eternal, necessary, immutable, omnipresent, and absolute. Denzinger, *Enchiridon*, 428 (355). Nothing exists which does not owe its existence to the Creator, and all creation is rooted in a single primordial event, the very genesis by which the world was constituted and time began. St. Augustine, *De Genesi adv. Man.* 1, 2, 4: PL 34, 175.

24. *Catechism of the Catholic Church*, 340–41.

25. *Gaudium et Spes* 36 § 1.

26. "Not only can faith and reason never be at odds with one another but the mutually support each other, for on the one hand right reason established the foundation of the faith and, illuminated by its light, develops the science of divine things; on the other hand, faith delivers reason from errors and protects it and furnishes it with knowledge of many kinds." Vatican Council I, *On Faith and Reason*, third session.

the Christian religion. As science functions in service to theology by broadening its vision of creation and enabling more accurate formulations, theology provides a critical evaluation of the theories and conclusions of science in relation to Christian revelation and works towards the integration of established scientific knowledge with orthodox teaching.

The physical universe provides the ultimate context of intelligibility within which theology, and accordingly, exotheology operates in conjunction with other disciplines and provides formulations which speak to the greater context of a potentially widely inhabited universe created and maintained by a divine being. Therefore, it relies upon dimensions of other theological disciplines, including contributions from scholars of other faith traditions. For exotheology, the disciplines of astronomy, cosmology, exobiology, and the fields encompassed by astrobiology are particularly vital in keeping theology on the forefront of new data that impacts Christian faith. Christian theology and science have fundamental roles in the unity of knowledge, and for centuries have focused on the uniqueness and centrality of human beings within creation; modern scientific discovery has revealed a vast universe which continues to provide opportunities for discovery for science and Christianity in the broadening of human knowledge. Therefore, science and theology remain colleagues as science discovers new realizations of divine creativity in the universe, where theology will be engaged and even challenged to provide new understandings of God's activity and beneficence.

A Thomistic Exotheology

Aquinas's teachings are foundationally important for Catholic theology, and especially relevant for this thesis are his fundamental emphasis on a non-conflict between faith and reason,[27] his support for natural theology, his incarnational theology, and his emphasis on the absolute freedom of the creator in creating and redeeming intelligent beings. Aquinas's principle that truth is to be accepted regardless of its source supports the use of the scientific data as it becomes available in the service of theology.[28]

27. Aquinas, *Summa Contra Gentiles*, Book 1, chaps. 1–7; *Catechism of the Catholic Church*, 159; "Though faith is above reason, there can never be any real discrepancy between faith and reason. Since the same God who reveals mysteries and infuses faith has bestowed the light of reason on the human mind, God cannot deny himself, nor can truth ever contradict truth" (*De Filius* 4:3017).

28. "Consequently, methodical research in all branches of knowledge, provided it is carried out in a truly scientific manner and does not override natural laws, can never conflict with the faith, because the things of this world and the things of faith derive

His natural theology and its relation to revelation for this thesis can be most appropriately described by Aquinas as the "book of nature" or God's works. God is known by reason alone, which is natural theology, and also by what has been revealed supernaturally.[29] The former remains an object of knowledge, as made available through natural reason, the latter remains an object of faith as it is revealed:

> Sacred doctrine essentially treats of God viewed as the highest cause, for it treats of Him not only so far as He can be known through creatures just as the philosophers knew Him—'That which is known of God is manifest in them.' (Rom. I. 19)—but also so far as He is known to Himself alone and revealed to others.[30]

> The existence of God and other truths about God, which can be known by natural reason, are not articles of faith, but are preambles to the articles; for faith presupposes natural knowledge, even as grace presupposes nature and perfection the perfectible.[31]

Catholic teaching holds that although the ultimate plans and operations of God remain mysterious to humans, this does not imply that this mystery is incomprehensible or unintelligible if it is revealed. Natural theology thus exists as a preamble, secondary to the articles of faith. In modernity natural theology operates as an interdisciplinary inquiry, providing new insights into the nature and modes of revelation, revealing what can be known by reason while in conversation with scripture and the deposit of faith.[32] Further, natural theology can reveal the reasonableness of faith.[33] In Catholic tradition, natural theology cannot be partitioned from divine revelation; rather it builds upon and reinforces what is revealed in special revelation,[34] and strives to understand divine purpose within a coherent framework, combining what

---

from the same God" (*Guadium et spes* 36 § 1).

29. *Catechism of the Catholic Church*, 50.

30. Aquinas, *Summa Theologica*, I.I, Q. 2, Art. 2.

31. Aquinas, *Summa Theologica*, I.I, Q. 2, Art. 2.

32. Paul revealed that "The invisible things of God are clearly seen, being understood from the creation of the world and through things that are made, both his eternal power and divinity" (Rom 1:20).

33. In contrast to the natural religion of deists Thomas Paine and Thomas Jefferson, who argued that God can be known through creation in a way which renders scripture unnecessary; pantheism represents another distortion where nature is not the expressive creation of a God but rather an extension of his being.

34. Hibbs, *Knowledge and Faith in Thomas Aquinas*, 603–22; Stump, *Aquinas*. 26–32.

is available through special and general revelation. For an exotheology which utilizes natural theology there must be an ultimate and intelligible context (according to this thesis, the cosmos) within which revealed and reasoned truths are contained,[35] and which the modern sciences form an integrated vision of reality combining both human and divine elements. This strength of natural theology provides for the formulation of the *varied* hypothesis as the best solution to address the concerns of Christian dogma and established tradition to provide this unified vision.

The Thomist incarnational teaching on the Redemption sees the Trinity, not Christ, as the source of grace for the angels and the first humans, and a Christocentric redemptive mode for all other humans.[36] This Thomist argument is expanded herein to accommodate a known greater creation, which suggests the possibility of models of redemptive modalities and economies of salvation of putative extraterrestrials at variance with our own as in the *varied* view. Other intelligent creatures accordingly might inherit a soteriology not terrestrial-bound nor Christ-centered; as God can operate in other non-human civilizations, allowing for diverse types of extraterrestrial economies of salvation. This is the most major development of Christian doctrine proposed in this thesis. Salvation for all intelligences is not construed as universally dependent on the person of Jesus of Nazareth, as presumed by, for example, the authors of the Nicene Creed. This departure is carefully argued for in the evaluation of '*inclusive*' and '*varied*' soteriologies in chapter 5. Nevertheless, I shall argue that this radical move is in full continuity with Thomistic theology.[37] Aquinas's incarnational theology is central to this thesis; his argument which allows for multiple incarnations of any person of the Trinity in rational beings expands notions of possibilities of divine activity in civilizations outside Earth, and which serves to support the *varied* soteriological view. As God is revealed and redeems on Earth through a gradual process of supernatural revelation which culminates in the mission of the incarnate Word, Jesus, on Earth, thus demonstrated is a definitive pedagogical pattern and trajectory of divine work which indicates potentialities of supernatural action in places outside the human sphere. Following this thought I shall show that revelation can vary considerably according to divine action elsewhere according to other histories.[38] As variations in biology, environment, and

---

35. Lonergan, *Method in Theology*.

36. Aquinas, *Summa Theologica*, III, Q. 70, Art. 4, ad. 4; III, Q. 3, Art. 5–7; Aquinas, *Commento al Corpus Paulinum*, #29.

37. "For God, in his omnipotent power, could have restored human nature in many other ways." See Aquinas, *Summa Theologica*, III, Q. 1, Art. 2.

38. *Catechism of the Catholic Church*, 53; Revelation is in the broad sense, knowledge of God deduced from observations of the natural world (general or natural revelation), (reason) and in the strict sense as divine events and utterances (special

histories are different, I suggest in this thesis, therefore, that the theological anthropology of creatures can vary; and predispositions to supernatural grace might be quite different, therefore, differences in divine relation are likely, resulting in variant soteriological economies.[39]

The Catholic teaching on the operations of grace, the supernatural gift to intellectual creatures for salvation and attained through salutary acts or states of holiness,[40] in extraterrestrial societies might have a multiplicity of divinely bestowed forms serving a range of purposes. I will argue that divine revelation, creaturely participation, the workings of grace, and creator-creator unity could manifest themselves differently; the operations of actual and sanctifying grace and creatures' reception to it within a plurality of supernatural religions might have great diversity. Each mode would act in accordance with the ultimate function of supernatural grace to enable the elevation to supernatural life and redemption. For this thesis, the Catholic teachings on the works of grace will be extrapolated to address the distinctiveness of non-human intelligences' life and relationship to forms and actions of divine grace. Accordingly, following Aquinas, as grace and nature are dialectical due to the necessary contingency of creatures, the nature of a subject would receive supernatural grace according to the particularities of its unique nature.[41] The provision of grace and knowledge of God is only possible through the workings of the Holy Spirit, who according to Christian faith has been at work from the beginning of creation to the culmination of all in its union with the creator. The Spirit as argued in this thesis rules, sanctifies, and animates a potentially diverse multitude

---

revelation). What knowledge of God that has been supernaturally revealed was not attainable through human reason, and God has revealed all that is necessary for human salvation. Pius XXII, *Humani Generis*, 561: Denzinger, *Enchiridon*, 3875.

39. *Catechism of the Catholic Church*, 1987, 1996; *Council of Trent*: Denzinger, *Enchiridon*, 1533–34.

40. Grace is distinguished into two types: actual grace, a temporary help to aid in holiness; and sanctifying grace or habitual grace, meaning creation of a state of holiness and justification. Actual grace consists of a passing influence on the soul and preordained for the end of one's eternal salvation. The relation between them is of action and state, not of actuality and potentiality. Grace serves to dignify humanity and makes it the image of God, as mankind can only claim his fundamental endowments as those qualities above nature are complete gift. 1 Cor 2:7–9; *Catechism of the Catholic Church*, 1998, 2000. Actual grace is that unmerited interior assistance from God, in virtue of the merits of Christ, conferred upon fallen man in order to strengthen, on the one hand, his infirmity resulting from sin and, on the other, to render him capable, by elevation to the supernatural order, of supernatural acts of the soul, so that he may attain justification, persevere in it to the end, and thus enter into everlasting life. The end purpose of all actual grace is directed to the possession of sanctifying grace, from which holiness and sonship of God solely depend. *The Catholic Encyclopedia*, 689–98.

41. Aquinas, *Summa Theologica*, I.I, Q. 1, Art. 8, ad. 2, *Gratia non tollit naturam, sed perficit*.

of intelligent and non-intelligent life. Although every action of a divine Person is attributed to the entire Trinity, the Spirit acting as an independent agent serves as God's manifest and powerful activity in beings and worlds who provides grace and gift (although its action and powers may manifest in different ways than with humans). Therefore, this thesis argues that an individual and communal journey of the spirit in unification with divinity is likely to be ubiquitous in extraterrestrial beings given a creaturely nature and a biological life-cycle.

In summary, this chapter has presented the credentials of a Thomistic approach within a Roman Catholic framework as a grounding for the investigation of the problem of possible extraterrestrial intelligences. The theological basis for the exploration in this thesis is as follows: The Catholic intellectual tradition has a long track-record of engagement with historical inquiries on extraterrestrial intelligence with scientific research, and is therefore, well-qualified theologically to provide new formulations in the present context in accordance with scripture and tradition. The new data in the space sciences has suggested challenges to long-held terrestrial Christological and soteriological doctrine, as well as the teaching of the fundamental uniqueness of humanity in creation, motivating theology to re-engage scripture, tradition, and science, in order to provide an informed account of God's works in a larger creation. This is argued in chapter 5 in which the *varied* view is evaluated and premised as superior to competing positions. Also, the Church's adoption of the modern, scientific methods of biblical criticism allows for a re-reading and development of scriptural interpretation in accord with modern scientific discoveries, providing the development of a biblical hermeneutic which provides for a soteriology of extraterrestrial intelligences. The Catholic approach, as described above supports the efforts of an exotheology engaged with the interdisciplinary field of astrobiology and other disciplines in developing its formulations on extraterrestrials and theology, as discussed in chapters 3 on theological developments, and chapter 4 on xenology and theological anthropology. The Church's established tradition of a development of doctrine, and as argued in this thesis, an evolution of soteriological formulations addressing the possibility of extraterrestrial life leads to the evolution of the *varied* view. Lastly, the recognition of the importance of the 'omni-properties' of divinity, with Aquinas, in considering extraterrestrial intelligence and divine work in a larger cosmos; and the theological engagement with Aquinas's Christology, soteriology, and teaching on natural theology, while preserving congruity with established teaching supports the conclusions of exotheology according to the *varied* view.

Chapter 2

Cosmology, Astrobiology, and Modern Astronomical Technologies

Section A: The New Cosmology

The scientific advances in cosmology and astrophysics have provided a clearer picture of the early history, development, and structure of the universe, as well as galaxy, star, and planet formation. Critical to our discussion will include, within the last few decades, the unprecedented and long-awaited discoveries of extra solar planets, which has led to the establishment of a new field—astrobiology, whose aim is to determine the conditions for life and assist in the development of space initiatives seeking extraterrestrial life forms. These environmental conditions and their relation to the development of extraterrestrial life, extrapolated from terrestrial studies, will figure prominently in the astrobiological section, which will be relevant to the later discussion on possible extraterrestrial biologies. The section will conclude with review of the telescopic advances which have revolutionized astronomy, primarily telescopes *Kepler* and *Hubble*, and discuss the most important soon to be launched program, the *James Webb Space Telescope*, capable of discerning extraterrestrial biosignatures on certain extrasolar planets.

Reliable historical summary works on astronomy include Morison,[1] Kutner,[2] and Carroll;[3] astrobiological works are Kolb,[4] Rothery and Gilmour,[5] Impey and Lunine.[6] As mentioned, the new scientific understandings of the

---

1. Morison, *Introduction to Astronomy and Cosmology*.
2. Kutner, *Astronomy: A Physical Perspective*.
3. Carroll, *An Introduction to Modern Astrophysics*.
4. Kolb, *Astrobiology: An Evolutionary Approach*.
5. Rothery and Gilmour, *An Introduction to Astrobiology*.
6. Impey and Lunine, *Frontiers of Astrobiology*.

twentieth and twenty-first centuries in cosmology have dramatically altered our inherited near-eastern Judeo-Christian anthropocentric view of the universe. The cosmological model of an early universe, existing in an extremely hot and dense state and exploding in a theorized Big Bang, followed by an expansion around ≈13.8 billion years and resulting in the creation of billions of galaxies with each containing around a hundred billion stars, now dominates our scientific worldview. Now considered primitive are the early geocentric attempts of the Ptolemaic, Copernican, and Galactocentric[7] universes designed to accommodate scriptural accounts of creation and maintain mankind's unique and central cosmic role.[8] The new universe, as illustrated by the *Hubble Ultra Deep Field* (HUDF) image, completed January 2004, reveals a universe approximately 13.8 billion light years in extent, composed of billions of evolving galaxies, nebulae, and stars, and having no discernable center. The *Kepler* mission, now in service since March 2009, and having studied only an infinitesimal fraction of the sky in the northern constellations of Cygnus, Lyra, and Draco, has confirmed the existence of over 2,662 extrasolar planets in more than 530,506 stellar systems, and as of 2020, has also detected 2,840 planet candidates yet to be confirmed.[9] There is presently a total of 4,276 exoplanets discovered via *Kepler* and other ground-based telescopes as of September 2020. Extrapolating from the *Kepler* data, in November 2013 astronomers reported there could be more than one hundred billion planets, and forty billion Earth-sized planets orbiting in the habitable zones of sun-like stars and red dwarfs within the Milky Way Galaxy, eleven billion of which may be orbiting sun-like stars.[10]

The exoplanet astronomical findings are the result of three new techniques used by *Kepler* in coordination with ground-based telescopes. These are the radial velocity method, the transit method, and the microlensing method. The radical velocity method is an indirect technique which looks for the gravitational influence of planets on their parent stars. This gravitational pull creates a very small but discernible wobble effect on the star which indicates the presence of a planet. Additionally, this wobble has another effect of splitting the light from the star into its spectral lines,

---

7. All geocentric models. Galactrocentrism was proposed in 1785 by William Herschel, whose observations indicated the Milky Way was a separate disk-shaped galaxy with the Sun in a central position. This was proven incorrect by Harlow Shapley in 1918.

8. See Appendix A: Historical Cosmology and Theological Timeline.

9. "NASA's Exoplanet Archive KOI Table," *NASA*.

10. Overbye, "Far-Off Planets Like the Earth Dot the Galaxy." *New York Times*; *17 Billion Earth-Size Alien Planets Inhabit Milky Way*; Petigura et al., "Prevalence of Earth-Size Planets."

COSMOLOGY, ASTROBIOLOGY, AND MODERN ASTRONOMICAL TECHNOLOGIES

and the resulting Doppler shift of these is used to measure the gravitation pull of a planet on the its star.[11] Use of this technique from radial velocity and combined with knowledge of the star's mass, astronomers can then calculate the radius of the planet. One difficulty with this method is as the orbital plane of a planet is unknown from our distance, only the minimum mass of the planet can be determined.[12] If a planetary tilt happens to be oriented ninety degrees from *Kepler*'s line of sight, this will render the Doppler measurement impossible.[13]

The transit method was first successfully used in 2003 to identify a planet five thousand light years away. This technique looks at a planet's orbit where it passes in front of its parent star, resulting in a slight dimming effect of its light. During this period which is termed an occultation, the atmosphere of the planet will absorb some radiation emitted by its star creating absorption lines which are then are detectable. These lines are able to indicate to astronomers the compositional makeup of a planet's atmosphere.[14] For the *Kepler* telescope, its instruments need only to discern a dip in light of eighty-four parts per million to detect a planet's presence. Once a transit is detected a planet's orbit can then be calculated from its period and mass of the star using Kepler's third law of planetary motion. The actual size of the planet can also be determined by drop in light and size of star. Therefore, from the orbit of the planet and the temperature of its parent star, the temperature of the planet can be known, which determines its habitability.[15] Astronomers have determined that stars hosting planets have to be less than about 1.5 times the mass of our Sun to provide the lifespan required for a stable planetary environment for life to form. As *Kepler* measures stars continuously, at least three planetary transits are required to verify a planet's presence, and then Earth based telescopes are utilized to review those findings, which are later confirmed by computer simulation programs to rule out other phenomena which could mimic the presence of a planet.[16]

The microlensing method takes advantage of the fact that a path of light can be bent by the gravitational field around a massive body such as a star or planet. This method detects planets over a wide range of mass and further from their stars. Microlensing is a technique where the light from distant stars produces a temporary brightening due to presence of mass

---

11. Wilkinson, "Searching for Another Earth," 417.
12. Wilkinson, "Searching for Another Earth," 417.
13. Wilkinson, "Searching for Another Earth," 418.
14. Wilkinson, "Searching for Another Earth," 418.
15. Wilkinson, "Searching for Another Earth," 418.
16. Wilkinson, "Searching for Another Earth," 418.

between star and observer. It is not as sensitive technique as radial velocity or transit methods where planets have to be massive or close to their star.[17] When the gravitational field of a host star combines with the gravitational pull of a planet, this acts like a lens, which magnifies the light of the background star. With proper alignment of the background and lensing star, a planet can brighten the effect of the background star. This technique was used to find nineteen exoplanets as of late 2019.[18] An important finding using these methods is that rocky planets appear to be plentiful in our Galaxy, with current calculations indicating that twenty-five percent of Sun-like stars have an Earth-sized planet in a habitable zone, there could be ten billion potentially habitable Earths in our Galaxy.[19]

The new data provide us with unprecedented evidence that planetary systems are a common and necessary corollary of stellar formation, and consequently, greatly increase the possibility of an eventual discovery of other simple and complex life forms, and potential contact with non-terrestrial biological intelligences in the future. Biological and geological evidence on Earth shows that life is highly adaptable and extremely diversified, and that species evolve in various environmental varieties of temperature, atmospheric pressure, access to sunlight, available oxygen, and other variables. Therefore, most scientists and astrophysicists have accepted the thesis of a universe where biological evolution follows stellar evolution, when conditions allow.

Before discussing contemporary Christian theology in this new setting, it will first be necessary to outline the fundamentals of what new cosmological and astrophysical findings have determined as the most plausible picture of the creation of the universe, our galaxy, and solar system so as to better establish the true place and role occupied by humanity within the larger cosmological context. It is currently understood by most cosmologists that after the Big Bang and during its initial stages of expansion, the universe first fragmented into super-massive globes of gas, some of which aggregated to form large protogalaxies and clusters of protogalaxies.[20] The collapsing globes subsequently fragmented into clusters of stars, which then separated into isolated stars. These fragmentation and clustering processes appear to be of comparable importance in the structural makeup of the universe. Most theories of the origin of galaxies start with large areas of varying density in the very early universe. Protogalaxies began as large

---

17. Wilkinson, "Searching for Another Earth," 420.
18. "Extrasolar Planet Detected by Gravitational Microlensing," NASA.
19. Wilkinson, "Searching for Another Earth," 420.
20. Eggen et al., "Evidence from the Motion of Old Stars," 748.

concentrations of hydrogen and helium gas, with helium derived from the bonding of hydrogen atoms within the reactions of the first generation of stars. These concentrations began with masses ranging from millions to trillions of solar masses, which continued to expand (although more slowly than the universe' expansion) resulting in a widening of their separation in space. At some stage, each protogalactic globe ceased to expand and began to collapse, with early population II stars beginning to form in the central region of protogalaxies where the density was highest. The brightest of these first generation stars evolved rapidly and erupted as supernovae, ejecting gas enriched with heavy chemical elements.[21] The infalling gases of the protogalaxies were, therefore, steadily enriched with these heavier elements, which were incorporated into the material for the creation of the next generation of stars. Meanwhile, these entire globes of gas began to collapse freely under the influence of their own gravity, and gradually began an infalling rotational movement. This collapse lasted hundreds of millions of years, and in some cases perhaps billions of years,[22] while stars formed continually in the central regions and the brightest lasted only millions of years. The inside of protogalaxies formed halos, and the outside of the protogalaxies fell and formed the nucleus and disk of newborn galaxies. The rotation, speed, time, and gravity of infalling gas determined whether the galaxy was of spiral, elliptical, or irregular form. The rate of star formation depended on the molecular hydrogen gas;[23] as the higher the density, the faster stars formed. If a protogalaxy had higher density (due to its separating at an earlier stage in the expanding universe) then stars formed more quickly and no infalling gas survived to form a disk or nucleus.[24] By determining the rate at which radioactive elements of atoms from stars and gas decay into their daughter elements, and measuring their relative quantities, it has become possible to determine the age of the Earth, the Solar System, and many galaxies within our view. As a result, astrophysicists have determined that our solar system is approximately 4.6 billion years old[25] and the Milky Way galaxy approximately thirteen billion years old,[26] with the universe an estimated age of ≈ 13.8 billion years.[27]

21. Heger and Woosley, "The Nucleosynthetic Signature of Population III," 532–43.
22. Noguchi, "Early Evolution of Disk Galaxies," 77–95.
23. Rana and Wilkinson, "Molecular Hydrogen," 323–24.
24. Baugh and Frenk, "How Are Galaxies Made?"
25. Bouvier and Wadhwa, "The Age of the Solar System."
26. Cayrel et al., "Measurement of Stellar Age," 691; Cowan et al., "The Chemical Composition," 861.
27. Planck Collaboration, "Planck 2013 results"; Bennett et al., "Nine-year Wilkinson Microwave Anisotropy Probe (WMAP) Observations."

Using nucleochronology, astronomers have determined that the light elements of hydrogen, helium and deuterium were produced by protons and neutrons in the Big Bang while the universe was still young, dense, and very hot. Most of the other, heavier chemical elements were made much later in subsequent generations of stars and were ejected into space via supernova explosions.[28] As noted, the first stars were composed of the lightest element, hydrogen. The principal elements for supporting life, that is, oxygen, carbon, nitrogen, and potassium, were produced only after the first generation of stars, composed primarily of hydrogen and some helium; as a result of their thermodynamic and thermonuclear evolution and culminating in supernovae, these stars expelled these heavier elements into space, which were then made available as material for the second generation of stars, of which our solar system was formed.[29] These stars, which were more stable and had a longer lifespan given their mass, provided orbiting planetismals a steady source of energy in the form of light and heat which may have provided the proper environment for the stimulus and development of the elementary forms of life on a range of planets. Astrobiologists have proposed the fundamental environmental conditions for the possibility of life to develop on a given planet. Potential planets in a solar system must have sufficient mass to maintain a gravitational field capable of retaining a gaseous atmosphere, but be small enough to cool off in a reasonably short period of time, in order to create the conditions suitable for life.[30] The distance of a planet to its parent star must be within its habitable zone, that is, not so close that it will receive an excessive amount of heat (which would inhibit the coalescence of molecules) but not so distant that lower temperatures would inhibit the ability for the necessary chemical compounds to interact within the medium of liquid water.[31] It was initially believed that for a planet harboring the conditions for life to develop, it could not belong to a binary or multi-stellar system, as due to great variations in gravitational forces these systems cannot provide a stable planetary orbit. However, it has now been established that these environments may also be conducive to life and that planets would not necessarily experience turbulent orbital periods.[32] Given the example of our Earth, we understand that oxygen was initially provided by plants which grew underwater, and released into the

28. Whittet, *Dust in the Galactic Environment*, 45–46.
29. Krebs and Hillebrant, "The Interaction of Supernova Shockfronts," 411–19; Allen, "Supernova Effects."
30. Wurchterl, "Planet Formation," 67–96.
31. Kasting et al., "Habitable Zones around Main Sequence Stars," 108–28.
32. Quintana and Lissauer, *Terrestrial Planet Formation in Binary Star Systems*; Wiegert and Holman, "The Stability of Planets in the Alpha Centauri System," 1445–50.

atmosphere where some of it was converted to ozone which protected the atmosphere, whereby allowing for a protective environment which enabled vegetation to grow on land, which in turn provided more oxygen for land life forms to develop.[33] The first simple life forms, plant and animal, by means of their biochemical processes supplied the biosphere with other important elements such as carbon, hydrogen, nitrogen, and phosphorus for the later and more complex life forms. The research and study of biological compounds in interstellar space and on the surface of heavenly bodies such as asteroids, comets, satellites, and planets has also resulted in some important findings. Using radio frequency observations and infrared spectroscopy, over one hundred different types of molecules have been discovered in interstellar space, including water, carbon monoxide, carbon dioxide, ammonia, ethanol, the amino acid glycine, methanol, formaldehyde, and various carbon, silicon, and nitrogen compounds, as well as a number of amino acids.[34] Extraterrestrial amino acids have been found inside carbonaceous chondrites, a primitive form of meteorite. The structural symmetries of their molecules differ from those of the same type on Earth, confirming their extraterrestrial origin.[35] In addition, organic molecules found on the surface of meteoric residues and observed on comets are identical to those that form the basis for terrestrial life, and have caused scientists to consider their possible role in pre-biotic processes, as in panspermia,[36] and their origins from biological processes already in existence. However, no nucleic acids, or DNA, have been discovered outside Earth as of yet.[37]

The Science of Astrobiology

Astrobiology is an interdisciplinary study encompassing the fields of geology, chemistry, biology, molecular biology, ecology, geography, astronomy, physics, and planetary science in the research and study of the origins, evolution, distribution, and future of life—intelligent or otherwise within and outside Earth. In 1982, the international scientific community granted astrobiology's research official status, being established as Commission

33. Crowe et al., "Atmospheric Oxygenation Three Billion Years Ago," 535–38.
34. Kwok and Zhang, "Mixed Aromatic-Aliphatic Organic Nanoparticles," 80–83.
35. Matson, "Meteorite That Fell."
36. The hypothesis is that life on Earth (or other planet) originated from microorganisms or chemical precursors of life present in interstellar space transported or distributed by space dust, meteoroids, asteroids, comets, planetoids, or unintentionally by spacecraft, and able to initiate life upon reaching a suitable environment.
37. Martins et al., "Extraterrestrial Nucleobases," 130–36.

number fifty one of the International Astronomical Union. This commission undertakes theoretical and experimental reconstruction of the processes understood to be responsible for life's origin on our planet so as to better understand its occurrence on a cosmic scale. Published information includes proceedings of international conferences, to include Papigiannis,[38] Shostak,[39] Cosmovici et al.,[40] Grady,[41] Goldsmith, Owen,[42] Dick and Strick,[43] and Meech et al.[44] The discipline of astrobiology occupies a central and comprehensive role in the human quest for possible intelligent extraterrestrial life, in conjunction with SETI radio telescopes and the *Pioneer* and *Voyager* probes, launched with a partial role of seeking extraterrestrial contact. Through examination of those environmental conditions conducive to the existence of extraterrestrial life as well as models drawn from terrestrial physical and cultural anthropology, astrobiology has through extrapolation begun to provide data on possibilities pertaining to non-intelligent as well as intelligent extraterrestrial biology, psychology, and sociology, which is valuable for theologians. Historian Steven Dick asserts that most astronomers and origin of life researchers accept life's cosmic abundance as the most likely scenario and the working hypothesis of those in the growing hybrid fields of bioastronomy and astrobiology.[45]

Astrobiologists have calculated the mathematical constants of galactic, stellar, and planetoid formation necessary to determine the physiochemical conditions for the development of environments capable of supporting life. There is still debate whether the emergence of life is a chance event, as described by Jacques Monod, or a necessary consequence of the proper conditions, as suggested by Christian de Duve.[46] Astrobiology has followed the assumption that extraterrestrial life forms will be based on carbon chemistries, as on Earth, as carbon allows for the building of long and complex molecules more than any other, although silicon-based chemistries are hypothesized.[47] According to current evolutionary theory, the period of time on Earth from the simplest forms of life to the appearance of synapsids was

---

38. Papigiannis, *The Search for Extraterrestrial Life*.
39. Shostak, *Progress in Search for Extraterrestrial Life*.
40. Cosmovici et al., 161.
41. Grady, *Astrobiology*.
42. Goldsmith and Owen, *The Search for Life in the Universe*.
43. Dick and Strick, *The Living Universe*.
44. Meech et al., "Commission 51."
45. Dick, *Many Worlds*, 191.
46. De Duve, *Vital Dust*.
47. "Polycyclic Aromatic Hydrocarbons," *Astrobiology Magazine*.

not more than 3.5 billion years.[48] Liquid water was probably critical to extensive biological evolution as it provided a fluid medium in which material could move and aggregate. Microscopic cellular life must have arisen about a billion years since the formation of Earth as fossils of methane-producing bacteria have been found in 3.4 billion year old rocks from South Africa in 1977.[49] The precise dating of the earliest life forms on Earth remains controversial. One review of the data lists the earliest fossil stromatolites, colonies of blue-green algae, at 2.7 billion years old from Canada to Zimbabwe.[50] The earliest probable evidence is stromatolites from Western Australia, which are between 3.4 and 3.5 billion years old; and the oldest possible evidence is a 3.7 billion year old rock containing carbon isotopes of possible biological origin from western Greenland.[51] Life is unlikely to have evolved much before 4.1 billion years ago due to intense early meteoritic bombardment and the possible magma ocean covering much of the Earth's crust. Paleontologist Stephen Gould has indicated that primitive life seems to have arisen "as soon as it could."[52] For the first two billion years, most life remained in the ocean where liquid water provided a supporting and protective environment, with early life forms consisting of mostly soft-bodied organisms which rarely produced fossils, so their development is difficult to trace. The flourishing of stromatolites and early plant forms 2.0 to 2.5 billion years ago helped boost oxygen production.[53] Therefore, the atmosphere evolved towards toward being dominated by oxygen compounds. Biologists consider that the evidence of the adaptability of life forms and the rapid proliferation of advanced species once they evolved is indicated in the geological time scale: the fossil record of Earth demonstrates that it experienced about a one billion year evolution from nonliving organic chemicals to small organisms, and then a much more rapid evolution to species with conscious intelligence.[54]

Crucial to the understanding of the environmental conditions for life is to consider the number of geological and astronomical processes that could have caused massive climate change and affected the course of biological evolution: planetary convection causing plate tectonic crustal splitting, resulting in landmass drifting, (such as the separation and isolation

---

48. Reisz and Fröbisch, "The Oldest Caseid Synapsid."
49. Gould, "The Great Dying," 33–42.
50. Schopf and Kudryavtsev, "Evidence of Archean Life," 141–55.
51. Witze, "Claims of Earth's Oldest Fossils Tantalize Researchers."
52. Gould, "An Unsung Single-Celled Hero," 22.
53. Walker, "Evolution of the Atmosphere."
54. Hartmann, *Astronomy*, 619.

of Australia, allowing for different species to evolve there), changing sea levels, ocean currents, wind patterns, and seasonal extremes. Volcanic eruptions could have spewed dust into the atmosphere resulting in a dimming of sunlight, which could cause extreme climate change. Slight changes in the planet's orbit and tip of the planetary axis plane of the ecliptic caused by gravitational forces could result in major climactic changes such as ice ages. Changes in the Sun's radiation could result in climate change, and, from great distances, irradiation from a nearby supernovae are also capable of causing climate change or directly affecting organisms. An asteroidal or cometary impact, or atmospheric explosion could have damaged the ozone layer, exposing organisms to increased radiation.[55] Any of these factors may have contributed to the five great extinction events[56] evidenced in Earth's evolutionary history, including one, the Permian-Triassic (colloquially known as the Great Dying) which is estimated to have occurred about 252 mya,[57] causing the extinction of up to ninety-six percent of all marine species[58] and seventy percent of all terrestrial vertebrates,[59] and known to have caused the mass extinction of insects present at that time.[60]

The great variety of ancient and modern species on Earth and the diversity of environments in which life flourishes suggest that given time, and with the proper conditions necessary, life on other planets could have evolved to fit a wide range of environments. Humans can survive certain variations in body temperature, and the environmental range covered by all terrestrial species is even greater. Some extremophile microorganisms live in Antarctic pools that remain liquid at -20°C as a result of dissolved calcium salts. Other thermophiles live in Yellowstone Hot Springs at temperatures of 140°C.[61] Habitable pressures range over a factor of one thousand: bacteria exist at altitudes of atmospheric pressure of only 0.2 atm,[62] while more advanced organisms live at ocean depths with pressures in hundreds of atmospheres.[63] Although still debated, the current upper temperature

55. Hartmann, *Astronomy*, 623.

56. Ordovician-silurian, 440 mya; Devonian, 365 mya; Permian-triassic, 252 mya; Triassic-jurassic, 210 mya; Cretaceous-tertiary, 65.5 mya.

57. Mya = Million years ago. Shu-zhong Shen, et al. "Calibrating the End-Permian Mass Extinction," 1367–72.

58. Benton, *When Life Nearly Died*.

59. Sahney and Benton, "Recovery from the Most Profound Mass Extinction of All Time," 759–65.

60. Labandeira and Sepkoski, "Insect Diversity in the Fossil Record." 310–15.

61. Beal, "Thermophiles."

62. Atm = Atmopheric pressure.

63. Hartmann, *Astronomy*, 627.

limit for life is 250°C, given that amino acids begin to break down at higher temperatures, as demonstrated by the decreased concentration of total hydrolysable amino acids when calcareous sediments were heated to determine their thermal stability in seafloor hydrothermal systems.[64] In October of 2011, based on data from the Infrared Space Observatory and the Spitzer Space Telescope, observed within infrared emissions in stars and interstellar space were "amorphous organic solids with a mixed aromatic-aliphatic structure" which could be created in the region around dying stars and then distributed into space.[65] This led many astrobiologists to consider that these organic solids could have provided some of the essentials for the formation of other organic compounds in young planets.[66] In September 2012, NASA scientists reported that polycyclic aromatic hydrocarbons (PAHs) subjected to interstellar medium conditions are transformed through hydrogenation, oxygenation, and hydroxylation to more complex organic compounds, which can lead to the formation of amino acids and nucleotides (the materials of proteins and DNA).[67] Also, in August of 2012, astronomers at Copenhagen University detected the specific sugar molecule glycolaldehyde in the gas surrounding protostellar binary IRAS 16293-2422, four hundred light years from Earth.[68] Glycolaldehyde is a possible precursor of RNA, which may well have been the genetic material which preceded DNA,[69] and similar in function to DNA. This finding suggests that complex organic molecules may form in stellar systems prior to the formation of planets, and this may have an effect on the development of other inorganic and organic compounds during the early formation of planets.[70] According to astrobiologists Bjorn Carey and Michael Mautner, more than twenty percent of the carbon in the universe may be associated with PAHs, possible starting materials for the formation of life and which appear to have been formed shortly after the Big Bang, are widespread throughout the universe, and are associated with new stars and exoplanets.[71]

64. Ito et al., "Thermal Stability of Amino Acids," 177–88.
65. Kwok and Zhang, "Mixed Aromatic-Aliphatic Organic Nanoparticles," 80–83.
66. Chow, "Discovery"; Kwok and Zhang, "Mixed Aromatic-Aliphatic Organic Nanoparticles," 80–83.
67. Gudipati and Yang, "In-Situ Probing of Radiation-Induced Processing."
68. Jorgensen et al., "Detection of the Simplest Sugar," 757.
69. Atkins et al., *The RNA World*.
70. Jorgensen et al., "Detection of the Simplest Sugar," 757.
71. Mautner, "Planetary Bioresources and Astroecology," 72–86.

Modern Telescopic Discoveries and Near Future Missions

As mentioned in the introduction, unprecedented findings in the detection and study of extrasolar planets and solar systems outside our own have transformed astronomy. The *Kepler* telescope, designed specifically for the detection of extrasolar planets has discovered thousands of planets and planet candidates in the small region within the constellation of Cygnus. *Kepler*'s observational technology is able to easily discern the presence and properties of larger gas giant planets, as well as smaller, rocky, Earth-like planets, and has begun to indicate the prevalence of these smaller bodies orbiting many parent stars.[72] *Kepler* is designed to discern planets by the transit method, thereby only detecting planets orbiting those stars whose orbital plane is in our line of sight from Earth. This suggests the likelihood of many more planets which exist, but which *Kepler* cannot see. Based on *Kepler*'s mission data, astronomers have stated there could be about 40 billion Earth-sized planets orbiting sun-like stars and red dwarfs in the Milky Way Galaxy, of which "at least five hundred million" are in the habitable zone.[73] Using the *Kepler* data, astronomers at NASA's Jet Propulsion Laboratory reported that about 1.4 to 2.7 percent of all Sun-like stars are expected to have Earthlike planets "within the habitable zones of their stars."[74] This translates to, conservatively, two billion Earth analogs in our own Milky Way galaxy alone. The *Hubble* telescope has revealed hundreds of billions galaxies within the observable portion of the universe, potentially yielding more than one sextillion Earthlike planets, that is, if all galaxies have similar numbers of planets to the Milky Way.[75] In January 2012, an international team of astronomers speculated that each star in the Milky Way Galaxy may host "on average . . . at least 1.6 planets," suggesting that over one hundred sixty billion star-bound planets may exist in our galaxy alone.[76] Of those planets discovered thus far, 207 are similar in size to Earth, 680 are super-Earth-size, 1,181 are Neptune-size, 203 are Jupiter-size and fifty-five are larger than Jupiter. Moreover, forty-eight planet candidates were found in the habitable zones of surveyed stars. The *Kepler* team estimated that 5.4 percent of all stars host Earth-size planet candidates, and that seventeen percent of all stars have multiple planets. These authors conclude:

---

72. Chow, "5 Rocky Planets Revealed."
73. Petigura et al., "Prevalence of Earth-Size Planets."
74. Choi, "New Estimate for Alien Earths."
75. "Hubble Reveals," NASA.
76. Wall, "Super-Earth Alien Planet May Be Habitable for Life."

A clear trend toward smaller planets at longer orbital periods is evident with each new catalog release. This suggests that Earth-size planets in the habitable zone are forthcoming if, indeed, such planets are abundant.[77]

The *Kepler* catalog database now holds over two hundred Earth-size planet candidates and over nine hundred that are smaller than twice the Earth's size, which makes for a 197 percent increase in these types of planet candidates, with planets larger than two Earth radii increasing by about fifty-two percent. Thirty planets in the habitable zone are near Earth in size, and the fraction of host stars with multiple candidates has grown from seventeen to twenty percent. The single most remarkable finding of the new research on extrasolar planets is that an enormous variety of systems exist—a diverse range of often-bizarre environments that is considerably broader than had usually been imagined before the first one was discovered.[78] The new results have driven important refinements to models of planetary formation and evolution. Apart from the *Kepler* findings, the *Hubble* telescope, in service since 1990, has provided unprecedented and detailed visible light images of deep space and time, and many of its observations have led to breakthroughs in astrophysics, including an accurate determination of the rate of expansion of the universe. One of the most important of those imaged, the *Hubble Deep Field* (HDF) image covers an area 2.5 arch meters across, about one 24-millionth of the whole sky, equivalent to the size of a grain of sand held at arm's length. The field is so small that very few stars in our galaxy lie within it, and almost all of the three thousand objects in the image are galaxies, the youngest and most distant known. By revealing such large numbers of very young galaxies, the HDF has become a landmark image. Three years after the HDF observations were taken, a region in the south celestial hemisphere was imaged in a similar way and named the *Hubble Deep Field South*. The similarities between the two regions strengthened the belief that the universe is uniform over large scales and provided further confirmation of the cosmological principle that the Earth occupies an ordinary region in the universe and that at its largest scales, the universe is homogeneous. In 2004 a deeper image, known as the *Hubble Ultra Deep Field* (HUDF), was constructed from a total of eleven days of observations. The HUDF image was at the time the most sensitive astronomical image ever made at visible wavelengths, showing an estimated ten thousand galaxies, until the Hubble Extreme Deep Field was released in 2012. Images from the Extreme Deep Field, or XDF, were released on September 2012 to

77. "1,901 New Kepler Candidates," NASA.
78. "NASA's Kepler Releases New Catalog—2,321 Planet Candidates," NASA.

a number of media agencies. Images in the XDF show about 5,500 galaxies within its smaller field of view, now believed to have formed in the first five hundred million years following the Big Bang.[79]

Hubble Ultra Deep Field Image (HUDF).

The *Transiting Exoplanet Survey Satellite* (TESS) was launched on April 18, 2018 and designed to search for exoplanets utilizing the transit method in a field encompassing four hundred times that of *Kepler*. TESS is designated to survey the brightest stars nearest Earth for a period of two years and will study the size, mass, density, and orbit of smaller planets hosted by these stars. As of August 2020, TESS has found sixty-six planets and nearly 2,100 planet candidates. The *James Webb Space Telescope* (JWST), three times larger than *Hubble*, is scheduled for launch in October 2021 and will be able to identify those planets with a small or intermediate mass, and through infrared spectrometry, determine the makeup of their atmospheres and hence their capacity for sustaining life. This will include imaging star-forming clusters, studying debris disks around stars, direct imaging of planets, and spectroscopic examination of planetary transits.

79. "Hubble Goes to the eXtreme," Hubble Site.

COSMOLOGY, ASTROBIOLOGY, AND MODERN ASTRONOMICAL TECHNOLOGIES   27

The JWST's unprecedented resolution and sensitivity from long-wavelength visible to the mid-infrared are designed to fulfill its two main scientific goals—studying the birth and evolution of galaxies, and the formation of stars and planets.[80] From these recent astronomical studies, it has become evident that planet formation as a consequence of stellar evolution is a relatively common phenomenon. Future projects will utilize interferometric radio telescopes in orbit around the Earth or on the far side of the moon where interference from Earth will no longer inhibit the ability of ground-based telescopes to discern extrasolar planets and their conditions. In Chile, The *Giant Magellan Telescope*, planned for the Las Campanas Observatory in southern Atacama, will enable images of the universe ten times sharper than *Hubble*; it will also be utilized in the continuing search for exoplanets, and is set to begin operation in 2023. Scheduled to become operational in 2024, the *European Extremely Large Telescope* (EELT) also in Chile, is a revolutionary new ground-based telescope, having the aim of observing the universe in much greater detail than even the *Hubble Space Telescope*. It's mirror of approximately forty two meters will allow *direct imaging* of larger extrasolar planets as well as the study of their atmospheres.[81] By means of examining elements present in the atmosphere, especially those of nitrogen dioxide and ozone, these possible signs of industrial pollution and Earth-like environments can indicate the presence of intelligent civilizations.[82] It will also perform "stellar archaeology"—measuring the properties of the first stars and galaxies, as well as probing the nature of dark matter and dark energy. The *Wide Field Infrared Survey Telescope* (WFIRST), designed by NASA and set for launch in the mid-2020s, will use a 2.4 meter wide field telescope with a 288-megapixel multi-band near-infrared camera, allowing for images comparable in sharpness to *Hubble*, but with a hundred times larger degree field of view. WFIRST primary mission will center on studying the expansion of the universe, large-scale cosmic structure, measuring dark energy, the consistency of general relativity, and the curvature of space-time. Secondly, it will contain a high contrast coronagraph which uses starlight suppression technology in order to detect planets only 0.1 parsecs from their host stars. These new telescopic technologies and resulting discoveries are leading a new era in cosmological understanding, providing new information valuable to astrophysics and the burgeoning field of astrobiology, and, as mission data continues, the possible detection of microbial, simple plant or animal life, or even intelligent life in the

80. Dressler and Frogel, "GSMT AND JWST."
81. "The European Extremely Large Telescope," ESO.
82. Rosbury and Yan, "High Resolution Transmission Spectrum," 255–66.

next few decades. Following this summary of the modern scientific astronomical and cosmological advances, we now consider the work of two key organizations where important research has been undertaken in the search and study of extraterrestrial life and its import for Christian theology—the National Aeronautic and Space administration (NASA) and the Search for Extraterrestrial Intelligence (SETI).

Section B: Science and Extraterrestrial Intelligence

At the beginning of the space age, the National Aeronautic and Space Administration (NASA) sponsored an important and exhaustive study performed by the Brookings Institution in 1960 entitled, "Proposed Studies on the Implications of Peaceful Space Activities for Human Affairs," prepared in collaboration with the Committee on Long-Range Studies of NASA, and submitted to the House Committee on Science and Astronautics of the United States House of Representatives in the 87th United States Congress. This study, now known as the *Brookings Report* has become well-known for its most often cited brief section titled, "The Implications of a Discovery of Extraterrestrial Life." In this section, the report acknowledged the (then) recent new efforts in detecting extraterrestrial messages via radio telescope by SETI, and conceded the real possibilities of the existence of semi-intelligent life as well as the likelihood of advanced intelligent life:

> It is conceivable that there is semi-intelligent life is some part of our solar system or highly intelligent life which is not technologically oriented, and many cosmologists and astronomers think it very likely that there is intelligent life in many other solar systems. While face to face meetings with it will not occur within the next twenty years (unless the technology is more advanced than ours, qualifying it to visit Earth), artifacts left at some point in time by these life forms might possibly be discovered through our space activities on the Moon, Mars, or Venus.[83]

In considering the potential impact to human society in an encounter with other life forms in space, the report referenced the resulting social consequences within historical exchanges between disparate human cultures, referring to impacts on indigenous peoples during national expansions of European colonialism and imperialism:

> Anthropological files contain many examples of societies, sure of their place in the universe, which have disintegrated when

---

83. Michael, "Proposed Studies," 182–83.

they have had to associate with previously unfamiliar societies espousing different ideas and different ways; others that survived such an experience usually did so by paying the price of changes in values and attitudes and behavior.[84]

More specifically, since the report considered within its scope the possibility of radio contact with intelligent extraterrestrials, the forecasted consequences here focus primarily on the psychological impact:

> It has been speculated, of all groups, scientists and engineers might be the most devastated by the discovery of relatively superior creatures, since these professions are most clearly associated with the mastery of nature, rather than with the understanding and expression of man. Advanced understanding of nature might vitiate all our theories at the very least, if not also require a culture and perhaps a brain inaccessible to Earth scientists.[85]

Much can be added to this statement given the inherent variables of the mode and means of contact, and whether an advanced extraterrestrial civilization would assume a cooperative, ambivalent, or other position with Earth leaders and scientists. Importantly, the report discussed religious reactions to a discovery/contact with an extraterrestrial life form or artifacts, considered in relation to those denominations with fundamentalist or literalist theologies:

> The positions of the major American religious denominations, the Christian sects, and the Eastern religions on the matter of extraterrestrial life need elucidation. Consider the following: 'The Fundamentalist (and anti-science) sects are growing apace around the world . . . For them, the discovery of other life—rather than any other space product—would be electrifying . . . some scattered studies need to be made both in their home centers and churches and their missions, in relation to attitudes about space activities and extraterrestrial life.[86]

The *Brookings Report* concludes with recommendations for further research, with a caution that given our limited knowledge of human behavior under even an approximation of unprecedented circumstances such

---

84. Michael, "Proposed Studies," 215.
85. Michael, "Proposed Studies," 103, n.34.
86. Michael, "Proposed Studies," 102, n.34. The Peters ETI (Extraterrestrial intelligence) study, referenced in the Introduction, was one such study, and clearly indicated fundamentalist doctrines to be the most challenged by discovery and/or contact with an extraterrestrial race with superior technological, cultural, and spiritual development.

as discovery/contact, study of historical analogs, as well as contemporary sociological studies should be conducted:

> Continuing studies to determine emotional and intellectual understanding and attitudes—and successive alterations of them if any—regarding the possibility and consequences of discovering intelligent extraterrestrial life.[87]

> Historical and empirical studies of the behavior of peoples and their leaders when confronted with dramatic an unfamiliar events or social pressures. Such studies might help to provide programs for meeting and adjusting to the implications of such a discovery. Questions one might wish to answer by such studies would include: How might such information, under what circumstances, be presented to or withheld form the public for what ends? What might be the role of the discovering scientists and other decision makers regarding release of the fact of discovery?[88]

Since the publishing of the *Brooking Report* in 1960, little if any of the recommended research on the religious impact to human society has been produced. A handful of scientists and astronomers have however, considered the probabilities that such intelligent civilizations exist in our own galaxy, as well as endeavored to contact intelligent extraterrestrials using radio frequency technology. The following year after *Brookings*, in 1961 astronomer Francis Drake developed what is now termed the "Drake's equation," as an attempt to formulate the probability of encountering intelligent extraterrestrial life within our galaxy. His formula consisted of a computation of statistical probabilities, and when multiplied, provide an estimate of the number of possible galaxies capable of communicating with humans. The formula proposed by Drake is $N = R^* \cdot fp \cdot ne \cdot fl \cdot fi \cdot fc \cdot L$. N indicates the number of advanced civilizations in the Milky Way galaxy whose electromagnetic emissions are detectable. R^* indicates the rate of formation of the central stars with adequate energetic properties for the development of intelligent life; fp the fraction of those that could have associated planets; ne the number of those planets with conditions suitable for life to develop; fl the fraction of suitable planets in which life actually appears; fi the fraction of those planets in which intelligent life appears; and fc the fraction that could develop the technology to indicate their presence to other civilizations. The last factor, L, regulates the "average life" of a technological civilization on a

---

87. Michael, "Proposed Studies," 216.
88. Michael, "Proposed Studies," 216.

planet capable and interested in emitting signals of presence. The estimates for N are very diverse. According to Drake's original calculation that resulted in an approximate value of N=100,000, other scientists have proposed a value of N = 100, and still others have speculated there be only one active technologically advanced civilization for every 300 galaxies.[89] As some have correctly remarked, the equation lacks a realistic model of scientific basis in which to perform the proper calculation. As of yet astronomers are just beginning to figure the fraction of stars which possess planets, based on the discovery of exoplanets gleaned primarily from *Kepler* data, but remain without a proper model of the parameters governing planetary formation from rotating star clouds. Latest calculations of the equation from NASA and the European Space Agency indicate R, the rate of star formation in our galaxy of 1.3 to 5 per year;[90] fp, those systems having associated planets, and analysis of microlensing surveys indicate this value approaching 1,[91] meaning on average, most stars host planets. Based on new Kepler readings, it has been estimated there are forty billion Earth-sized planets in the galaxy orbiting the habitable zone of their host stars.[92] Given there are about one hundred billion stars in the galaxy, $fp \cdot ne$ has been assigned a value of about 0.4. Therefore, only the first two of its variables (R and fp)[93] can be established observationally at this time, with the remaining conjectured and fall beyond our current technology and observational abilities.

Drake later became one of founding members of SETI institute (The Search for Extraterrestrial Intelligence), begun in 1984, which searches for radio signals from potential technologically advanced alien civilizations via radio telescopes in the centimeter and decimeter wavelengths. The SETI program utilizes the neutral hydrogen electromagnetic radiation spectral line at 21 cm (1420 MHz), chosen as a reference point for other technological civilizations given hydrogen's intensity and wide diffusion throughout the universe. SETI received its conceptualization from the suggestions of Cocconi and Morrison,[94] in the hopes of receiving a signal from an intelligent alien civilization outside our solar system. The hypothesis among radio astronomers is many technological civilizations may populate our galaxy and

89. *Sydney Morning Herald*, "Life on Other Planets?"

90. Robitaille and Whitney, "The Present-Day Star Formation Rate," Vol. 710 (2010), L11-L15.

91. Palmer, "Exoplanets Are Around Every Star."

92. Petigura et al., "Prevalence of Earth-Size Planets," 19273-78.

93. The Kepler telescope observational data indicate that one out of ten stars hosts a planet; therefore, *fp* would be the first variable to be determined (at 0.1). Cf. Batalha, "Exploring Exoplanet Populations with NASA's Kepler Mission," 12647-54.

94. Cocconi and Morrison, "Searching for Interstellar Communications," 844-46.

the universe, and would have a communication network through which they may initiate and receive messages from other civilizations with similar radio telescope technology. Begun at the dawn of radio-astronomy, and formerly in collaboration with NASA, it now operates on a budget sustained by researchers and private funding sources. Presently, terrestrially produced radio waves beginning in the early twentieth century from commercial, industrial, and government broadcasting have unwittingly been sent into interstellar space to a spherical distance of approximately eighty light years. Having at present received no messages, and given the distance transmitted and time for a signal to travel back to Earth, SETI researchers assume there may be no intelligent life (or ability or desire to respond) within a distance of forty light years. Radio transmissions in binary code were sent from the *Aricebo* radio telescope in Puerto Rico towards the center of the galactic globular cluster M13 in 1974, decodable in a black and white image containing information on Earth and human biology. No responding message has been received to date. Future envisioned SETI projects in this century include interferometric radio telescopes in orbit around Earth and on the dark side of the moon, in order to block Earth signal-noise and increase the power of resolution and sensitivity for receiving extraterrestrial signals.

The SETI Academy Committee of the International Academy of Astronautics dedicates some of its work to the study of the social and cultural consequences of contact with intelligent extraterrestrial civilizations, and creates protocols if communication is established. Noted research is this area has been conducted by Vakoch,[95] Harrison and Dick,[96] Tough,[97] and Michaud.[98] However, these studies have not seriously nor systematically considered the particular religious implications of discovery/contact. It is noteworthy that SETI's contingency plans only include a contact scenario entailing messages received by long-distance messaging through radio telescopes. The possibility of a meeting or discovery of any other contact mode or scenario is categorically dismissed by almost all SETI researchers given several anthropocentric assumptions: the vast distances required for an extraterrestrial race to travel to our solar system would pose the same difficulties as our current understandings of physics and space travel dictate; limited extraterrestrial motivations for journeys to remote locations such as our Earth; and the biological characteristics of potential visitors. Extraterrestrial

---

95. Vakoch, "Reactions to Receipt of a Message," 737–44.

96. Harrison and Dick, "Contact," 227–57. Dick's *Many Worlds* provides an important contribution to some of the fundamental religious concerns in a contact event.

97. Tough, *When SETI Succeeds*.

98. Michaud, *Contact with Alien Civilizations*.

races thousands, tens of thousands, or hundreds of thousands of years more advanced could use methods of travel completely foreign to us given our current understanding of physics, just as our modern space travel technologies would have been inconceivable to humans even one hundred years ago. Our notions of extraterrestrial motivations for contact could, in many cases, be psychological projections of our own needs and fears based upon our own biology, social and cultural structures, and human history. That is not to say speculative scenario analysis is not valuable, however given that we know little other than human conjectures and extrapolations we cannot assume a specific mode of contact and agenda. Extraterrestrial beings (especially those who may initiate contact) could have life expectancies far advanced from that of humans, allowing for lengthy excursions into space, and/or use generations space vessels, capable of traversing large spatial distances which allow for varying types of contact with other races.

Earlier speculation on the potentialities of the existence of extraterrestrials in the late 1940s led to the question "Where are they?" by Italian physicist Enrico Fermi at Los Alamos National Laboratory in 1950, when discussing the lack of evidence given the high probabilities of advanced civilizations in our galaxy. Theorists in support of Drake have argued that if any one extraterrestrial technological civilization began deploying slower than light vessels that would settle on host planets and develop colonies, and in turn a few generations later send out other vessels to other habitable places within the galaxy, that civilization could have colonized all the habitable planets in the galaxy. Therefore, where is the evidence of this colonization? Many solutions to the problem of lack of evidence have been offered, and most recently, Steven Webb compiled seventy-five potential answers to the question.[99] These can be summed up in three main categories: intelligent civilizations are rare, where the Drake values for $R^* \cdot f \cdot n_e \cdot f_l \cdot f_i$ have much lower values than have been argued by proponents; there exist intelligent civilizations, however they may be too distant to communicate or are unable or uninterested in communicating with humans; or the lifespan of advanced civilizations is shorter than projected, either by self-destruction, by natural events, or other unknown reasons. There remains the possibility that alien intelligences prefer to interact with other societies furtively, where humans are unable to detect their presence. Spatially, if there were a million such civilizations in our galaxy alone, they would be separated by an average of one hundred light years, clearly providing challenges for communication among certain species.

---

99. Webb, *If the Universe Is Teeming with Aliens . . . Where is Everybody?*

We now turn to an examination of the history of philosophical and scientific explorations of possible extraterrestrial life, from the rudimentary telescopic observations of Kepler, Brahe, Galileo, to the most recent and proposed future space missions of NASA. Within the chronology of the development of observational techniques, the earlier philosophical and metaphysical conjecture of the early scientific age gradually was replaced with scientific rigor and discipline.

Section C: Ancient and Early Medieval Philosophy on the Plurality of Worlds

This section will consider the historical thought pertaining to the concept of a plurality of worlds [hereafter just 'plurality' and later equated with the possibility of extraterrestrial beings] from Greek antiquity to the beginning of the twentieth century—beginning as a philosophical concept, to a medieval metaphysical construct and theological possibility, to scientific probability in the late seventeenth century. As indicated in chapter 1, the Roman Catholic Church has been prominent in considering this question since the Middle Ages. Upon entrance into the scientific age, the subject of other worlds containing intelligent extraterrestrial life continued as a matter of philosophical inquiry as well as an object of scientific endeavor through advances in astronomy and space research, as well as major advances in cosmology due to greatly enhanced observational technologies. The focal point of this discussion will be to understand the philosophical, metaphysical, theological, and scientific arguments and evidences first for a plurality of worlds; and later in the history, the possibility of extraterrestrial life. It will discuss how these are relevant to the twenty-first theological discussion of intelligent extraterrestrials, presented in chapter 3.

The question of other worlds and their possible inhabitants is not a twentieth-century idea as many may assume, but rather extends back to ancient Hellenistic philosophy, received its Christian formulations in the medieval period, and later support by the sciences at the beginning of the seventeenth century. The plurality of worlds debate (inhabited or uninhabited) has been well documented by Dick,[100] Crowe,[101] Darling,[102] and Angelo.[103] The ontology of the early Greek atomistic philosophy argued a

---

100. Dick, *Plurality of Worlds*; Dick, *The Biological Universe*.

101. Crowe, *The Extraterrestrial Life Debate, 1750–1900*; Crowe, *The Extraterrestrial Life Debate, Antiquity to 1915: A Source Book*.

102. Darling, *The Extraterrestrial Encyclopedia*.

103. Angelo, *The Extraterrestrial Encyclopedia*.

plurality, developed first by Leucippus, (fl. 480 BC), and his student Democritus (c. 460–370 BC), two of the most radical innovators of philosophical and early scientific thought, consisted of a universe with an infinite quantity of atoms, each infinitesimal, indestructible, and eternal, too small to be seen and in perpetual motion. Given the infinite expanse of the universe, these atoms were likewise capable of an infinite number of combinations, which Leucippus believed must result in an infinity of worlds. Diogenes Laertius, a compiler in the early Christian era, made an early reference to the earliest atomist: "Leucippus holds that the whole is infinite . . . part of it is full and part void . . . Hence arise innumerable worlds, and are resolved again into these elements."[104] Another teacher of atomist thought, Epicurus (341–270 BC), modified and presented the atomist argument in its final classical form, and was first to consider the possibility of the existence of intelligent life outside Earth. In his "Letter to Herodotus" he wrote:

> There are infinite worlds both like and unlike this world of ours. For the atoms being infinite in number, as was already proved, are borne on far out into space. For those atoms which are of such nature that a world could be created by them or made by them, have not been used up either on one world or a limited number of worlds, nor again on worlds which are alike, or on those which are different from these. So that there nowhere exists an obstacle to the infinite number of worlds.[105]

In his study of ancient atomism, British historian Cyril Bailey concluded that the infinity of worlds concept was "practically a direct deduction from the infinity of the universe."[106] He considered this a classical Epicurean inference, where a hypothesis is considered valid given the absence of data to the contrary. Lucretius (99–55 BC), a later Roman expositor of Greek atomism, argued even more strongly the existence of these worlds as a necessary principle of atomist thought:

> Now since there is illimitable space in every direction, and since seeds innumerable in number and unfathomable in sum are flying about in many ways driven in everlasting movement, it cannot by any means be thought likely that this is the only round Earth and sky that has been made, that all those bodies of matter without have nothing to do: especially since this world was made by nature . . . wherefore again and again I say you

---

104. Kirk and Raven, *The Presocratic Philosophers*, 403, 409–10. Kirk and Raven date his floruit at around 440–435 BC.
105. Bailey, *Epicurus*, 25.
106. Bailey, *The Greek Atomists and Epicurus*, 361.

must confess that there are other assemblages of matter in other places, such as this which the air holds in greedy embrace.[107]

Furthermore, Lucretius asserted that if atoms of matter can potentially be affected by the motion of others, all matter would realize its potentialities in an infinite variety of forms; it could thus produce within the infinity of matter an infinite cosmos. This argument can be seen as an early form of Lovejoy's "Principle of Plenitude": "[W]hen abundant matter is ready, when space is to hand, and no thing and no cause hinders, things must assuredly be done and completed."[108] This new argument or principle—that if something can potentially affected, then it must be so affected[109]—was subsequently invoked in theistic terms[110] and proved a foundationally positive argument throughout the remaining history of the plurality of worlds debate. However, the atomists were not preoccupied with the possibility of extrasolar life, rather they wished to develop a system which accounted for observation and which also supported a definitive cosmology and cosmogony. The reemergence of the plenitude concept in the form of Epicurean atomism in the seventeenth century, as one of the possible responses to Copernicanism demonstrated once again its utility, and later was proven the basis for atomic physics. Being the first system to necessitate the idea of a plurality within its cosmogony, it became in important tradition by which later thinkers would consider the plurality. However, in antiquity atomism did not gain a large cohort of believers, as it was largely overshadowed by Aristotle's system throughout the Middle Ages, and its credibility was further weakened in theological circles due to its strong atheistic tendencies. Aristotle in *De Caelo* and those following his thought argued against a plurality of worlds, envisioning a heavenly sphere characterized by eternity, immutability, and incorruptibility—thus making them uninhabitable places for creatures such as extraterrestrials. In *De Caelo*, Aristotle rejected atomism and asserted his theory of the four primary elements of Earth, air, fire, and water—originally borrowed from Empedocles.[111] He theorized that if one accepts that all the elements of another world are of the same form and nature as ours, it would be impossible for them to exist due to his principle of natural place:

> The same rule must apply to all, since all alike exhibit formal identity with each other but numerical individuality. My

107. *De rerum natura*, Book II, 1067–69.
108. *De rerum natura*, Book II, 1067–69.
109. Lovejoy, *The Great Chain of Being*.
110. The notion "that God does not waste space," or that the full power of the Creator must be realized in his creation.
111. Sarabji, *Simplicius*, 157–59.

meaning is this, that if the relation of particles in this world to each other and their relation to those in another world are the same, then any given particle from this world will not behave otherwise towards the particles in another world, than towards those in its own, but similarly; for in form they do not differ one another at all.[112]

In other words, regardless of its distance from the particles of our world, every particle of the same form must tend towards the center of our Earth, which is the center of the world, and thus eliminate the possibility of other worlds. For Aristotle, the idea of a space with even two worlds leads directly to a contradiction, as the elements of Earth would move violently from one world, and naturally with respect to the other. He illustrated:

> It must be natural therefore, for the particles of Earth in another world to move towards the center of this one also, and for the fire in that world to move towards the circumference of this. This is impossible, for if it were to happen the Earth would have to move upwards in its own world and the fire to the center; and similarly Earth from our own world would have to move naturally away from the center, as it made its way to the center of the other, owning to the assumed situation of the world relative to each other. Either, in fact, we must deny that the simple bodies of the several worlds have the same natures, or if we admit it we must, as I have said, make the center and the circumference one for all; and this means that there cannot be more than one.[113]

As the atomists believed they had naturally deduced the necessity of a plurality of worlds as a result of an infinity of atoms, a void, and random motions, Aristotle's theory of elements and his assumptions of natural motion and natural place provided the basis for his rejection of an infinity of worlds, let alone a plurality. Aristotle supported his denial of a plurality from what he called his "arguments of first philosophy," taken from his *Metaphysics* that a plurality of worlds would necessitate a plurality of first movers.[114] To illustrate his conception of the universe, Aristotle adopted the Pythagorean two-sphere universe devised in 540 BC which consisted of the Earthly realm as a central sphere containing the four elements. The celestial realm, that of the area between the sublunar sphere and the outer sphere of stars, was the realm of the ethereal (a fifth element) containing the Moon

---

112. *On the Heavens*, I, 8, 227a 1.
113. *On the Heavens*, I, 8, 276b 12.
114. Cornford, *Plato's Cosmology*, 277–78.

and planets Mercury, Venus, Sun, Mars, Jupiter, and Saturn, each attached to translucent spheres which rotated about the Earth.

Peter Apian's *Cosmographia*. Renaissance woodcut illustrating the Ptolemaic system, 1524.

The Aristotelian rational cosmic system of the antiquity dominated the Mediterranean world from the fourth century BC to the third century AD, combining philosophical, scientific, and ethical principles. After its translation and presentation to the Latin West in the Middle Ages, and in combination with the scriptures synthesized by Aquinas, Aristotelian thought comprised the core of knowledge available to philosophers and theologians. The cosmology of *De caelo* went essentially unchallenged for a century following its introduction, until Paris Godefroid of Fontain, Henry of Ghent, and Richard of Middleton argued for the possibility of a plurality, each world determining its own natural motions.[115] Christianity in the Middle Ages did not implicitly reject the notion of a plurality of worlds, according to natural theology, in that it allowed for the possibility for God to create *ex nihilo* according to his unfathomable will. Indeed the Church, while embracing the Aristotelian geocentrism insisted that while God had the power to produce many worlds, he did not do so—to say otherwise was to deny God's omnipotence.

In 1543, shortly before his death, Polish astronomer-priest Nicolaus Copernicus's life's work, *Revolutions of the Celestial Spheres*, was printed. Copernicus was dismayed at the loss of the Platonic ideal of perfect circular motion, which had been accepted by Aristotle, but abandoned when Ptolemy introduced equants. His radical heliocentric theory positioned a motionless Sun near the center of the universe, with Earth and other planets rotating at uniform velocities in circular paths, modified by epicycles to explain planetary retrogression. Further, he reintroduced other innovations of the Ptolemaic system, including three distinct motions of the Earth: daily rotation, annual revolution, and annual tilting of its axis; including calculations minimizing the Earth's distance from the Sun compared to the stars.[116] Heliocentrism was initially not accepted by most astronomers, requiring further epicycles to explain the motion of the heavenly bodies, and hence did not match the precision achieved by Ptolemy. Copernicus's methodology included the compilation of millennia-old Greek astronomical thought and proposed a heliocentric universe, transformed into the infinite and centerless Cartesian universe in the seventeenth century, and later into the Newtonian universe in the eighteenth century. However, his primary aim was not to define the universe as a possible abode for extraterrestrials, but rather to provide an accounting for observed phenomena by way of a system based on consistent principles and hypotheses and terrestrial and planetary motion. At the beginning of the seventeenth century we come to the end of the purely philosophical speculation on a plurality of worlds with the advent of the first crude telescopes developed in the Netherlands and improved by Galileo and others.

History of Scientific Searches for Extraterrestrial Life

The development of the optical telescope for astronomical observation during the seventeenth century led to a resurgence in the theme of a plurality of worlds, given the multitudes of new stars which were revealed.

---

115. Duhem, *Le systéme du monde*, vol. IX, 373–78.
116. Kuhn, *The Copernican Revolution*.

Johannes Kepler, (1571–1630), German mathematician and astronomer, using the accurate observations of Tycho Brahe, gave credence to Copernican theory and allowed him to develop his laws of planetary motion and hence a more accurate model of how the solar system functioned. Kepler reviewed the first lunar and stellar observational reports of Galileo Galilei (1564–1642), published in his *Siderius Nuncius* in 1610. Within it, Kepler saw details of the Moon, the Jovian moons, and a multitude of stars. In response, Kepler speculated regarding our relation to possible inhabitants on the moon and elsewhere in the universe:

> Well, then someone may say, if there are globes in the heaven similar to our Earth, do we vie with them over who occupies a better portion of the universe? For if their globes are nobler, we are not the noblest of rational creatures. Then how can all things be for man's sake? How can we be the masters of God's handiwork?[117]

Kepler, in his *Dissertatio cum nuncio sidereo* conjectured the possibility of extrasolar inhabitants, including those on the Moon as well as the visible planets and their moons, although he insisted on the primacy of Earth's inhabitants.[118] Later, in his posthumous *Somnium* (1634), Kepler detailed his belief in inhabitants on his favorite astronomical object, the Moon, as well as the other planets and their satellites visible at that time. In his defense of Copernicanism, Kepler argued a lunar observer would consider the Moon motionless among the stars just as terrestrial observers do—calling into question the accuracy of observation. It was his belief that the Copernican system implied the existence and motion of other planets similar to Earth, and that the doctrine of heliocentrism would be accepted only when a plurality of worlds was accepted.[119] He postulated a physical system incorporating a plurality of worlds stemming from heliocentrism where the Earth, Moon, planets, and stars shared in common elements, motion, and purpose. For his part, while Galileo was reticent to consider the idea of other inhabitants, he realized the heliocentric system put the Earth in a position possibly similar to other planets of the solar system, which later brought him into conflict with Church authorities due to the theological implications related to the special creation of Earth as the center of God's attention.[120]

---

117. Rosen, *Kepler's Conversation with Galileo's Sidereal Messenger*, 43.
118. Rosen, *Conversation*, 42.
119. Dick, *Plurality of Worlds*, 8.
120. McMullin, "Galileo on Science and Scripture," 271–347.

In the latter half of the seventeenth century, Bernard le Bovier Fontenelle (1657–1757), a French poet, dramatist, and later member of the French Academy, wrote *Entretiens sur la pluralité des mondes* (1686). A critically important work that attracted a large audience, *Entretiens* tied the existence of innumerable planets to the image of Cartesian vortices, or planetary motion as the result of circulating bands composed of atom-sized globules and debris, whereupon were lodged planets which rotated around their host suns.[121]

Fontenelle expresses his concept of a plurality tied to these vortices:

> If the fix'd Stars are so many Suns, and our Sun the center of a Vortex that turns round him, why may not every fix'd star be the center of a Vortex that turns round the fix'd star? Our Sun enlightens the Planets; why may not every fix'd Star have Planets to which they give light?[122]

The frontispiece to the first and other subsequent editions of the *Entretiens* provided an illustration depicting a multitude of planets orbiting fixed stars, showcasing Fontenelle's belief in a plurality of planetary systems.

---

121. Descartes, *Principles of Philosophy*, 42–93.
122. Fontenelle, *Entretiens sur la pluralite des mondes*, 125.

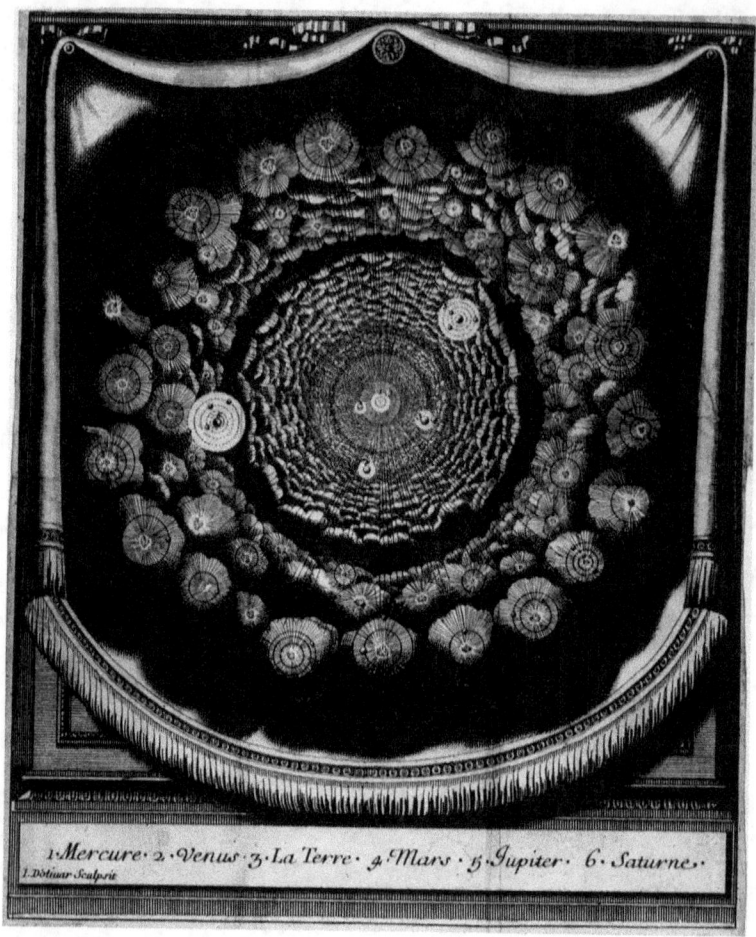

Frontispiece of Fontenelle's *Entretiens*.

Nonetheless, he made efforts to avoid conflict with Christian doctrine by ignoring religious issues, instead concentrating his arguments on varying planetary conditions and their comparisons to Earth. Fontenelle's colloquial promotion of the Cartesian vortices, joined to the new Copernican astronomy, and cast within the expanse of the plurality of worlds concept had high appeal to the masses, and popularized the idea of a plurality in France and England. A second important writing of the period to consider the plurality was the posthumous work of prominent Dutch mathematician and astronomer Christiaan Huygens (1629–1695). His *Kosmotheoros, sive de terres coelestibus earumque ornatu conjecturae* (1698), was translated into several languages. Huygens is best known for his accurate explanations of the rings of Saturn and his early (albeit later to be determined inaccurate) non-periodic theory of light.[123] Huygens was educated in Cartesian thought all his life but never accepted the Cartesian cosmology to the extent Fontenelle did:

> All that I shall do more is to add somewhat of my opinion concerning the world, as it is a place for the reception of the Suns or fix'd Stars, every one of which I have show'd may have their planetary systems about them. I am of the opinion that every Sun is surrounded with a whirlpool or vortex of matter in very swift motion; tho not in the least like Cartes's either in their bulk or manner of motion.[124]

In *Kosmotheoros* the idea of a plurality did not depend entirely upon the plausibility of the Cartesian vortices. Huygens formulated his evidence, "[T]o reason from what we see and are sure of, to what we cannot . . . wherein from the Nature and Circumstances of that Planet which we see before our eyes, we may guess at those that are farther distant from us."[125] Huygens indicated our Sun as merely a star in close proximity, and explained with his limited scientific understanding (what we now term ecospheres or habitable zones around each star) considered how heat, water, and air were necessary for life on any given planet. From these variables, he extrapolated the possibility of extraterrestrials on the visible planets based on these judgments of planetary habitability. Using Earth as a model, he argued for the existence of extraterrestrials from three points: the uniformity and diversity of nature, the Copernican likenesses between the planets and Earth, and similar to Cusa, Bruno, and Huygens as we will see in the next section, the concept of plenitude.[126]

In 1687, Isaac Newton published his first edition of *Principia Mathematica*, which abandoned the Cartesian vortex cosmology in favor of his natural philosophy, in which he insisted that God had produced the kinetics of the solar system, although he clung to the basics of atomist thought (revived by Walter Charlton)[127] as regards the infinity of atoms.[128] This can be highlighted in his letters of 1692–1693 with English theologian Richard Bentley, where he argued a theistic atomist cosmogony: that an infinity of matter with an

123. Leiter and Leiter, *A to Z of Physicists*, 108.
124. Huygens, *The Celestial Worlds Discover'd*, 149–57.
125. Huygens, *The Celestial Worlds Discover'd*, 18–19; Huygens, *Oeuvres*, 699.
126. Huygens, *The Celestial Worlds Discover'd*, 21–23; Huygens, *Oeuvres*, exp. 703–7.
127. Charleton, *Physiologia Epicuro-Gasssendo-Charltoniana*.
128. Isaac Newton, *Le Traite 'De l'infini' de Jean Mair*.

active God in the universe would make such a thing possible.[129] There are several important direct references to Newton's advocacy of extraterrestrials: A passage from "General Scholium" of his 1713 edition of *Principia* contains the hypothetical statement, "If the fixed stars are the centers of other like systems, these, being formed by the like wise counsel, must all be subject to the dominion of One."[130] A second passage, published by David Brewster in his 1855 biography of Newton related to the final judgment:

> [Christ] ... will give up his kingdom to the Father, and carry the blessed to the place he is now preparing for them, and send the rest to places suitable to their merits. For in God's house (which is the universe,) are many mansions, and he governs them by agents which can pass through the heavens from one mansion to another. For if all places to which we have access are filled with living creatures, why should all these immense spaces of the heavens above the clouds be incapable of inhabitants?[131]

A third reference comes from a record made by John Conduitt, husband of Newton's niece, two years prior to Newton's death. In conversation, Newton conjectured that distant stars can serve as suns for host planets containing extraterrestrial inhabitants, given the providence of the Creator, and that these beings may be demigods, angels, or other type of life.[132]

For Newton, the creation of our solar system, as well as that of other stars systems was impossible without an active God, and the concept of other world systems were a non-integral part of the principles of his mechanistic philosophy, of which he remained an uncommitted believer. Following Newton, the continuing expansion of astronomical understanding during the latter seventeenth century stimulated further speculation on extraterrestrial life by several well-known astronomers and scientists. The first scientist of repute in the eighteenth century to support publicly the idea of a plurality was William Herschel (1738–1822), a British astronomer and composer, who is best known for his formulation on the structural characteristics of the galaxy by means of his study of the distribution of stars using astronomical spectrophotometry, the discovery of

---

129. Mott, *Sir Isaac Newton's Mathematical Principles*, 544.

130. Mott, *Sir Isaac Newton's Mathematical Principles*, 544.

131. Brewster, *Memoirs of Life, Writings, and Discoveries of Sir Isaac Newton*, vol. II, 354. Crowe, in his *The Extraterrestrial Life Debate, 1750–1900*, 25, points out that Brewster did not indicate that in Newton's manuscript the portion of the statement following the words "many mansions" is not italicized and crossed-out, and in place is written, "We are also to enter into societies by Baptism & laying on of hands & to commemorate the death of X in our assemblies by breaking of bread."

132. Turnor, *Collections for the History*, 172–73.

Uranus, and of infrared radiation. He was highly devoted to the plurality of worlds concept, and defended his convictions in several papers published between 1783 and 1784. He commented that Mars and the Moon were not unlike the Earth in climate, and even speculated that the Sun was populated, as he believed that its surface was cool and its heat produced only upon sunlight entering Earth's atmosphere.[133] Further, he postulated that every distant star hosted a multitude of populated planets:

> [S]ince stars appear to be suns, and suns, according to popular opinion, are bodies that serve to enlighten, warm, and sustain a system of planets, we may have an idea of numberless globes that serve for the habituation of living creatures.[134]

Herschel's support of a plurality of worlds has been termed purely metaphysical,[135] and he answered objections to his odd belief in solar inhabitants by means of analogical comparisons between the surfaces of the moon and Earth, each having their own stable habitats. More accurately, and eventually more scientific, he defended the plurality theory according to the adaptability of creatures to their specific environments, and the uniformity and diversity of nature (the latter in accord with Huygens).[136] Hershel's main supportive argument for his belief in a plurality of worlds was likely his attraction to James Ferguson's[137] claims of lunar life, which he sought to detect directly. His ambiguous early observations compelled him to build larger and more accurate telescopes in efforts to confirm his pluralist hypotheses.

Following Herschel, the major scientific figure in support of inhabited worlds was French Astronomer Camille Flammarion (1842–1925), who main work, *La pluralite des mondes habites* was published in 1862. Exposed to the thought of Darwin and Lamarck and spiritualist churches sprouting throughout Europe as a youth, he later abandoned his Catholic beliefs due to its supposed irreconcilability with Copernicanism,[138] becoming a devotee of trans-planetary reincarnation as promoted by Jean Reynaud in his *Terre et ciel* (1854).[139] Reynaud's doctrine of indefinite perfectibility, that is, souls passing

---

133. Basalla, *Civilized Life in the Universe*, 52.
134. Crowe, *The Extraterrestrial Life Debate, 1750–1900*, 66.
135. Holden, *Sir William Herschel*, 149.
136. Crowe, *The Extraterrestrial Life Debate, 1750–1900*, 66.
137. Scottish astronomer (1710–1776), contemporary of Herschel who wrote *Astronomy Explained upon Sir Isaac Newton's Principles*. Ferguson espoused a strong pious pluralism with an inhabited solar system and universe.
138. Flammarion, *Mémoires*, 168–88.
139. Reynaud, *Terre Et Ciel*, 48–49. Reynaud sought to establish a religious system,

upon death from the terrestrial to higher states, from planet to planet for eternity, left Flammarion "obsessed by life after death, and on other worlds, and [he] seemed to see no distinction between the two."[140] Following the thought of Reynaud, Flammarion's Darwinism was applied on a cosmic scale, where he argued "all planets would attain life in due time."[141]

Flammarion Engraving.

His passionate belief in a plurality was nurtured by authors as Fontanelle, Huygens, and Brewster, and his books were extraordinarily popular on both sides of the Atlantic. His *Le pluralite* reprinted thirty-three times. His most popular work, *Astronomie populaire* sold in excess of one hundred thousand copies, where he professed his evolutionary ideas and the likelihood of the superiority of inhabitants on celestial bodies over those of Earth.[142] Arguing for a plurality of worlds inhabited by a variety of beings

---

compatible with Catholicism, accommodating a pluralist and reincarnationist theology. It was rejected by a council of bishops in 1857.

140. Herrick, *Scientific Mythologies*, 56.
141. Herrick, *Scientific Mythologies*, 56.
142. Herrick, *Scientific Mythologies*, 57.

in body and immaterial spirits in evolution, he can be considered a double-pluralist: "Plurality of worlds; plurality of existences: these are two terms which complement and illuminate each other."[143] Flammarion is most noted in the development of the idea of plurality of worlds for his metaphysical presuppositions rather than scientific observations, plenitude arguments, or use of natural theology. He was heavily criticized by W. H. M. Christie, Astronomer Royal of England, in his tendency "to mix fact and fancy" and his "hasty inferences from doubtful observations."[144] However, it can be argued that his linking of metaphysical ideas to a "cosmic" Darwinism supported the idea of varied planetary conditions resulting in diverse extraterrestrial forms, (both adopted by astrobiology), and final causality. He was also (somewhat paradoxically given his predilection for conjecture) a devoted advocate for the science of astronomy through his Flammarion societies.[145]

A contemporary of Flammarion, Italian astronomer and science historian Giovanni Schiaparelli (1835–1910) provoked controversy. A strong believer in a plurality, he reported visualizing *canali*, or "channels" on the Martian surface, drawn originally by Angelo Secchi (1818–1878), a Jesuit astronomer.[146] The English translation of *canali* as "canals" was picked up by the press, giving rise to a variety of hypotheses, speculation and folklore about the possibility of intelligent life on Mars.[147] Scientists Camille Flammarion and Percival Lowell, among others accepted the artificial canal hypothesis, and although Schiaparelli's claims were later disproved as the effect of pareidolia,[148] many works of fiction and nonfiction emerged from 1877 and thereafter, speculating on the possibilities of intelligent life there, including Edgar Rice Burroughs's Human-Martian hero John Carter. The writings of Schiaparelli, in *La via sul pianete Marte: tre scritti*

---

143. Flammarion, *La pluralite des mondes habites*, 324.

144. Crowe, *The Extraterrestrial Life Debate, 1750–1900*, 384.

145. Meetings of amateur astronomers and those interested in astronomical matters known to exist in French provinces, South America, and parts of Europe.

146. Secchi remarked on the possibility of extraterrestrial life, "It is absurd to claim that the worlds surrounding us are large, uninhabited worlds and that the meaning of the universe lies just in our small, inhabited planet." Poe, *To Believe or Not to Believe*, 264.

147. Washam, "Cosmic Errors."

148. The psychological phenomenon resulting from an ambiguous or arbitrary sound or image perceived as having order, definition and human significance. Regarding the martian "canals," Eugène Antoniadi at Meudon Observatory with a eighty-three cm telescope at the 1909 opposition showed no canals; clearer photos taken at the new Baillaud dome at Pic du Midi Observatory in 1909 also brought discredit to the canal hypothesis; the 1965 NASA Mariner 4 surveyor mission provided final confirmation of the absence of canals.

su Marte e I marziani, with Flammarion and others, created the cultural mindset whereby inhabitants of other worlds were termed "Martian."[149] The opinion of a non-astronomer on the question of the plurality, Alfred Wallace (1823–1913), British naturalist and parallel discoverer of natural selection, was published in his *Man's Place in the Universe: A Study of the Results of Scientific Research in Relation to the Unity or Plurality of Worlds* in 1903. In contrast to many of the aforementioned pluralists, he argued in favor of an anthropocentric universe, concluding Earth to be the single planet in the solar system capable of supporting life due to its unique habitat with liquid water—known today as the "Rare Earth" argument.[150] He supported this conclusion by criticizing the claims of Schiaparelli and Percival Lowell that Martians had constructed canals by performing his own research of the Martian climate and atmospheric conditions, and concluded by means of spectroscopic analysis that Lowell had overestimated Mars' surface temperature, rendering liquid water on the surface impossible, let alone any large-scale irrigation.[151]

The concept of the plurality of worlds and extraterrestrial inhabitants endured with the beginning of the twentieth century. It saw the theme of life in the cosmos, for millennia the subject of philosophical speculation (and during the previous century still rife with metaphysical presupposition and unreliable observational equipment and techniques), supplanted as a purely scientific endeavor by way of improvements in telescopic technologies, space research, and advances in astrophysics. However, the concept of the discovery of intelligent life beyond Earth continued to be served by the powerful motivators of anthropology, philosophy, and religion, however implicit. Eminent astronomer Paul Davies has remarked on these aspects as humanity ventured deeper into space in search of life:

> The powerful theme of alien beings acting as a conduit to the Ultimate—whether it appears in fiction or as a seriously intended cosmological theory—touches a deep chord in the human psyche. The attraction seems to be that by contacting superior beings in the sky, humans will be given access to privileged knowledge, and that the resulting broadening of our horizons will in some sense bring us a step closer to God. The search for alien beings can thus be seen as part of a long-standing religious quest as well as scientific project. This should not surprise us. Science began as an outgrowth of theology, and all

---

149. Sindoni, *Esistono gli extraterrestri?*
150. Morris, *Life's Solution*, ch. 5.
151. Slotten, *The Heretic in Darwin's Court*, 474.

scientists, whether atheists or theists, and whether or not they believe in the existence of alien beings, accept an essentially theological world view.[152]

One of the most notable figures of this new age of astronomy, Edwin Hubble (1889–1953), the discoverer of galaxies outside our own in 1923 within an expanding universe due to red shift, in 1929 confirmed Belgian Monsignor Georges Lemaître's Big Bang theory of the origin of the universe, which fundamentally altered the scientific perspective of the cosmos and the possibilities contained wherein.[153]

With the establishment of NASA in 1958 by President Dwight Eisenhower, early missions of Mercury, Gemini, and Apollo manned space flights culminated with the moon landing in 1969. Many following missions have been explicit searches for extraterrestrial life, simple or complex, on the planets of our solar system: The space probes *Pioneer* 10 and 11, launched in 1971 and 1973 became the first to venture outside our solar system,[154] containing a message on a plaque with the image of a human couple and some coded scientific data, in the hopes of a future interception by intelligent extraterrestrials. *Pioneer 11* lost contact with Earth in 2005 due to reduced battery power and its remote distance, and similarly, *Pioneer 10*'s last transmission was in 2003, although it continues its journey into space.[155] The *Voyager* probes, launched in 1977 to study the outer solar system, included a twelve-inch golden phonograph record containing pictographs embodying an interstellar message, Earth sounds, and a symbolic map for possible aliens to trace its terrestrial origin. As of August 2012, *Voyager 1* became the first man-made object to enter interstellar space,[156] and *Voyager 2* was expected to reach the same frontier by 2016.[157]

By the mid-century Mars became a primary object of interest for exploration by both the Soviets and the U.S., due to the earlier discovery of polar ice caps by Herschel and its proximity to Earth. It was first explored with the surveyor probe *Mariner 4* which showed the first close-up images of Mars, and then the *Mariner 6* and *Mariner 7* flyby missions from 1964–1971. *Mariner 9* achieved orbit above Mars along with Soviet orbiters *Mars 2* and *3*. The first successful soft surface landings came with *Viking 1* and 2 in 1976 and *Viking 3* in 1977, which provided the first

152. Davies, *Are We Alone?*, 137–38.
153. Moore, "Life in the Universe."
154. Fimmel et al., *Pioneer Odyssey*.
155. "The Pioneer Missions," NASA.
156. "Voyager Enters Interstellar Space," NASA.
157. Grant, "At Last, Voyager 1 Slips into Interstellar Space."

direct surface images, and included retrieval of data on any biosignatures, and meteorologic, seismic, and magnetic properties.[158] The later 1997 rover reconnaissance missions starting with the probe *Pathfinder* proved a successful airbag landing system, exploited in the subsequent rover missions.[159] Both the *Viking* and *Pathfinder* missions sought to discover evidence of extraterrestrial life forms, of which none were found. More recent initiatives for Mars explorative missions have included the *Mars Reconnaissance Orbiter*, also in 1997. This was a highly successful mission which has provided images of the entire Martian surface, and studied the atmosphere and interior.[160] In 2001, NASA's orbiting probe *Mars Odyssey* effectively used spectrometers and imagers to glean data regarding past or present water; however the *European Mars Express*, launched in 2003 lost contact with Earth upon entering the atmosphere. NASA's famously successful rovers *Sojourner*, *Spirit* and *Opportunity*, which landed on the surface in January 2004, confirmed the past existence of liquid water; and the *Phoenix* mission, which landed in May 2008 have provided further data on the Martian surface and geology. Water present on Mars in the remote past has been evidenced by large empty channels and riverbeds, and evidence of water ice was discovered by the *Phoenix* lander in the summer of 2008. In 2011 the *Mars Science Laboratory* mission was launched and delivered the *Curiosity* rover, as well as the *Opportunity* rover of the *Mars Exploration Rover* mission launched earlier. Both rovers provided unprecedented, detailed images of the planet surface as they search for evidence of habitability, taphonomy,[161] and organic carbon.[162] The MAVEN (Mars Atmosphere and Volatile Evolution) orbiter, launched in 2013, was designed to explore Mars' upper atmosphere and ionosphere. In 2016, the *ExoMars* mission consisted of an orbiter and lander, designed to monitor trace gases in the Martian atmosphere from a 400 km orbit. The lander delivered a science payload designed to take atmospheric measurements, as well as act as a relay for the 2020 *ExoMars* mission, currently under development as an astrobiological robotic mission by the European Space Agency in collaboration with the Russian Federal Space Agency.[163] NASA's most recent mission, the *InSight* (Interior Exploration using Seismic Investigations, Geodesy, and Heat Transport) mission landed on Mars in

---

158. Bianciardi et al., "Complexity Analysis," 14–26.
159. "Mars Pathfinder," NASA.
160. "Mars Global Surveyor," NASA.
161. The study of decaying organisms and their fossilization.
162. Grotzinger, "Introduction to Special Issue," 386–87.
163. "ExoMars," European Space Agency.

November 2018 and will conduct surface observations for one Martian year. Currently, Mars 2020 is a Mars rover mission forms part of NASA's Mars Exploration Program, which includes the rover *Perseverance* and the small robotic, coaxial helicopter *Ingenuity*. The mission will investigate signs of habitable conditions on Mars in the ancient past, as well as evidence of past microbial life, and water. An important part of this mission is to estimate possibilities for future colonies to collect CO_2 and derive oxygen from it, as well as rocket fuel.[164] To date, although it has been determined that the possibility for life forms is likely excluded from other planets of our solar system due to the prohibitive chemical and physical conditions, there is renewed interest in some of the larger satellites of the bigger planets. Images obtained from the *Pioneer* and *Voyager* probes in the 1970s and 80s, and from the *Galileo* mission, launched in 1989 (which released a probe to Jupiter in 1995) and the *Cassini-Huygens*, launched in 1997 which began its orbit of Saturn in 2004, have brought attention to the Jupiter satellite Europa due the discovery of water there, as well as to Saturn's satellites Enceladus and Titan.[165] Clearly, the significant resources brought forth to explore and understand the history of Mars, as well as other space missions, has important implications for our theme of humanity's understanding of extraterrestrial life and the implications of its discovery. A mere finding of independently evolved microbial life would have important meaning in determining if there is or is not a universal biology, and scientists could begin to consider the provenance of life a cosmic process rather than an improbable event. In the event of the discovery of intelligent life, the impact would be much farther reaching and expansive. Most unquestionably, terrestrial theological systems would require a shift, institutionally and individually, away from any remaining anthropomorphic notions regarding creation, to consider the larger context of God's creation of intelligences of which humanity may comprise a small part. In the next section, we will see how Christian theologians considered such a possibility, and ways the institution of the Roman Church reacted to these and other deep theological issues that appeared to threaten the long-standing geocentric faith.

---

164. "NASA Announces Mars 2020 Rover," NASA.

165. Cook, "Clay-Like Minerals Found on Icy Crust of Europa."

Chapter 3

Historical Literature on the Theology of Extraterrestrial Life

Section A: History of Theological Developments on Extraterrestrial Life: Antiquity to Nineteenth Century

Imago Mundi (Courtesy of British Museum).

HISTORICAL LITERATURE ON THE THEOLOGY OF EXTRATERRESTRIAL LIFE 53

The Old Testament writers inherited their cosmology from the civilizations of Ancient Near Eastern (ANE) peoples. One example of such an ANE cosmology is that depicted on a Babylonian clay tablet world-map known as the *Imago Mundi* from about 600 BC. Depicted is a flat, round world with Babylonia at center, encircling an *Earthly Ocean*, and eight outer regions beyond the waters, shown as equal triangles. These outer regions, according to Babylonian cosmology, were said to form bridges between the *Earthly Ocean* and the *Heavenly Ocean*, wherein are the various animal constellations, eighteen mentioned by name. Beneath Earth existed an underworld for the passage of the Sun, Moon, and planets. The later Hebraic cosmology of antiquity was composed of a flat, disc-shaped Earth floating on an ocean of water, supported by massive submerged pillars, preventing the land from sinking into the sub-Earth waters of chaos. Rising above the land, was the expanse of a great vault, containing within it the sky or firmament with the Sun, Moon, stars, and clouds, which rested on foundations beyond the limits of vision.[1] Surrounding the Earth were the original waters of chaos (or cosmic ocean), spoken of in Genesis which God separated from the land's oceans at creation.[2] The waters over the Earth existed above the firmament, such that the sky (or *raqia*) a solid inverted bowl, could be seen beyond the clouds-colored sapphire blue from the heavenly ocean above it. Rain, snow, and hail were supposedly kept in storehouses outside the *raqia*, having windows which released onto the Earth their contents (i.e., Noah's flood). Beyond the waters above the firmament was the realm of "heaven of heavens," or *shamayim*, the dwelling place for God and his angels.[3] The heavenly realm was believed to extend down to, and be coterminous with, the farthest edges of the Earth[4]—it was understood that people looking up from Earth could see the floor of heaven and what they believed the base of God's throne.[5] A hidden, morally neutral underworld *sheol*, the abode of the dead, existed below the landmass. However absent from this model was

---

1. Seely, "The Firmament and the Water Above," 227–40.
2. Gen 1:6–8.
3. Gen 1:2–8.
4. Deut 4:32.
5. Pennington, *Heaven and Earth in the Gospel of Matthew*, 42; Wright, *The Early History of Heaven*, 56–57; Also see Exod 24:9; Job 22:14; Ezek 1:26; cf. 1:28; 10:1; Prov 8:27. Exod 24:9 speak of a floor of heaven, composed of lapis-lazuli (bluish) brick as clear as sky under the feet of God; Ezek 1:26 describes the divine presence in a vision of "above the firmament which is over their (the cherubs) heads an appearance like lapis-lazuli stone as a throne," where on this throne was seated "the appearance of the likeness of the glory of Yahweh."

any place which could be inhabited by extraterrestrials apart from the heavens, which were populated by an angelic hierarchy and God. It was not until

Ancient Hebriac Cosmology (Used with permission by the artist: Michael Paukner).

after the Hellenistic period circa 330 BC that the Hebrews abandoned the older three-level cosmology and adopted the Aristotelian concept of a spherical Earth suspended in space at the center of a number of concentric circles containing the planets and stars.[6] As the early Christian Church had no formalized cosmology, it adopted the biblical narrative of a single creation by God of a distinct race on Earth, encapsulated within the dominant Greek cosmology. The earliest and most well-known theologian to support the idea of other worlds, later termed a *plurality of worlds*, was Origen of Alexandria (c. 185–254 AD). Essentially a Platonist, and in contrast to the inherited Hebraic cosmology and scriptural silence on the subject, he postulated God could create worlds other than our own:

> [B]ut that a diversity of worlds may exist with changes of no unimportant kind, so that the state of another world may be for some time unmistakable reasons better (than this), and for others worse, and for others again intermediate.[7]

Origen viewed the universe as populated with stars, planets, and angels—each alive, rational, spiritual beings ranging from high angels to fallen ones—which we cannot equate in our modern sense with extraterrestrials. With a belief in these outside intelligences, his Christian apocatastatic redemption was extended beyond the terrestrial and included all intelligences: "The altar was at Jerusalem, but the blood of the victim bathed the universe."[8] As this theological history will show, Origen stood as a remarkable figure in affirming within Christian tradition a theology adapted to a type of inhabited plurality before the thirteenth century, due to the early Church's objections to the atheistic atomist philosophy and its general acceptance of Aristotelian cosmology and philosophy. The foremost theologians following Origen in the third century were Hippolytus; in the fourth century Eusebius, bishop of Caesarea; and Theodoret, bishop of Cyprus in the fifth century; all of which rejected the plurality.[9] St. Augustine of Hippo (354–430 AD) in his work *City of God* also denied the idea of other worlds, basing his objection on the prevailing scriptural interpretation of humanity's special creation.[10] He equally objected to the Stoic doctrine of successive worlds, as well as the concept of simultaneous worlds and the principle of plenitude:

---

6. Aune, "Cosmology," 119.
7. Origen, *De Principiis*, Book II, Ch. III, Sec. 4.
8. Origen, *Homilies on Leviticus*, homily I, n° 3.
9. McColley, "The Seventeenth Century Doctrine," 393.
10. Augustine, *City of God*, Book XI, Ch. 5.

> For they imagine infinite spaces of time before the world during which God could not have been idle, in like manner they may conceive outside the world infinite realms of space, in which, if any one says that the Omnipotent cannot hold His hand from working, will it now follow that they must adopt Epicurus's dream of innumerable worlds?[11]

Similarly, Augustine also considered it heretical to believe in the existence of a population in the Antipodes, as it would compromise the unity of the human race, and thus universal redemption:

> As to the fable that there are Antipodes, that is to say, men on the opposite side of the Earth, where the Sun rises when its sets on us, men who walk with their feet opposite ours, there is no reason for believing it. Those who affirm it do not claim to possess any factual information; they merely conjecture that, since the Earth is suspended within the concavity of the heavens, and there is as much room on the one side of it as on the other, therefore, the part which is beneath cannot be void of human inhabitants. They fail to notice that, even should it be believed or demonstrated that the world is round or spherical in form, it does not follow that the part of the Earth opposite to us is not completely covered in water, or than any conjectured dry land there should be inhabited by men. For Scripture, which confirms the truth of its historical statements by the accomplishment of its prophecies, teaches not falsehood; and it is too absurd to say that some men might have set sail from this side and, traversing the immense expanse of ocean, have propagated there a race of human beings descended from that one first man.[12]

Continued rejection of the plurality concept within Christendom in the eighth century can be illustrated by Pope Zachary (741–752) in his letter to St. Boniface where he mentions Virgil of Salzburg, a priest who taught on the plurality of worlds. Zachary recommends his defrocking and expulsion from the Church in his *Quod alius mundus et alii homines sub terra sint, seu sol et luna* as punishment for heresy:[13]

> As for this perverse and abominable teaching, which he [Virgil] has proclaimed in opposition to God, and to his own soul's

---

11. Augustine, *City of God*, Book XI, Ch. 5. Also Book XII, Chs. 11–15, 19, and Book XIII, Ch. 16.

12. Augustine, *City of God*, Book XVI, Ch. 9.

13. Migne, "Epistola XI ad Bonifacium," 946–47; Dick, "Plurality of Worlds," 502–12.

detriment-if the report of his having spoken thus be true-that is, that there are another worlds and other men beneath the Earth, or even the sun and moon: ("Quod alius mundus, et alii homines sub terra sint, seu sol et luna") take counsel and then expel him from the church, stripped of his priestly dignity.[14]

Zachary's motivation for dispelling this idea was to counter the apparent difficulty in considering non-terrestrial humans created by God but not descended from Adam and free from original sin. This condemnation was in addition to censure Virgil's similar teaching on the existence of the Antipodes, the hypothetical inhabitants of the southern hemisphere believed in and taught by the pagans. It was these supposed peoples that the Church would not consider "descendants of Adam" until about the twelfth century,[15] and which as we will see in a later chapter, are highly relevant to our discussion on intelligent extraterrestrials and Christianity. Albertus Magnus (1193-1280), another prominent theologian initially interested in the plurality concept, famously questioned, "Since one of the most wondrous and noble questions in Nature is whether there is one world or many... it seems desirable for us to inquire about it."[16] However, he and his pupil, Thomas Aquinas (1225-1274) both ultimately decided against the idea, with Aquinas following the Aristotelian argument in his *Summa Theologica* that only one world existed. Aquinas derived his understanding of a world from Plato, which held that it would be more in accord with God's omnipotence that he created a single perfect world than many others comparatively imperfect, as division implied imperfection.[17] He makes his argument in the *Summa Theologica*:

> This world is called one by the unity of order whereby some things are ordered to others," and as it is axiomatic that "whatever things come from God, have relation of order to each other, and to God himself... hence it must be that all things should belong to one world.[18]

Aquinas's simple argument against other worlds can be made as follows: if God created other worlds, they would either be similar or dissimilar

---

14. On Virgil's life see, e.g., Krusch, *MGH SSrerMerov*, 6:517-20; Grosjean, "Virgile de Salzbourg en Irlande," 92-123; Kenney, *The Sources for the Early History of Ireland Ecclesiastical*, 2nd ed., 523-26.

15. Rainaud, *Le continent austral*, 133-34, 159-65. This date has been questioned; see Delhaye, *Le microcosmus de Godefroy de Saint-Victor*, 282-86.

16. Bless, *Discovering the Cosmos*, 686.

17. Plato, *Timaeus*, 31b.

18. Aquinas, *Summa Theologica*, I, Q. 47, Art. 3: "Whether there is only one world?"

to ours. If similar, they would be in vain and contrary to divine wisdom. If dissimilar, none would contain all things and thus none would be perfect, as an imperfect world is incompatible with a perfect Creator.[19] Of important note is the contemporary of Aquinas, the great Italian scholastic dogmatic theologian St. Bonaventure (1221–1274), who as a writer of the Franciscan school was more accepting of the idea of a plurality than the Dominican followers of Aquinas. In his speculative and mystical writings he asserted God could create other worlds: "He was able to make a hundred such worlds, one in a higher place than another, and, still more, one embracing all of them. And too God could make a time before this time and in it make a world."[20] Bonaventure, along with Francis Mayron and especially, William Vorlong, (discussed below) comprise the better-known dissenters against the Aristotelian cosmological monopoly against a plurality. Not three years after Aquinas's death, many questionable ideas held in the universities emerged as the wholesale adoption of Aristotelianism appeared to threaten certain Christian doctrines. For example, Aristotle's teaching on the eternity of the world conflicted with the day of judgment; and while conceding that the Earth is truly a sphere at the center of the universe, the Church denied the Aristotelian argument that God could not move the Earth if he willed, or could not create other worlds if he willed.[21] Ecclesiastical authorities grew alarmed, and in 1277 Pope John XXI directed Etienne Tempier, the bishop of Paris to investigate the intellectual controversies. Within three weeks the bishop condemned 219 propositions, with excommunication the penalty for holding even one of the heresies. The thirty-fourth article denounced Aquinas's earlier teaching that the First Mover could not create other worlds, while asserting that he did not do so, thereby protecting God's omnipotence and freedom. In making this declaration in view of Aristotelian physics, thinkers in the aftermath of the condemnation could conclude that God could create such worlds either by supernatural suspension or modification of the laws of nature as defined by Aristotle. Further, the condemnation gave rise to the idea that God could have created the world in a different manner but with the same observational effects, which implied that no particular hypothesis could be insisted upon. Historians Pierre Duhem[22] and Richard Dales[23] maintain the con-

---

19. Aquinas, *Expositio in Aristotelis libros*, Lectio XIX, 94.

20. Bonaventure, *Commentaria in quatuor libros sententiarum*, Lib. 1, Dist. 44, Art. 1, Quaest. 4 (*Opera Omnia* 1, 789); McColley and Miller, "Saint Bonaventure," 388–89.

21. Grant, *A Source Book in Medieval Science*, 48.

22. Duhem, *Etudes sur Leonard de Vinci*, 412.

23. Dales, *The Intellectual Life of Western Europe*, 550.

demnation had a positive effect in the early development of science by not relying on an uncritical acceptance of Aristotelian philosophical works.[24]

An early example of this greater freedom from an overly-strict Aristotelianism for Christian philosophers and theologians is the thought of Jean Buridan (1295–1358), rector of the University of Paris, in his questions on Aristotle's *De caelo et mundo* where he argued for God's ability to override limitations of natural law and create different effects, or other world(s) of a different form, not ordered to ours but according to their own elements and their own natural place. This formulation provided compatibility with Aristotle and the requirements of Christian theology's free and omnipotent God. William of Ockham (c. 1287–1347), an English Franciscan friar and scholastic theologian, also addressed the subject by relativising Aristotle's anti-pluralistic stance.[25] He writes in his commentaries on Distinction XLIV of *Peter Lombard's Sentences*:

> The reason for this is that God could produce an infinite number of individuals of the same species as those which now exist, Therefore, God could produce as many individuals and many more than are now produced, and of the same species; but he is not limited to producing them in this world, Therefore, he could produce them outside this world and make a world of them just as he made this world from those things which he produced here now.[26]

In even stronger dissension from Aristotle, Ockham justified how "natural place" for any object does not have to be a *unique* natural place, allowing for multiple worlds:

> I say that although all individuals of the same species can be naturally moved to the same place and number, yet if they should occupy the same place outside their natural place, at least successively, then it is not necessary that they always be moved to the same place in number naturally, but it is possible that they could be moved simultaneously to different places according to number.[27]

Nicole Oresme (1325–1382), who later became Bishop of Paris, composed a French commentary on the *De caelo*, taking even further Ockham's

---

24. Woods, *How the Catholic Church Built Western Civilization*, 91–92.

25. Dick, *Plurality of Worlds*, 23.

26. Ockham, *Opera Plurima*, Distinction XLIV. "*Utrum deus posse facere mundum meliore isto mundo,*" Section E.

27. Ockham, *Opera Plurima*, Section F.

argument against Aristotle's theory of natural place as well as Aquinas's argument that all things in the universe must have relation to each other: two worlds sufficiently separated do not necessarily have relation to each other, only to their own parts; and apart from God's specific act of creation of another world, natural laws remain.[28] This was a defining moment in the history of the plurality of worlds idea, that the motions of celestial bodies were governed by their individual place and surroundings, and a clear denunciation of Aristotle. However, while Buridan, Ockham, and Oresme were the first in the scholastic tradition to argue for a theological and natural possibility of a plurality, they ultimately appealed to the dominant medieval conservative religious authority by denying its reality.[29] An additional difficulty for late-thirteenth and fourteenth century thinkers in their consideration of a plurality of worlds was a hesitancy to contemplate the possibility of extraterrestrial inhabitants.[30] William Vorilong (c. 1390–1463), French philosopher and scholastic theologian, was the first to consider the possibility of extraterrestrial beings within the context of the Christian doctrines of the Incarnation and Redemption. He wrote, "Infinite worlds, more perfect than this one, lie hid in the mind of God,"[31] and, "It is possible that the species of each of these worlds are different from those of our world."[32] While these quotes do not suffice as definitive formulations, he did develop much more concrete opinions. While he did not hold that extraterrestrial inhabitants are complicit in original sin, he viewed Christ as a universal figure capable of redeeming all intelligent beings. In his commentary on the *Sentences of Peter Lombard* he postulated,

> If it be inquired whether men exist on that world, and whether they have sinned as Adam sinned, I answer no, for they would not exist in sin and did not spring from Adam. As to the question whether Christ by dying on this Earth could redeem the inhabitants of another world, I answer that he is able to do this even if the worlds were infinite, but it would not be fitting for Him to go unto another world that he must die again.[33]

Although this commentary reveals Vorilong as a true innovator in considering the foremost consequences for Christian theology given rational extraterrestrials unlike humans, he followed earlier scholastics in

28. Menut and Denomy, *Le livre du ciel et du monde*, 171–73.
29. Sylla, "Aristotelian Commentaries and Scientific Change," 37–83.
30. Dick, *Plurality of Worlds*, 43.
31. De Vaurouillon, *Quattuor*, Vorillonis Lib.1, Dist. xliv, 105.
32. De Vaurouillon, *Quattuor*, 107.
33. Brady, "'The Declaration seu Retractatio' of William of Varuouillon," 394.

concluding God did not create other worlds. However, Nicolas of Cusa (1401–1464), German philosopher, theologian, and astronomer, showed no fear of Church authorities when he conceived of a cosmology of an edgeless, homogenous universe without absolute center, thereby decentering Earth within creation. As he phrased, its "center is everywhere and its circumference nowhere."[34] A centerless universe would include as many centers of attraction as there are planets, a significant departure from Aristotle's theory of natural place. The introduction of this important concept of "centers of attraction" in the universe created a marked shift in the plurality of worlds tradition, later to be verified and systematized by Newton. In his *Of Learned Ignorance*, Cusa also developed philosophical formulations attempting to describe the relationship other worlds and their inhabitants would have with humans, and considered their physical and intellectual natures. His conclusions were expressed: "These possible inhabitants . . . wherever they are, do not have any proportion with the inhabitants of our world."[35] He asserted the proposition at the time that life could exist throughout the universe:

> Life, as it exists here on Earth in the form of men, animals and plants, is to be found, let us suppose, in a higher form in the solar and stellar regions. Rather than think that so many stars and parts of the heavens are uninhabited and that this Earth of ours alone is peopled-and that with beings, perhaps of an inferior type-we will suppose that in every region there are inhabitants, differing in nature by rank and all owing their origin to God, who is the center and circumference of all stellar regions.[36]

Also:

> It is convenient to regard the Earth as the center of the universe, although nothing in reality compels us to do so.[37]

Cusa's writings signaled a crucial shift in departure from Aristotle's natural philosophy, as for the first time he envisioned a hierarchy of living beings from beyond the Earth to include the entire universe:

> And we may make parallel surmise of other stellar areas that none of them lack inhabitants, as being each, like the world we

---

34. Cusa, *Of Learned Ignorance*, 111.

35. Cusa, *Of Learned Ignorance*, Book II, Ch. 12. For the Latin original see *Nicolai de Cusa de docta ignorantia*, eds. Ernest Hoffman and Raymund Klibansky.

36. Cusa, *Of Learned Ignorance*, 114–15.

37. Tyrone Lai, "Nicolas of Cusa and the Finite Universe," 161–67.

live in, a particular area of one universe, which contains as many such areas as there are uncountable stars.[38]

He further asserted that while Earth and its inhabitants were within this vast hierarchy of intelligent beings, and believed by virtue of the enlightened inhabitants of the Sun, humans retained their nobility compared to other inhabitants:

> [E]ven if inhabitants of another kind should exist in the other stars, it seems inconceivable that, in the line of nature, anything more noble and perfect could be found than the intellectual nature that exists here on this Earth and its region.[39]

Cusa is significant in the history of the plurality of worlds concept as being the first theologian to divinize the atomist and atheistic plenitude concept that infinite atomic matter in infinite space must result in an infinite number of combinations and possibilities: that the Creator must realize his omnipotence in all creation, and that where God creates there is meaning. In essence, God does not waste space. This teleological form of the "principle of plenitude," would provide much philosophical and theological support for a plurality of worlds and their inhabitants throughout the succeeding centuries. Arthur Lovejoy has described it thus:

> No genuine potentiality of being can remain unfulfilled, that the extent and abundance of the creation must be as great as the possibility of existence and commensurate with the productive capacity of a 'perfect' and inexhaustible 'Source,' and that the world is better, the more things it contains.[40]

Cusa's hierarchy of intelligent beings throughout the universe was an expanded form of the Platonic and Aristotelian (in his *Historia Animalium*) "Great Chain of Being" of antiquity.[41] In it, rather than God placed at the top of a hierarchy of heavenly beings, humans, and animal life, he was enthroned at the center of all places of creation, maintaining his rightful place among all extraterrestrial beings.

---

38. Cusa, *Of Learned Ignorance*, 116.
39. Cusa, *Of Learned Ignorance*, 115.
40. Lovejoy, *The Great Chain of Being*, Ch. IV, "The Principle of Plenitude and the New Cosmography," 99–143.
41. See template below.

HISTORICAL LITERATURE ON THE THEOLOGY OF EXTRATERRESTRIAL LIFE 63

1579 Drawing of The Great Chain of Being by Didacus Valades, *Rhetorica Christiana*.

Cusa saw no dogmatic incompatibility with this new vision. Further, he viewed the terrestrial Jesus of Nazareth, the divine Logos, as an omnipresent, immanent, and divinely transcendent reality in a universe in which Earth did not lose its primary importance despite losing its physical centrality in the cosmos.[42] He based his conjectures on a metaphysical system completely antithetical to Aristotelian natural philosophy, gleaned from his consideration of the hotness of the Sun, elements of the Moon, and weighty

42. Cusa, *Of Learned Ignorance,* Book III, 1–4.

elements of the Earth. Although Cusa's propositions for a plurality of worlds were far from scientific, as observed by Lovejoy, his metaphysical system included natural theology, the principle of plenitude and sufficient reason.[43] Even before Copernicus published his heliocentric proposal, Philip Melanchthon (1497–1560), a thoroughly biblical and less scientific theologian of the Protestant Reformation, was aware of the new cosmological considerations, and while agreeing with Vorilong on the inimitable nature of the Incarnation on Earth, warned of dangers in considering Christ's sacrifice applicable to sinful extraterrestrial inhabitants:

> [T]he Son of God is One; . . . Jesus Christ was born, died, and resurrected in this world. Nor does he manifest Himself elsewhere, nor elsewhere has he died or resurrected. Therefore, it must not be imagined that Christ died and was resurrected more often, nor must it be thought that in any other world without the knowledge of the Son of God, that men would be restored to eternal life.[44]

These new theological possibilities inherent to the concept of a potential non-geocentrism would become obvious to many only well after its acceptance in the seventeenth century, and Copernicus was careful to publish his thesis as he lay dying, expecting criticism of his ideas. In fact, an unauthorized forward to Copernicus's *De revolutionibus orbium caelestium* presented the heliocentric theory as convenient mathematical fiction in order not to offend Church censors. *De revolutionibus* was dedicated to Pope Paul III, in his hope that "My labors contribute somewhat even to the Commonwealth of the Church . . . for not long since the question of correcting the ecclesiastical calendar was debated."[45] Even before its publication, Copernicus's teaching on heliocentricity circulated in manuscript form and was known by astronomers and others, and was immediately recognized to contradict the prevailing biblical understandings. Certain scriptures were central to this conflict of ideas: "[T]he world also shall be stable, that it be not moved"; "[the Lord] who lay the foundation of the Earth, that it should not be removed forever" and "the sun also arises and the sun goes down, and hastens to his place where he rose."[46] Martin Luther, prior to the publication of *De revolutionibus*, cited scripture against the new cosmology and warned, "This fool wishes to reverse the entire science of astronomy; but sacred scripture tells us that Joshua commanded the Sun to stand still,

---

43. Lovejoy, *The Great Chain of Being*, 112–15.
44. Melanchthon, as translated and quoted by Dick, "Plurality of Worlds," 89 n. 4.
45. Hetherington, "Cosmology, Religious, and Philosophical Aspects," 143.
46. 1 Chr 16:30; Ps 104:5; Eccl 1:5.

and not the Earth."⁴⁷ However, it was not until the Galileo episode in 1633 that heliocentrism was formally condemned by the Catholic Church. Prior to the Counter-Reformation, the Catholic Church was more relaxed in its biblical interpretation and willing to accept the Copernican theory being taught in some universities, and indeed the theory was incorporated in the new calendar promulgated by Pope Gregory XIII in 1582.⁴⁸

In 1584 Giordano Bruno, an Italian Dominican friar, philosopher, mathematician, and astronomer, arrived in Oxford and began preaching with missionary zeal of an infinite homogenous universe containing worlds inhabited by extraterrestrials—going well beyond Copernicanism and Ptolemaism to transform every celestial body into another populated world. Bruno proclaimed that the Sun is merely another star, among countless other stars with infinite numbers of inhabited worlds populated with intelligent beings. Similar to Cusa, he asserted a homogenous universe, and proposed a Christianized version of the Epicurean universe, that is, a universe of infinite extent, populated with numberless planetary systems, all teeming with life. This type of an infinite universe was Bruno's and not an established notion of heliocentricity.⁴⁹ The following quote sums up his position on the plurality question:

> God is infinite, so His universe must be too. Thus the greatness of God magnified and the greatness of His kingdom made manifest; He is glorified not in one but in countless Suns; not in a single Earth, but in a thousand, I say, an infinity of worlds.⁵⁰

Bruno stated it was only he who "has penetrated into the heavens, past the frontiers of the world, [and] shattered the fantastic walls of the spheres (Aristotelian and Ptolemaic)."⁵¹ His works, known as the six "Italian dialogs," which included the principal cosmological writing *De l'infinito Universo et Mondi*, published in 1584, in fact make no mention of Copernicus. Bruno's cosmological views were his own, stemming from another work entitled *De la causa, principio et uno* (also 1584) in which he expounded his metaphysical principle of a unity of all creation, juxtaposed to the dichotomic Earthly/celestial realms of Aristotle which he repudiated,⁵² and the idea that God

---

47. Luther, *Luther's Works*, 30, 42. Ref. Josh 10:13.

48. Hetherington, "Cosmology, Religious, and Philosophical Aspects," 143.

49. Copernicus, "Let us leave to the Physicists the question whether the universe is finite or not, holding only to this, that the Earth is finite and spherical." *De revolutionibus*, Book I, Sec. 8.

50. Bruno, *De l'infinito*, 1584.

51. Aquilecchia, *Le cena de le ceneri*, 98–99.

52. Bruno, *De l'infinito*, Books VI-VIII.

is compelled to realize his omnipotence through the plenitude of nature.[53] According to Bruno, each star was a sun around which revolved any number of worlds, imperceptible to terrestrials due to their great distance. *De l'infinito* was the product of Bruno's years in England. In this work as well as his later Latin work *De immenso et innumerabilibus*, he systematically refuted the arguments of Aristotelian cosmology.[54]

> Concerning this question, [of Aristotle whether beyond this world there lieth another] you know that his interpretation of this word *world* (mondi) is different from ours. For we join world to world and star to star in this vast ethereal bosom, as is seemly and hath been understood by all those wise men who have believed in innumerable and infinite worlds. But he applieth the name world to an aggregate of all those ranged elements and fantastic spheres reaching to the convex surface of that primum mobile . . . It will be well and expedient to overthrow his arguments insofar as they conflict with our judgment, and to ignore those which do not so conflict.[55]

Here, Bruno provided a new connotation to the term *world*, extended to mean that of an infinite homogeneous universe with innumerable celestial bodies, and because the universe was infinite it must be populated with "imperfect" intelligent beings.[56] His principle of unity, which he perceived as more natural and perfect than plurality (in the Platonic sense), was expanded beyond the Platonic and Aristotelian models, creating a vision of the universe in which the Earth and other planets became their *own* centers of attraction, each with its own system within a universe of infinite diversity.[57] Bruno, a singular thinker, sought to fundamentally deconstruct the Aristotelian cosmology that had dominated the scene within most of Europe. After the publication of his Latin works from 1589 to 1590, Bruno was unable to obtain a permanent teaching position in the universities, and, having worn out his welcome due to his lack of tact and diplomacy in intellectual debates, moved to Venice, serving as a tutor for Giovanni Mocenigo on mnemonics in 1592. When after a few months Bruno announced his intent to leave Venice, Mocenigo, unhappy with his teachings, had him turned over to the Venetian Inquisition, where he underwent a prolonged interrogation and trial. After an extended imprisonment, he was

53. Bruno, *De la causa*.
54. Bruno, *De l'infinito*, Book II; *De immenso*, Book II.
55. Singer, *Giordano Bruno*, 329.
56. Bruno, *De l'infinito*; *De immenso*, Books VI–VII, 306, 310, 314, 370–71.
57. Dick, "The Origins of the Extraterrestrial Life Debate," 4–6.

burned at the stake in 1600 for heresy in claiming, among charges of holding opinions contrary to the Catholic faith pertaining to the Trinity, the divinity of Christ and the Incarnation, to include the belief in a plurality of worlds throughout the universe. Since the time of Bruno's execution, the Vatican has made some statements after the discovery of some of lost trial documents. In 1942, Cardinal Giovanni Mercati, who made this discovery, related his position that the Church was justified in condemning Bruno. On the four hundredth anniversary of Bruno's death in 2000, Cardinal Angelo Sodano declared Bruno's death to be a "sad episode," while defending his inquisitors in their "desire to serve freedom and promote the common good and did everything possible to save his life."[58] Bruno's opposition to the prevailing Aristotelian geocentric cosmology, and more so certain heretical beliefs regarding the true presence and the divinity of Christ sealed his fate, not necessarily his belief in other worlds. Additionally, the Roman Church saw Bruno's views and rhetoric dangerous as it sought to reestablish authority after the Reformation. Hegel in his *Lectures on the History of Philosophy* remarked that Bruno represented a rejection of all Catholic beliefs which rested on mere authority.[59] Bruno and Cusa both demonstrated, albeit Bruno more systematically, the possibility of planetary systems containing inhabitants using a metaphysical system synthesizing several elements: the principle of cosmological unity and individual centers of attraction, the *a priori* atomistic principle of a plenitude of nature, a natural theology supporting God's omnipotent use of creation, and sufficient reason. Unfortunately for Bruno, it would take four hundred years before his beliefs in the plurality of worlds were confirmed by science.

The plurality concept and its implications for theology remained highly problematic at the end of the sixteenth century, as it questioned the centrality of humanity in God's plan, and more seriously, the accuracy and validity of God's communication through the scriptures. This grave concern was centered on the belief of a single, unique, and unrepeatable incarnation of the divinity on one planet, to one singular species that was the sole focus of God's attention, to which speculations of a plurality, hosting certain extraterrestrial beings was heresy. In 1600 there existed no official Catholic position on the Copernican system.[60] It was the telescopic findings of Italian physicist and astronomer Galileo Galilei (1564–1642), and his later confrontation with the Church Inquisition that led to the eventual censure of heliocentrism. Galileo in 1610 published his small book, *Sidereus*

---

58. Salvestrini, *Bibliografia di Giordano Bruno*.
59. Merati, "*Il Sommario del Processo di Giordano Bruno*," 101.
60. McNulty, "Bruno at Oxford," 300–5.

Nuncius, describing his observations of the Venusian and Galilean moon phases, which contradicted Ptolemaic geocentrism and affirmed Copernican theory. Aristotelian philosophers at the Italian universities, who with the Church had tightened their hold on orthodoxy as a result of the Counter-Reformation council of Trent (1545–1563) clashed with Galileo over several of his ideas. In 1616 Galileo was subjected to a Church inquiry after a letter he wrote to one of his former students defending heliocentrism was brought to the attention of the Inquisition. In 1616 the Inquisition submitted questions regarding the motion of the Earth and the stability of the Sun to ecclesiastical judges, who found the idea of the Earth's motion erroneous and the Sun's stability heretical,[61] and the Commissary General of the Inquisition ordered Galileo not to hold, teach, or defend the teachings. However due to rumors that he had abjured, Galileo requested and was provided by the Commissary General an affidavit stating he was under no restriction in holding or defending his positions other than those applying to all Catholics. Later, Pope Urban VIII, a friend of Galileo who granted him six audiences in 1624, encouraged him to write on heliocentrism, and to include the pope's own orthodox view. In response, he published his *Dialog on the Two Chief World Systems* in 1632, arranged as a conversation between a Copernican scientist, *Salviati*, impartial scholar *Sagredo*, and an Aristotelian named *Simplicio*. When Urban VIII read the *Dialog*, to his embarrassment identifying with *Simplicio* whose name had the connotation of "simpleton" (and was losing the argument), Galileo was summoned and charged with contravening the 1616 order, despite his producing the signed and dated affidavit in his defense.[62] He was charged with suspicion of heresy, and of holding belief in Copernicanism after 1616, was ordered to abjure, curse, detest his errors and heresies, and spent the last nine years of his life under house arrest.[63] His *Dialog* was banned, and for many decades afterwards any discussion of heliocentric cosmology was considered heretical to the Roman Catholic establishment. The motivations of Urban VIII and Galileo in the condemnation have been the subject of much historical study, some contending Galileo acted innocently and Urban VIII was misled by advisors opposed to Galileo,[64] others defending the Church's actions given the current

---

61. Langford, *Galileo, Science, and the Church*, 56–57.
62. Pantin "New Philosophy and Old Prejudices," 237–62.
63. Finocchiaro, *Galileo on the World Systems*, 47.
64. Sobel, *Galileo's Daughter*, 223–25. Sobel contends Urban VIII had succumbed to the influence of court intrigue and preoccupation with affairs of state, his friendship with Galileo taking second place to concerns over personal persecution. Amid claims by a Spanish Cardinal that Urban was a poor defender of Church doctrine, and Galileo's case presented to the pope by those opposed to him and his book, the situation did not favor a positive outcome for the astronomer.

limited scientific knowledge. It is important to note that a second charge against Galileo was the idea of "other worlds," in which he ultimately was found innocent. He was very reticent to make any claims of plural worlds, as noted in his exchanges with Kepler. It would be more accurate to say that he espoused one world with many systems, rather than a plurality.[65]

Anglican Bishop John Wilkins, a Copernican and anti-Aristotelian, and one of the founders of the Royal Society,[66] sought to reconcile his belief in the plurality of worlds with Christian doctrine in his 1638 *Discovery of a World in the Moone*. Writing the *Discovery* within the framework of the new cosmology, he made his case for a world in the Moon by following an ordered set of thirteen propositions from the general to the specific, notable among which were a) that the strangeness of the idea did not disprove it, b) that possible inhabitants cannot contradict reason or faith, (only Aristotle), and c) the Moon has seas, land, and atmosphere.[67] Although Wilkins avoided doctrinal formulations such as original sin, he confidently stated one cannot argue against the plurality from the "negative authority of Scripture," a position considered heretical in antiquity by what he termed "ignorance of the period."[68] By advancing the plausibility of the Moon as "another Earth," Wilkins argued the Copernican system implied the possibility of a solar system with other worlds. *Discovery* was convincing to many other philosophers and observers, and a Moon "world" became a critical part in the idea of a plurality concept for empiricist thinkers. By the mid-seventeenth century the plurality of worlds concept had gained more natural philosophical and theological viability due to the growing dissatisfaction with Aristotelian cosmology, commented upon and critiqued throughout the Middle Ages, and a rising consensus of the validity of heliocentrism for several reasons. The influence of Bruno's advocacy of a plurality and extraterrestrial life within his "unity" of infinite diversity; the metaphysical notion of the plenitude advocated by Bruno and Cusa, attached with it the notion of a "Great Chain of Being" of which humanity was a small part; and newer empirical observations of the Moon by Galileo and Kepler. These observations revealed not the perfectly polished sphere as Aristotle taught, but rather conditions similar to Earth and hence another world, further undermining the Aristotelian view for more scientifically-minded thinkers as Wilkins.

65. According to Campanella in *Apologia pro Galileo*, 1622.
66. Shapire, *John Wilkins*, 32–25.
67. Wilkins, *The Discovery of the World in the Moone*.
68. Wilkins, *The Discovery of the World in the Moone*, citing Baronius, *Ann. Eccl.*, who was excommunicated for his belief in the plurality of worlds.

The second half of the seventeenth century saw Bernard le Bovier de Fontenelle's *Entretiens sur la pluralité des mondes*, written within the Cartesian tradition, placed on the Roman Church's Index of Prohibited Books in 1687 for its support of extraterrestrial beings. He had reasoned, not unlike Vorilong, that inhabitants of other worlds did not descend from Adam, were not in fact humans,[69] and thus did not partake in his sin. Despite the Church's censure, *Entretiens* was immensely popular in Protestant Europe, eventually with many editions in nine languages. A few years later, Richard Bentley's 1693 *A Confutation of Atheism from the Origin and Frame of the World*, showed his agreement with Wilkins on the negative authority of the scriptures.[70] He echoed Fontenelle in proposing that rational extraterrestrials did not necessarily take the form of humans, with some higher in natural perfections, others inferior to humans, rendering moot (in his mind) the question regarding their relation to the human family and related implications to the Incarnation and Redemption. For Bentley, the primary force supporting the plausibility of a plurality and extraterrestrial inhabitants was the force of the principle of plenitude: "[Heavenly bodies] were formed for the sake of intelligent minds . . . why may not all other planets be created . . . for their own inhabitants which have life and understanding?"[71]

In contrast to the Roman Church's anti-pluralist stance, many European Protestants in the beginning of the Enlightenment accepted a belief of a cosmos containing extraterrestrial life, as a form of reconciliation of science and religion, as well as an enhancement of a natural religious perspective. This prevailing view can be attributed to three main factors: Huygen's *Kosmotheoros*, and more so Fontenelle's *Entretiens*, which were both highly influential among natural philosophers and theologians; the principle of plenitude continued to demonstrate its strength in many European countries; and many astronomers supported the concept (Lambert, Herschel, Schröter, Bode, Laplace, Lalande), although much was based upon conjecture rather than science.[72] It was in this positive environment that Emanuel Swedenborg (1688–1772), scientist turned mystic, revealed his revelations,[73] which included those of highly spiritually evolved extraterrestrials on our moon and other planets, and which recognized Jesus Christ by means of spiritual intermediaries. He argued for the existence

69. Fontenelle, *Conversations on the Plurality of Worlds*, 14.

70. Bentley, *A Confutation of Atheism*, 358. The negative authority of scripture refers to the absence of arguments to a contrary position. In this case, that of extraterrestrial intelligences.

71. Cohen, *Isaac Newton's Papers and Letters on Natural Philosophy*, 358.

72. Crowe, *The Extraterrestrial Life Debate, 1750–1900*, 81, 161.

73. Swedenborg, *Earths in Our Solar System*.

of these otherworldly beings through the philosophical analogy between terrestrial and extraterrestrial environments, and from plenitude. His mystical excursions to other planets conveniently included only those unknown to science at the time and given the nature of his revelations, they were least likely to receive invalidation in his time. His visions, which he intended to be taken literally, became a portion of the basis for the New Jerusalem or Swedenborgian Church. It was to those who found a satisfaction in the mystical and open-minded embracing of extraterrestrials that deist Thomas Paine (1737–1809), in his *Age of Reason* of 1794 caused serious doubt. In it, Paine questioned the validity of the Christian notion of the divine Incarnation and Redemption within a universe populated by a multitude of intelligent beings:

> From whence . . . could arise the . . . strange conceit that the Almighty . . . should . . . come to die in our world because, they say, one man and one woman had eaten an apple! And, on the other hand, are we to suppose that every world in the boundless creation had an Eve, an apple, a serpent, and a redeemer? In this case, the person who is irreverently called the Son of God, and sometimes God himself, would have nothing else to do than to travel from world to world, in an endless succession of death, with scarcely a momentary interval of life.[74]

Paine's book was very popular in Britain and America, and generated more than fifty published responses opposing his objections to Christianity.[75] His chief argument involved his disbelief in a revealed religion with an incarnated God (he considered Christianity a "pious fraud"),[76] and instead advanced a natural religion, whereby if the Christian religion claims title to a God-human redemptive figure for a sentient race, it cannot be compatible with intelligent extraterrestrial life. Paine had turned the argument on its head, arguing the implausibility of Christianity on the basis of a plurality. Astronomers of that time, Thomas Wright (1711–1786), Johann Lambert (1728–1777), and William Herschel (1738–1822), all pluralists, disagreed with Paine's assessment on the basis of plenitude and metaphysical arguments. However, most notable among Paine's challengers were the theological works of Scottish minister and theologian Thomas Chalmers's (1780–1847) *Astronomical Discourses* (1817); Thomas Dick's the *Christian Philosopher* (1823); and Timothy Dwight's (1752–1817) *Theology Explained and Defended in a Series of Sermons* (1818). Most significant of these for

---

74. Paine, *The Age of Reason*, 704.
75. Crowe, *The Extraterrestrial Life Debate, 1750–1900*, 213.
76. Paine, *The Age of Reason*, Part 1.

our discussion was Chalmers's seven discourses (five of which deal with the plurality question), which provided several arguments in support of Christianity and pluralism using natural theology. Chalmers reasoned the likelihood of extraterrestrial existence due to the vastness of the universe and the basic similarity of other planets in composition to Earth, citing "other mansions" mentioned by Christ;[77] that it was impossible to assert the moral status of otherworldly beings or their salvation history;[78] and proclaimed unfounded the idea that creation was intended exclusively for Earth. Therefore, he disputed the idea that human beings were the center of God's attention and hence the Fall not was a universal event, which rendered Christ's multiple deaths unnecessary. Conversely, he stated it incompatible with God's omnipotence and omniscience to ignore a small Earth, noting God's attention to the lilies of the field.[79] Chalmers provided a more comprehensive Christian theology to incorporate extraterrestrials when conservative theologians were wrestling with the serious implications raised by Paine; however for much of the public, scientists, and Christians, belief in the plurality persisted.[80]

In 1853 William Whewell (1794–1866), British scientist and Anglican priest, published an anonymous book entitled *Of the Plurality of Worlds: An Essay*, arguing against the natural philosophy and theology of Chalmers. His main position concerned the new geological assertions that humankind appeared only recently compared to the extreme age of the Earth, and provided evidence against the plenitude argument, that God would not be wasting space on other heavenly bodies due to the lack of intelligent life, analogous to Earth where no life existed for long geological periods, concluding that not all divine power must be actualized. His essay initiated a bitter debate with Scottish physicist Sir David Brewster and other intellectuals, as well as many common people over the plurality question in Victorian England. The debate was extensive; however the significant elements in relation to extraterrestrials were Whewell's often impressive scientific bases against a plurality. He indicated that variable stars and gravitation around double stars would not be conducive to planets capable of supporting life, the same being true of eclipsing binary star Algol, and similarly dismissed the outer planets as abodes of life given their low temperatures or low density, as well as the inner planets due to their close proximity to the Sun's heat. He regarded the new geological findings regarding the extreme age of the Earth

---

77. Chalmers, *Discourses on the Christian Revelation*, 14, 17; John 14:2.
78. Chalmers, *Discourses on the Christian Revelation*, 119.
79. Matt 18:12.
80. Todhunter and Whewell, *An Account of His Writings*, 185.

and the relative recent appearance of humanity as evidence of the likelihood of non-intelligent life in the cosmos, which raised questions regarding the non-contemporaneous existence of extraterrestrial intelligence. His primary theological position was anthropomorphic and geocentric—that the uniqueness of the Incarnation for humanity necessitates Christianity's incompatibility with other worlds and beings. Though Whewell agreed that there may exist other inhabitants, they were not created in the image of God; while under his jurisdiction as creatures, they could not enjoy a special relationship with the Creator. In response, Brewster supported the traditional belief of the time that the Incarnation applied to all people, past, present, and future, including the people of the antipodes as well as those populating the cosmos, claiming the atonement of inhabitants of other worlds was simply a logical progression.[81] Further, Brewster claimed Whewell's conception of God omnipotence was inadequate, and that the plurality was necessary to complete the Christian faith. Although Brewster had less scientific support and exhibited deficient biblical exegesis for his position, the debate served to strengthen the possibility of the plurality among the religious minded. In the long-term, the exchange of Whewell and Brewster will be seen (in retrospect to later theologians) as a beginning of framing the central Christian doctrines of Creation, Incarnation, Original Sin, Sanctification, and the Redemption in respect of non-terrestrial beings.

In America, a second figure emerged after Swedenborg to experience visions involving alien beings and a theology incorporating extraterrestrials. Ellen G. White (1827–1915), a prophetess of the Seventh-Day Adventist Church was former member of the Millerite movement.[82] In 1846 she related a theology of extraterrestrials (based upon her own revelations) detailing that among the vast number of intelligences in the universe, only the inhabitants of Earth sinned—thus God's Incarnation within human society was a one-time event unrepeated throughout the universe:

> It was the marvel of all the universe that Christ should humble himself to save fallen man. That he who has passed from star to star, from world to world, superintending all . . . [should take] upon himself human nature, was a mystery which the sinless intelligences of other worlds desired to understand.[83]

---

81. Crowe, *The Extraterrestrial Life Debate, 1750–1900*, 304–5.

82. William Miller founded the Millerites, who shared in the belief in the Second Advent of Jesus Christ in 1844. The movement suffered defeat at the when in October 22, 1844 the day of the prophecy went unfulfilled, and most followers left disillusioned, known as the 'Great Disappointment.' Some former members went on to found new apocalyptic movements, including White.

83. White, *The Story of Patriarchs and Prophets*, 69–70.

White's treatment of the doctrine of original sin sees some resolution here, following the earlier formulation of Whewell. Shortly after White, in 1851, Joseph Smith's Church of Latter-Day Saints endorsed a teaching where other worlds and extraterrestrials were given a principal theological role. Their scriptures[84] advocated a universe of many inhabited worlds of human beings; that some worlds have ceased to exist and others have yet to be created; that all humans, terrestrial and extraterrestrial, are subject to the kingdom of the same Jesus; and that God himself dwells on a specific star or planet known as Kolob. Further, Mormonism scripturally and theologically addressed Augustine's concern regarding the salvation of the people of the Antipodes, teaching that Christ, after his resurrection appeared to the peoples of the New World and spread the gospel.[85]

French theologian Monseigneur de Montignez between 1865 and 1866 wrote nine essays which contended with extraterrestrial intelligence and the doctrines of the Incarnation and Redemption. In his fourth essay he provided a simplistic and poetic expression: "The blood which flowed out at Calvary has gushed out on the universality of creation,"[86] echoing Origen's "The altar was at Jerusalem, but the blood of the victim bathed the universe."[87] Further development of the Christological thought with extraterrestrials was provided by Catholic Neo-Scholastic German theologian Joseph Pohle, who published in 1884 *Die Sternenwelten und ihre Bewohne*, where he employed metaphysical arguments in conjunction with empirical astronomical data to make several claims in support of extraterrestrial beings. His first and second were plenitude arguments, wherein God created the universe for his own glory, and that the highest form of that glorification comes from the creation of intelligent beings. In his third argument, using Aquinas and Secchi as sources, he asserted a universe populated with intelligent beings is more perfect than one consisting of "unadorned deserted wastelands."[88] Fourth, he inferred given the creation of a diversity of lifeforms on Earth, that God would have acted similarly on other planets; and finally, God would create other worlds and extraterrestrial intelligences better disposed to worship him due to the great evil of humanity.[89] Pohle treated briefly but importantly Christ's relationship to supposed extraterrestrials:

---

84. The *Book of Mormon, Doctrine and Covenants*, and *The Pearl of Great Price*.
85. Augustine, *The City of God*, Book XVI, Ch. 9.
86. Crowe, *The Extraterrestrial Life Debate, 1750–1900*, 103.
87. Crowe, *The Extraterrestrial Life Debate, 1750–1900*, 181.
88. Pohle, *Die Sternenwelt und ihre Bewohner*, 416; See Gummersbach, "Pohle, Joseph," *Lexikon für Theologie und Kirche*, 2nd ed., vol. 8, 578.
89. Pohle, *Die Sternenwelt und ihre Bewohner*, 427–29.

Concerning the dogma of the Redemption of fallen men through the God-man Christ, it is not necessary to assume as probable also the fall of species on other celestial bodies. No reason . . . obliges us to think others as evil as ourselves. However even if the evil of sin had gained its pernicious entry into those worlds, so would it not follow from it that also there an Incarnation and Redemption would have to take place. God has at his disposal many other means to remit a sin that weighs either on the individual or on an entire species.[90]

Pohle's treatment of the central doctrines impacted by possible intelligent extraterrestrials constituted the most extensive to that historical point, providing a more precise definition to other possible modes of grace and sin, and non-competing economies of salvation for extraterrestrial races. Another Roman Catholic theologian, Januarius De Concilio (1836–1898), priest and professor, argued from a position of theological geocentrism in favour of extraterrestrial life from the principle of plenitude, that "[God] must, if He would follow the requirements of wisdom, draw from the given forces to be created all the possible good in view of the end," claiming that extraterrestrials exist as an intermediate species between humanity and angels, are created in and through Christ and attain their eternal end, and are redeemed by Christ's Earthly sacrifice,[91] dissenting with Pohle by the inclusion of other intelligent races within the Christian redemptive framework:

God, in His infinite goodness, wanted to make the universe an infinite expression of Himself, at least by union. His divine Word, the infinite expression of His grandeur, came to reside in the universe by uniting to Himself the human nature in the body of His own personality, and thus He divinized the whole universe, inasmuch as human nature represented all its existing species.[92]

Following Pohle, theologians of the twentieth century continued to consider the issue of extraterrestrials within the Christian dogmatic structure, albeit in much smaller numbers due to the advent of planetary exploratory missions, which brought increasing evidence substantiating the absence of life outside Earth in our solar system, and lack of any indications of extrasolar planets which may contain lifeforms. Modern thinkers of the latter twentieth and twenty-first centuries, writing during the height of the space age, would have more significance for our contemporary discussion on the implications for Christian theology and intelligent extraterrestrials.

90. Pohle, *Die Sternenwelt und ihre Bewohner*, 457–58.
91. De Concilio, *Harmony between Science and Revelation*, 207, 215.
92. Weintraub, *Religions and Extraterrestrial Life*, 94.

Conclusion

By 1916 the plurality of worlds and the question of extraterrestrial life had produced, by Michael Crowe's calculation, about 140 books, not including thousands of essays, papers, and reviews by not unlearned thinkers. Perhaps three-quarters of astronomers and possibly almost half of the most prominent Western intellectuals of the eighteenth and nineteenth centuries took part in the discussion of the possibility of extraterrestrial life,[93] the overwhelming majority which served as its advocates. Crowe[94] and Dick[95] have pointed out that the history of the debate, beginning in antiquity and an important subject to philosophers, theologians, and scientists, and throughout virtually every century to the present, repudiates the view that the question of other worlds and extraterrestrial life are a mere product of the space age. Often, the plurality question appeared as a necessary outgrowth of cosmological systems and others' attempts to validate or invalidate them with recourse to metaphysical arguments, appeals to religious doctrine, or scientific observations. In every instance, the plurality concept showed remarkable resilience and adaptability to fit new philosophical and theological environments that either threatened or supported its plausibility, largely due to it being an unfalsifiable hypothesis given limitations in observational technology to confirm or deny its reality. Astronomers lacking empirical evidence designed metaphysical conjectures to support their advocacy of extraterrestrials, and as negative evidence about life in the solar system grew in the nineteenth century, pluralists modified their opinions rather than discard the idea. From a scientific viewpoint, early scientific-minded thinkers misconstrued what later has been determined as the necessary conditions of life, many assuming that a mere atmosphere sufficed to support lifeforms. These conditions have now been quantified with much more precision, and remain the subject of intense research among astrobiologists with discoveries of extremophiles, habitable zones, and the possibilities of extraterrestrial organic chemistries based on carbon, and conceivably, silicon. In addition to these, the strong historical advocacy of plurality and of intelligent extraterrestrials within the broad range of thinkers and periods can be considered in no small part a religious quest to confirm the existence of other beings, existing beyond the "heavenly realm" of older cosmologies. These beings, by their non-existence, served to confirm humanity's special and solitary creation, and unique relationship with

---

93. Crowe, *The Extraterrestrial Life Debate, 1750–1900*, 547.
94. Crowe, *The Extraterrestrial Life Debate, 1750–1900*, 547.
95. Dick, *Plurality of Worlds*, 1.

God in support of biblical and orthodox theology; and by their possible existence, affirmed the ancient longing to know that we are not alone in the universe, and that humanity coexists with beings who may serve as helpers or mediators more spiritually advanced, carrying divine messages to bring us greater understanding of ourselves and the Creator.

Review of the aforementioned thinkers who considered the plurality of worlds concept from antiquity to the end of the nineteenth century indicates a varying but nonetheless consistent conceptual pattern in support of a plurality of worlds, and later, extraterrestrial inhabitants. Preeminent within this array, as we have seen, is the argument from the teleological and metaphysical "Principle of Plenitude." Plato's first postulated powerful ontology of a "Great Chain of Being," the earliest known form of the plenitude, which entailed the "necessarily complete transition of all the ideal possibilities into actuality," became an enduring and powerful force for the plurality.[96] The early Greek atomists were first to argue a plurality from the idea of infinite potentialities of infinite matter, and can be credited with the concept of the plenitude, albeit in embryonic, atheistic form. The notion that God must actualize his potential power, that there exists a divine purpose to all creation where God achieves the greatest results by the least possible means, and of a natural or theistic universe thus compelled to contain all possible forms of existence, became a definitive teleological Christian form for later theistic thinkers. The plenitude concept received greater recognition and theological refinement with the condemnation of 1277, when in an effort to protect God's freedom and omnipotence against false and heretical propositions, the Church affirmed that God could create other worlds if he so willed in a direct declaration against Aristotelian physics and cosmology, while denying that he did so. This was reinforced by developments in natural theology led by growing distaste on the negative authority of scripture; increased support for heliocentrism; clearer observations of the moon that did not reveal a perfectly polished sphere as Aristotle claimed; and analogical and empirical arguments which enriched the image of God in the seventeenth century as active throughout the universe in creating other worlds and inhabitants. As a result, the ontological form of the plenitude concept was further strengthened in the eighteenth century by the addition of the metaphysical notion of cosmic uniformitarianism.[97] In the late nineteenth century, Pohle's thesis exemplified the Christian synthesis of the plenitude argument: that the Creator's greatness and an

---

96. Lovejoy, *The Great Chain of Being*, 50.

97. Also known as the Doctrine of Uniformity, it refers to invariance in foundational scientific principles in nature, in physical laws governing matter, time, and space, and as a constancy in causality.

incomprehensibly vast universe are *a priori* compatible with the genesis of life (specifically intelligent life), of each extraterrestrial civilization having its own, unique relationship with the Creator; and that God's creation is completely fulfilled in the contemplation of rational beings.

The plenitude concept in modernity has received much attention in the cosmic discoveries of other galaxies, quasars, black holes, dark matter and others, and most importantly in the detection of exoplanets. The question of the plurality of worlds, debated since antiquity, has been answered in the affirmative by science. However, the intimately linked and much-debated subject of extraterrestrial life remains a profound question, as do the major religious and theological consequences in the event of its discovery or contact with extraterrestrials. As extraterrestrial life has not yet been scientifically proven to exist (or not exist), the modern debate continues to employ scientific, as well as philosophical, metaphysical, and religious arguments. Given that humanity lacks verifiable information confirming the *exact* conditions and genesis of life on Earth, and while the argument over the twentieth and twenty first century has become more science-centered, the possibility of extraterrestrial life and thus the veracity of the plenitude concept continues to be debated along with juxtaposed ideologies of chance and necessity.

Until the general acceptance of Copernicanism within the intellectual world of the seventeenth and eighteenth centuries in western Europe,[98] and the removal of Copernicus's *De revolutionibus* from the Catholic Church's list of forbidden books in 1758,[99] the cosmology of Judeo-Christian tradition remained geocentric and anthropocentric as well as universal and transcendental, with the scope of God's creation within the original context of the formation of scripture limited to a Ptolemaic universe composed of a spherical Earth encapsulated by a finite sphere of fixed stars, itself surrounded by an infinite void.[100] Wholly missing was any conception of a cosmos and deity that would allow for a theology that included the existence of extraterrestrial life forms, apart from Giordano Bruno's teaching of a populated homogenous edgeless universe in the latter sixteenth century. Copernican theory by the seventeenth century and well into the twentieth had generated a passionate search for other solar systems, albeit a difficult one due to formidable limitations of observational technology, lending itself to ambiguous results with

---

98. Haldane, *Descartes*, 292.

99. Finochiario, *Retrying Galileo*.

100. Fraser, *The Cosmos*, 14. The basic tenets of Greek geocentrism were well established by the time of Aristotle, and the Ptolemaic system was further developed and standardized by Hellenistic astronomer Claudius Ptolemaeus in the 2nd century AD, and superseded only until the advent of Copernicanism.

competing interpretations. Now, with highly sophisticated astronomical equipment and a century of careful observation, the premise of Copernican theory had been affirmed and expanded from heliocentrism, to Wallace's galactocentrism, to our modern sense of a centerless homogeneous universe composed of like elements and subject to consistent and definitive laws based on Newtonian and Einsteinian physics.

Early scientists of the telescopic era, inheriting the arguments in support of a plurality of worlds of the Middle Ages, often sought to confirm their predilection for the existence of extraterrestrial inhabitants by means of conclusions drawn on often imprecise observations, while building upon the older arguments their metaphysical presumptions. In the seventeenth century, metaphysical concepts such as the plenitude became significant only when consistent evidence from observation and better established physical theories were accepted. These metaphysical presumptions of centuries past and their influence on astronomy in the search for extraterrestrial life have now been fully supplanted by modern physics, including relativity and quantum theory, and the disciplines encompassed by the field of astrobiology.

The absence of scriptural references to extraterrestrial life, its emphasis on humanity's special creation and Earth's central role in the cosmos and the scriptural account of one incarnated divine being were, until only the early eighteenth century provided *prima facie* evidence of a single world populated by intelligent beings created in the image of God. The serious implications for Christianity in relation to systematic and biblical theology—specifically the doctrines of the Incarnation and Redemption in consideration of intelligent extraterrestrials, were positioned in direct confrontation with established orthodoxy, as best expressed in Paine's *Age of Reason*. The majority of Christian theologians prior to the thirteenth century, with the exception of Origen and William of Ockham, saw the plurality as wholly incompatible with doctrine, and due to fear of Church censors few of its advocates are notable until the seventeenth to nineteenth centuries with the gradual acceptance of Copernican cosmology.

The Reformation imposed a solely biblical perspective until the Enlightenment, where liberal theologians began to consider a more philosophical perspective of an intelligible universe and a plurality of worlds. During the first decades of the twentieth century, theological interest in intelligent extraterrestrials waned in comparison to the more intensive debate of the eighteenth and nineteenth centuries due to the advent of the space exploration, including rocketry and satellite capabilities, telescopic and radio technology, and planetary probes which brought unprecedented knowledge of our local system and the composition of the universe, which have laid to rest the scriptural, theological, metaphysical, and pseudo-scientific objections against the existence of other worlds from the pre-scientific era. We now know they exist, via the *Kepler* telescope and other Earth-based telescopes, and in great numbers. However there remain arguments against complex extraterrestrial life, premised on some of the old Whewellian arguments. In planetary astronomy and astrobiology, the Rare Earth hypothesis argues an improbable combination of fortuitous geological and astrophysical events is required for an Earth-type habitat to form, hence humanity is either alone or excessively rare in the universe.[101] However, new data on exoplanets continues to emerge at a rapid pace and this issue may be closer to resolution in the near future. A successful SETI message or exchange, as well as future telescopic missions designed to spectrographically detect signs of industrialization in extraterrestrial atmospheres would eliminate these doubts. The theological objections to extraterrestrials from the beginning of the space age have loosened according to popular opinion (see aforementioned Peters and Alexander surveys in the introduction); but have not been resolved theologically for two main reasons: we have not had contact with an extraterrestrial civilization which could provide the appropriate amount of credible information needed to devise the corresponding theological response; and in anticipation of possible future extraterrestrial discovery/contact, Christian theology has not deliberately and systematically examined the implications and possible adjustments/clarifications that might be necessary.

Section B. The Plurality of Worlds and Christian Theological Formulations

This section will provide a new classification of the historical theological approaches to the consideration of intelligent extraterrestrials in conjunction with the doctrines of the Incarnation and Redemption. Acceptance of the plurality of worlds concept and its later, closely associated outgrowth of the postulate of the existence of intelligent extraterrestrial inhabitants of other worlds experienced major obstacles within the framework of Church authority and Christian theology before the early eighteenth century. Subsequent development within Protestant natural theology led to supportive opinions where the plurality was embraced as evidence of God's omnipotence and magnanimity among a variety of philosophical and religious systems, most notably evangelicals. Meanwhile, three new groups fully incorporated the plurality and extraterrestrials into their theological canons: the Swedenborgians, Seventh-day Adventists, and Mormons—each claiming new and

101. See Morris, *Life's Solution*.

divinely inspired mystical revelations affirming Christ's redemption of humans and cosmic inhabitants. From antiquity to the early twentieth century, the plurality had evolved from Greek philosophical atheistic atomism, to metaphysical and philosophical possibility, to Christian theological heterodoxy, to early scientific hypothetical probability, to alternative religious orthodoxy. Despite the Roman Church's and Protestantism's eventual acceptance of heliocentrism and relaxation of natural theology with reference to extraterrestrial beings, the theological integration of extraterrestrial intelligences with the directly associated doctrines of the Incarnation and Redemption of Christ had evaded final or satisfactory resolution in the minds of many thinkers. The majority of plurality advocates up to 1900 (and following, to be discussed in next section) formulated incarnational and soteriological theologies comprising several consistent historical distinctions. This thesis will propose that each historical approach can be classified according to four basic forms. These can be termed the *inclusive, exclusive, multiple*, and *varied* types or positions.[102] The *inclusive* type[103] are those holding one incarnation of Christ on Earth within a cosmos of fallen, sinful or sinless extraterrestrials—where the salvific effects of Christ's work on Earth encompass all intelligent free beings universally. In this scenario, extraterrestrials may or may not have knowledge of their redemption won on Earth, some writers, therefore, postulate a future space-faring humanity with an imperative to evangelize extraterrestrial civilizations. Major figures espousing this view were Vorilong, Brewster, Henry More, George Adams, and Montignez. Second, an *exclusive* type[104] are those holding one exclusive incarnation of Christ on Earth for

---

102. Taxonomy is mine.

103. Prominent *inclusive type* soteriological proponents include: William de Vorilong, also known as Vaurouillon, Vorilongus, Vaurillon, or Vorrilon; see Brady, *William of Vaurouillon*, 291–315; Campanella, *Apologia pro Galileo*; More, *Divine Dialogs*, 523–36; Porteus, "On the Christian Doctrine of Redemption," 59–86; Adams, *Lectures*, vol. IV, 244; Brockes, *Irdisches Vergnügen in Gott*, vol. I, 435; Fuller, *The Gospel Its Own Witness*, 270–83; Nares, *An Attempt*, 18; de Maistre, *Soirées de Saint-Pétersbourg*, II, 319–20; Sir William Rowan Hamilton, according to Graves's *Life of Sir William Rowan Hamilton*, 383; Noble, *Astronomical Doctrine of a Plurality of Worlds*, 33–48; Chalmers, *Astronomical Discourses*, Discourse II 69–70, Discourse III 89–90, Discourse IV 130, 134–35; Birks, *Modern Astronomy*, 61–62; Miller, *Geology Versus Astronomy*, 33; Crampton, *Testimony of the Heavens to Their Creator*, 19–23; Flammarion, *La pluralite des mondes habites*, 340ff; Ebrard, *Der Glaube*; Ebrard, *Apologetik*, 355–56; de Concilio, *Harmony between Science and Revelation*, 215; Brewster, *More Worlds than One*; Powell, *The Unity of Worlds and of Nature*, 291; de Montiqnuez, *Theorie chretinne sur la pluralite des mondes*, essay 10, 274–75; Courbet, *Cosmos*, 4th ser., 28:273–76; Ortolan, *Astronomie et Theologiei ou l'erreur géocentrique*, 320–21.

104. Prominent *exclusive type* soteriological proponents include: Dwight, *Theology Explained and Defended*, V, 509; Friedrich Gottlied Klopstock in Wöhlert, *Das Weltbild in Klopstocks's Messias*, 36–37; Kurtz, *Die Astronomie und die Bibel*; Kurtz,

sinful humans among multitudes of unfallen, sinless extraterrestrials or in some instances, fallen, sinful ones, with redemption limited to humanity, unnecessary for sinless extraterrestrials, or unavailable to those with sin.[105] Advocates for this type included William of Ockham, Whewell, and Campanella. Third, a *multiple* type[106] are those holding that Christ, in the Creator's free offer of redemption, as the Logos incarnates and takes the form of each fallen and sinful intelligent extraterrestrial species throughout the universe. Prominent figures holding this view include Baden Power, who attempted to find a middle ground between Brewster and Whewell, Abbe Jean Terrasson, and R. M. Johuan. Fourth, a *varied* type[107] are those maintaining the position that God manifests *or* makes himself known within extraterrestrial societies in a variety of modes according to his own desires and ways, *not necessarily* by incarnation. Advocates of this type include Fontenelle, Bentley, John Keill, Chalmers, and Pohle. In the next chapter, each of these views will be evaluated in turn with discussion of twentieth century theologians from the beginning of the space age to the present. Specific attention will be given to exploring the *varied* type. Following this summary of the work in the history of the pursuits of other planets and extraterrestrial life before the twentieth century, we now turn to consider the relevant modern theologians and scientific opinion.

Section C: Theological Thought on Extraterrestrials in the Twentieth and Twenty-First Centuries

The preceding sections on the historical and contemporary scientific and theological thought provide the physical and conceptual setting for a serious consideration of intelligent extraterrestrials within the context of twenty-first century Christian religion. The discovery of multitudes of galaxies by means of more advanced telescopes in the early twentieth century such a the 1897 *Yerkes* 100 cm refractor in Wisconsin, the 1917 2.5 meter *Wilson*

The Bible and Astronomy, 509; Burque, *Pluralité des mondes*, 246–61; Maunder, *Initia doctrinae physicae*, 1:221; Leitch, *God's Glory in the Heavens*, vol. IX, 461–62.

105. Christian faith's enumeration of God's attributes renders incoherent a theology of extraterrestrials where the divinity does not at some historical point have provision for the redemption of an intelligent creation.

106. Prominent *multiple type* incarnationists include: Haye, *Religion Philosophi*; Jouan, *La question de l'habitabilité des mondes*; Terrasson, *Traité de l'infini créé*; Pohle, *Die Sternenwelten und ihre Bewohner*, 457–58.

107. Prominent among those holding the *varied type* position include: Keill, *Introductio ad veram astronomiam*; de Fontenelle, *Entretiens sur la pluralite des mondes*, in Marsack, *The Achievement of Bernard le Bovier de Fontenelle*, 125.; Chalmers, *Discourses on the Christian Revelation*, 14, 17; John 14:2.; Bentley, *A Confutation of Atheism*, 358.

reflector telescope in California, and in 1949 the 200 inch *Hale* telescope at Palomar in California led the way to the later, more advanced telescopes *Hubble* and *Kepler*, which greatly expanded our view of our place in the universe by more accurately determining the age and extent of the cosmos, as well as the existence of exoplanets. In the last one hundred years, we have realized the true vastness of an expanding universe, with better understanding of the formation of galaxies, stars, and planets due to advances in astrophysics, increased knowledge of habitable extraterrestrial environments from astrobiology, and confirmed data on multitudes of exoplanets from space and Earth-based telescopes. Each has contributed to provide an unprecedented awareness of humanity's place within an immense cosmos, demonstrated the limitations of philosophical and theological anthropocentrism, and enhanced the potentialities of discovering extraterrestrial habitats and life. Research performed by key organizations as NASA's astrobiology institute and SETI have indicated the vital importance of future studies of the possibility of extraterrestrial life, and its impact on science, culture, and religious structures. Some of these have been performed and contingency plans have been developed, however these only consider radio message transmissions and pertain strictly to scientists—no formal governmental protocols have been established to date. Among those studies which feature significant social science research, little or no consideration or input has been provided by theologians in forecasting consequences for religious dogmas or religious institutions.

In regards the thought of Christian theologians in our modern era, this chapter will demonstrate that of the twentieth and twenty-first thinkers who have considered the aforementioned doctrinal positions with regard to intelligent extraterrestrials, most have merely re-presented variations of the medieval and renaissance arguments, while none has proposed a comprehensive, systematized formulation to accommodate intelligent extraterrestrials within established Trinitarian and Christological doctrine. For this thesis, specific attention will be given to the *varied* type incarnational position, which historically has had only a handful of proponents and the least theological development (in fact generally only a few sentences or paragraphs in each source). As the discussion has shown, the Roman Catholic Church since the seventeenth century has not considered the possibility of intelligent extraterrestrials antithetical to the Christian scriptures or theology; however precise definitions regarding extraterrestrial relation to the Christian religion remain incomplete, unformulated, and unresolved to the present time.

Exclusivist

The *exclusive* view represents the "classical" approach by theologians, most prominent in pre-scientific worldviews, and exemplified best by the thought of Augustine, Aquinas, and Whewell. By the middle of the twentieth century, this perspective had lost favor among the few theologians, astronomers, and philosophers who considered intelligent extraterrestrials in a rapidly advancing scientific era, as evidenced by the works of Mascall, Raible, and McHugh (discussed below). Among the handful of modern theologians who have published on the subject, this perspective has few proponents. This approach took one of the following forms: a) most noted among pre-scientific age theologians is the position that the Incarnation of God in Christ is a one-time unrepeatable event on Earth for the edification and salvation of humans in a universe where no other intelligences exist; b) among more modern theologians, that the Incarnation is for humans only as extraterrestrial beings, if they exist, are without sin and need no saviour; or c) the Incarnation is exclusively for humans, and alien beings, if they are sinful are without redemption due to their non-relationship with Christ. As indicated earlier, this hypothesis did not represent any real effort to reconcile Christianity with the existence of intelligent extraterrestrials, and typically relied upon philosophical, theological, biblical, or scientific "arguments of absence" against the existence of extraterrestrials. Joseph Breig in his 1960 article "Man Stands Alone" was the earliest writer of the period defending the exclusivist position by claiming a singular divine act in all creation: "[That there was only] one Incarnation, one mother of God, one race into which God has poured His image and likeness."[108] Likewise, Jesuit Paul Steidl in his 1979 book *The Earth, the Stars, and the Bible*, rather than contend with potential relations between humans and a second genesis of intelligence, provided a popular, simplistic exclusivist argument by merely dismissing any possibility of extraterrestrial existence:

> He is moving his home to Earth permanently in a wonderful marriage of heaven and Earth. What does this mean for our question of life on other planets? It shows God's ultimate eschatological plan is Earth-centered. In the end, God, the Lord of the Universe lives on Earth. Does this mean that intelligent races on other planets will come up to planet Earth to worship God just as the Gentiles came up to Jerusalem to worship Israel's God?

---

108. Breig, "Man Stands Alone," 294–95.

Again, the simpler solution is to reject the notion that there is life on other planets.[109]

Although Steidl makes an oblique reference to a possible inclusivist approach of incorporating non-human intelligences under the headship of the earthly Incarnation of Christ, his 'simpler' solution is a refusal to consider the actual implications by affirming a single intelligent creation. Brian Hebblethwaite also defended the exclusivist position based on the denial of extraterrestrial existence. In his article "The Impossibility of Multiple Incarnations" he rejects the idea of other incarnations of Christ, arguing that if the earthly, human Jesus is the same person as the divine God the Son in the hypostatic union, then other incarnations must also be in that form; for the Second Person to have another simultaneous form would be nonsensical. He writes, "One individual subject cannot, without contradiction, be thought capable of becoming a series of individuals, or, a fortiori, a coexistent community of persons . . . [therefore] only one human subject can be the incarnate, human form of that one divine life."[110] As there could only be one Incarnation of the divine Son in a finite *personal* form, according to Hebblethwaite, it would make more sense to suppose that humanity is the sole instantiation of finite personal life in the universe, and as extraterrestrials are without access to Christ's redemption they cannot receive salvation; they must not exist. In his argument, Hebblethwaite rejects the possibility of any further human as well as extraterrestrial incarnations. However, as the Second Person of the Trinity is a divine, infinite person, multiple thinkers discussed below will maintain that an unlimited, divine being can unite itself with a variety of beings simultaneously and over vast expanses of space and time, manifesting in any manner desired; to argue a human incarnation poses limits to God's ability to incarnate would be considered incompatible with God's omni-properties. In accord with Hebblethwaite and Breig, Aristotelian-Thomist Marie George, in *Christianity and Extraterrestrials? A Catholic Perspective* also rejects the idea of the existence of extraterrestrials by referencing Aquinas's denial of extraterrestrial beings in her advocacy of human Christocentric exclusivity. She indicates

---

109. Steidl, *The Earth, the Stars, and the Bible*, 230–32.

110. Hebblethwaite, "The Impossibility of Multiple Incarnations," 323–34. William Letich in his 1862, *God's Glory in the Heavens*, argues similarly, stating in God's Incarnation on Earth within the context of a vast cosmos: "He will forever bear his human nature." Lutheran Johannes Brenz observed that the communication idomatum in the person of Christ requires us to think of the humanity of Christ as forever conjoined with his divinity. "For if the deity of Christ is anywhere without his humanity, there are two persons, not one." Rahner, *De personali unione duarum naturam in Christo*, 3–4, (quoted in Jenson, *Systematic Theology: Volume 1*, 203, n40).

that lacking our scientific perspective, Aquinas did not consider intelligent inhabitants of other planets as he believed Earth analogs did not exist in the first place; if there were extraterrestrials in the universe, he believed they may exist as animated celestial bodies containing an intellectual substance (as philosophers of his time believed).[111] As such, he did not envision our modern notion of the biological extraterrestrial. Given our present knowledge confirms celestial bodies are not life forms, George maintains the exclusivity of the Incarnation for humans amidst a non-existence of outside intelligences.[112] The exclusivist view makes no attempt to incorporate intelligences outside Earth into divine earthly revelation, and reveals the difficulty of holding an exclusive argument for a singular divine election of humans, forcing the question as to whether it is more plausible (and responsible) for Christian theologians to deny the existence of extraterrestrials and/or God's involvement with other intelligences and state the Incarnation of God on Earth was the only instance of divine activity within an inconceivably vast universe of ≈13.8 billion light years; or to acknowledge and provide some accounting of the actual theological possibility of other, outside divinely created creatures. As discussed below, inclusivist thinkers would consider more fully divine activity among an array of other putative beings within established Christian doctrine.

Inclusivist

The *inclusivist* view represents the first modification of the exclusivist "classical" approach to the consideration of putative intelligent extraterrestrials within an orthodox Christological framework. It extends the salvation won by Christ on Calvary to intelligent beings throughout the cosmos while maintaining the singularity and uniqueness of the Christ event for humans on Earth. Inclusivists generally deem original sin as universal and extended throughout the universe if it be inhabited by intelligence, with varying views on the locale, impetus, extent, and nature of that sin throughout space, time,

---

111. Aquinas, *Summa Contra Gentiles*, II, 94; Aquinas, *Quaestio Disputata de Anima*, 3. Aquinas derived his understanding of celestial spheres as possessing a soul and intelligence. In his *Treatise on Separate Substances*, he states, "between us and the highest God, there exist only a two-fold order of intellectual substances, namely, the separate substances which are the ends of the heavenly motions; and the souls of the spheres, which move through appetite and desire." (Inter nos et summum Deum non ponitur nisi duplex ordo intellectualium substantiarum, scilicet substantiae separatae quae sunt fines coelestium motuum, et animae orbium quae sunt moventes per appetitum et desiderium).

112. George, *Christianity and Extraterrestrials?*

personages, and civilizations. The Pauline hymns in Colossians 1:15–20, Ephesians 1:20–23; and references to Hebrews 2:7–9; and Romans 6:10, according to inclusive thinkers, reveal a Christocentric universe, and are often cited as scriptural justification for a "Cosmic Christ" in domination of all creation, earthly and otherwise. For some thinkers, such as Jürgen Moltmann the doctrine of original sin may not be universal or cosmic in scope, however the redemption wrought by Christ is. As we have seen, by the fifteenth century theologians considered the possibility of other worlds real, and whether Christ in his terrestrial sacrifice could achieve redemption for intelligences on other worlds. Christ could not suffer and die again on another world, as inclusivists would cite Hebrews 9:25–26 that Christ "suffered once and for all."[113] It is worth noting that a majority of Roman Catholic and Anglican theologians (discussed below) who have speculated on the Christological implications of intelligent extraterrestrials in the twentieth and twenty-first century occupy the inclusive axis, such as Milne, Teilhard de Chardin, and Consolmagno, each having abandoned the rigidity of the pre-scientific classical solution offered by the likes of Aristotle, Augustine, Aquinas, and modern exclusivists discussed above.

British astrophysicist Edward A. Milne's solution to the problem of Paine's disdain for the notion of a single and unique "planet hopping" suffering messiah was to insist on the singular and unrepeatable divine act on Earth within a universe populated by other possible beings. In his 1952 *Modern Cosmology and the Christian Idea of God* Milne continued the tradition of systematic avoidance of the possibility of potential divine acts or absence of such acts with other intelligent races, or consideration of their ultimate spiritual relationship and destiny:

> God's most notable intervention in the actual historical process, according to the Christian outlook, was the Incarnation. Was this a unique event, or has it been re-enacted on each of a countless number of planets? The Christian would recoil in horror from such a conclusion. We cannot imagine a Son of God suffering vicariously on each of a myriad of planets. The Christian would avoid this conclusion by the definite supposition that our planet is in fact unique. What then of the possible denizens of other planets, if the Incarnation occurred only on our own? We are in deep waters here in a sea of great mysteries.[114]

Milne's assertion of a single divine intervention in one species within the universe, while maintaining the existence of outside divinely created

---

113. McColley and Miller, "Saint Bonaventure," 388.
114. Milne, *Modern Cosmology*, 153.

intelligent beings but without possessing a means to redemption locally, created a quandary where salvation of extraterrestrials is unavailable until the creation of a remote humanity, spatially and temporally, and the advent of a single, terrestrial saviour. By insisting on the uniqueness and unrepreatability of the Christian message, salvation of extraterrestrials, according to Milne, could only be accomplished through the efforts of humans spreading the gospel throughout space. Given the isolation of humanity in the cosmos and by the use of the then new radio telescopic technology, humans could make known to other civilizations our salvation history and the specialness of the Incarnation so as to motivate these others to accept a Christian religious faith:

> In that case there would be no difficulty in the uniqueness of the historical event of the Incarnation. For knowledge of it would be capable of being transmitted by signals to other planets and the re-enactment of the tragedy of the crucifixion in other planets would be unnecessary.[115]

Milne failed to acknowledge that radio waves, which travel at the speed of light, diffuse and scatter given the vast distances in interstellar space, and hence would be unable to reach most of the supposed inhabitants of the universe. Anglican theologian E. L. Mascall, in his book *Christian Theology and Natural Science* criticized Milne's cosmic evangelization scheme in requiring humans to carry out messages to other distant civilizations, regarding it as nonsensical, adding that God was fully capable of making himself known to all creatures. Mascall was open to other ways that God could redeem a creature beyond Incarnation, as discussed below among *multiple* thinkers.

A more comprehensive accounting of the mechanisms of redemption available to putative intelligently created extraterrestrials was developed in the works of Jesuit priest and paleontologist Pierre Teilhard de Chardin, who held both inclusive and multiple incarnational views. Several of his works were suppressed by the Roman Church during his lifetime in the early twentieth century, and only later acknowledged. The central teaching of his principal work, *The Phenomenon of Man*, argued for an evolution of the universe from stages of inorganic matter, to the organic, to life, and to the human; with the divine purpose to advance matter from an inanimate stage to a conscious stage. Within this framework on cosmic timescales, Teilhard de Chardin proposed the historical, Earthly Jesus as the new summit and purpose of divine creation, the Logos fulfilling the goal of divine creation by uniting the human with the divine, made present and revealed

---

115. Milne, *Modern Cosmology*, 153.

in every creative evolutionary process.[116] He also postulated that evil was "universal" in the cosmos and existed in all realms and epochs, while not explaining, if universal in scope within creation, its definitive and ultimate origin. Teilhard de Chardin maintained a Christocentric universe, while emphasizing the action of a third "cosmic" nature of the Word in Christ whose duty was the work of recapitulating in him all creation and all beings which belong to it.[117] However, Teilhard de Chardin struggled to demonstrate how this recapitulation and resulting redemption of all creatures throughout the cosmos is accomplished:

> The hypothesis of a special revelation, in some millions of centuries to come, teaching the inhabitants of the system of Andromeda that the Word was incarnate on Earth is just ridiculous. All that I can entertain is the possibility of a multi-aspect redemption which would be realized on all the stars. There were worlds before our own, and there will be worlds after it . . . Unless we introduce a relativity into time we should have to admit, surely, that Christ has still to be incarnate in some as yet unformed star? There are times when one almost despairs of being able to disentangle Catholic dogmas from the geocentrism in the framework of which they were born.[118]

Teilhard de Chardin argued for a redemption available to all creatures, and that humanity cannot be considered the sole location for salvation from spiritual death: "The idea of an Earth chosen arbitrarily from countless others as the focus of the redemption is not one that I can accept." He encountered the same predicament articulated by Paine in maintaining a Christocentric universe, while the Christian redemption remained inapplicable to extraterrestrials. To resolve his dilemma, he realized he must argue for a Christ incarnated on other planets, or maintain that the Christian redemption only concerned humans. He was uncomfortable with either solution, and concluded that Christ on Earth was one of a multiplicity of incarnations among a variety of beings in the universe.[119] A later inclu-

---

116. Teilhard de Chardin, *The Phenomenon of Man*.

117. Teilhard de Chardin, *The Heart of the Matter*, 93; Teilhard de Chardin devotes only a rudimentary essay to the topic. He holds the centrality of Christ in the cosmos in the "strong sense" and incorporates this third dimension of nature of the Word, (*La multiplicite des mondes habites*, 282). In this way he avoids the problem of anthropocentrism but introduces a concept foreign to the long-accepted teaching of the dual nature of Christ.

118. Teilhard de Chardin, *Hymn of the Universe*, 44.

119. Lyons provided a comparison of Teilhard de Chardin's evolutionary theology to Origen, who presented Christ's redemptive worked as a transcendent action which

sivist, Cistercian Roch Kereszty in an appendix titled, "Christ and Possible other Universes and Extraterrestrial Beings" to his 1991 book *Jesus Christ: Fundamentals of Theology*, did not examine the issue of universal redemption as thoroughly as Teilhard de Chardin, conceding the possibility of extraterrestrial beings as "theologically possible but not more probable than their non-existence." He argued that intelligent extraterrestrial beings would have been created "in the image of the Son of God . . . because they are endowed with intellect and freedom . . . in some sort of communion with God the Father through the Son."[120] However, Kereszty provided no formulations on this "communion" between divinity and creature according to his Roman Catholic faith. Similarly, Brother Guy Consolmagno, Jesuit Vatican astronomer in *Astronomer: Adventures of a Vatican Scientist* makes the claim of a cosmic Christological reality, however like Kereszty, he provides no articulations of the relationship between extraterrestrials and the Christ as understood by humans. "It is not just humankind, but the whole of creation, that was transformed and elevated by the existence of Christ . . . finding any sort of life off planet Earth, either bacteria or extraterrestrials, would pose no problem for religion."[121] This assertion is problematic given the historical paucity of exotheological research, current deficit of knowledge of actual extraterrestrial life forms, and absence of comprehensive social studies on religious responses to contact/discovery. Consolmagno does not discuss the fundamental question of original sin for extraterrestrials, or explain how the message of Christ's redemption for all would be known universally:

> St. Paul's hymns in Colossians 1:15–20 and 1 Ephesians make it clear that the resurrection of Christ applies to all creation. It is the definitive salvation event for the cosmos. Another bit of biblical evidence is the opening of Johns Gospel, who tells us that the Word (which is to say, the Incarnation of God) was

---

gradually through time takes effect in every realm of creation but which, nevertheless, needs to find corporeal expression in a particular place on a particular occasion. Teilhard de Chardin, on the other hand, looks at the Redemption from within the creative process. It is like a feedback control, supplying a compensating correction to the process in order to bring it to a successful conclusion. "Christ's redemption is but a single activity; nevertheless, on the supposition that the universe contains a plurality of inhabited worlds, its presence must be multiplied throughout those worlds . . . Such a multiplied presence presupposes a multiplicity of Incarnations on the part of Christ . . . New knowledge about the physical cosmos leads to new suggestions about what a fully cosmic redemption entails." Lyons, *The Cosmic Christ*, 214.

120. Kereszty, *Jesus Christ*, 376–81.

121. Consolmagno, *Brother Astronomer*, 150–52.

present from the Beginning; it is part and parcel of the woof and weave of the universe.[122]

Consolmagno affirms salvation for all beings through a preexistent cosmic Christ of all creation, and admits "ET's may not be aware of the idea of an Incarnation, or they may have their own experience of the matter. Their experience may be so alien from ours . . . that we will never be able to share, nor they share in our experience."[123] This projected geocentralized and universalized Christology is shared by Fr. George Coyne, SJ, a colleague of Consolmagno, who also suggests a single Incarnation on Earth as applicable to all created intelligences:

> God chose a very specific way to redeem human beings . . . There is deeply embedded in Christian theology . . . especially in St. Paul and St. John the evangelist, the notion of the universality of God's redemption and even the notion that all creation, even the inanimate, participates in some way in his redemption.[124]

Episcopal priest and physicist George Murphy in his article, "Cosmology and Christology" argues similarly to Hebblethwaite, in regard to a Second Person of the Trinity limited to one incarnation given the uniqueness of the hypostatic union. He thus describes the Logos as the universe's 'pattern maker,' whereby Jesus as Logos cannot be just one aspect of the divine nature, one actualized pattern among many potential patterns, or he could not be regarded as the fullness of God incarnate. Therefore, Murphy concludes his nature, to include divine and human only, must be the foundation for the whole universe and for all other potential universes, whereby multiple incarnations are eliminated.[125] This is a reversed form of the proposition made by Pittenger, (discussed below with *multiple* theologians) whereby the Earthly Incarnation of the divinity provides a full, but nonetheless corporeal form of an infinite, inexhaustible *prosōpon* as corporeality entails limitedness; thus his fullness cannot be exhausted within the human form. Murphy's position that God's fullness is limited according to a particular form is suspect; God cannot be held to limitations consonant with the forms he chooses to manifest among beings, however in the revealing of his 'omniproperties' through creatures he chooses to incarnate he may limit as he designs. In fact, incarnation necessarily entails limitedness, however an infinite being cannot be limited in the actuality of its being, despite any

---

122. Weintraub, *Religions and Extraterrestrial Life*, 104.
123. Weintraub, *Religions and Extraterrestrial Life*, 104.
124. Coyne, "The Evolution of Intelligent Life," 177–88.
125. Murphy, "Cosmology and Christology," 109–11.

appearance. These arguments represent some of the subjects of divine *representation* and *actualization* in creatures of the inclusivist position, and will be explored in more detail in chapter 5.

Another foremost theologian, the late Wolfhart Pannenberg conceded the possibility of extraterrestrial intelligence and like Consolmagno, saw no conflict with regard to traditional Christology, stating "The as yet problematic and vague possibility of their existence in no way affects the credibility of Christian teaching . . . The turning of the Father to each of his creatures . . . is always mediated through the Son." He continued with the classic argument from geocentric anthropocentrism, "The Logos who works throughout the universe became a man and thus gave to humanity and its history a key function in giving to all creation its unity and destiny."[126] Pannenberg envisioned, as did Milne and other inclusivists, that the salvation history of the Earth and special role of human beings served as a focal point to unite with other societies, and in this way communicate and effect the salvation of Christ throughout the universe. However, he did not provide specific formulations on how this salvation is accomplished given the remoteness of humans and Earth. In considering extraterrestrials in his book *Jesus Christ for today's world*, Jürgen Moltmann argues for a cosmic Christology combining elements of the exclusivist and inclusivist position. He claimed since an evolution of intelligent life occurred only once in the universe on Earth, other living beings may be found in various evolutionary stages, but will be without divine presence; thereby eliminating the need to reconcile Christ's redemption with divinely created intelligences. He does, however include these other living beings as part of the divine plan, wherein "The fellowship of all created beings goes ahead of their differentiations and the specific forms given to them."[127] His Christology proclaims a Christ

> [who died] so as to reconcile everything in heaven and on Earth . . . and to bring peace to the whole creation . . . The transition of Christ . . . has cosmic meaning . . . through this transition resurrection has become the universal 'law' of creation, not merely for human beings, but for animals, plants, stones and all cosmic life systems as well.[128]

By means of Christ, he espouses a cosmic Christocentrism; therefore, resurrection has become a new, never-before transition to a higher plane of existence of creation for all living beings and matter; however he makes no attempt to explain how this universal reconciliation is accomplished

126. Pannenberg, *Systematic Theology*, vol. 2., 21, 34–35, 74, 76.
127. Moltmann, *Jesus Christ for Today's World*, 255–58.
128. Moltmann, *Jesus Christ for Today's World*, 255–58 .

theologically or along vast cosmic timescales or distances. This approach does not in reality present an effort to reconcile the Earthly Incarnation and Redemption of Christ with intelligent extraterrestrials, as Moltmann has removed their possibility from the equation, thereby eliminating any formulation necessary to accommodate them within traditional theology. A few years later, John Jefferson Davis wrote in his article "The Search for Extraterrestrial Intelligence and the Christian Doctrine of Redemption," where along with Moltmann and Pannenberg, affirmed a similar hypothesis regarding the cosmic centrality of Christ within the material universe and his kingship over all creatures, while maintaining that the uniqueness of the hypostatic union in Christ is not an expression of anthropocentrism, but rather a consequence of a coherent Christocentrism. That is, the Incarnation of the Word constitutes the greatest self-communication of God to creation, even among all other possible intelligent creatures, as other possible incarnations of the Word would entail a non-Christocentric universe. Davis claims that Colossians 1:15–20 has received inadequate attention, as it portrays redemption as being cosmic in scope, using repeatedly the words "all things, and everything" in support of the "cosmic" view. He set aside the issue of the possibility of unfallen extraterrestrials, those who may have maintained original righteousness;[129] and argues from the view of the biblical centrality of Christ over the entire material universe and headship over all possible creatures. Davis contends that the best theological position is to assert the uniqueness of the hypostatic union, which could occur only once within the earthly economy of salvation. The result for humanity is, therefore, not merely a geochristocentrism where we hold the headship of Christ, the God-man in a "strong sense," but rather that the Incarnation represents God's greatest, unrepeatable, once-for-all universal self-communication to creation amongst all possible creatures in the universe.

The late Monsignor Corrado Balducci, theologian, demonologist, and parapsychologist who studied the subject of UFOs and extraterrestrials and discussed the subject on Italian television, briefly explored the Christological implications of extraterrestrials, again proclaiming a cosmic Christocentrism. In reference to Colossians 1:16–17 in an interview he stated,

> If Christ is the center and head of all creation, no world exists which doesn't refer to Christ, as everything is under the influence of the divine Word and His glory . . . [Extraterrestrial] existence might very well be correlated with the salvation through Christ. Therefore, there exists no world which is not related to

---

129. Davis, "Search for Extraterrestrial Intelligence," 21–34.

Him. As the Word incarnate, he has, as the Bible confirms, an influence on every possible inhabited planet.[130]

Like other inclusivists before him, Balducci did not elaborate on this redemptive correlation or manner of Christic influence throughout the universe. Similar to Consolmagno, Australian Gerald O'Collins in a *National Catholic Reporter* article in 2004 expressed religious and theological optimism in the event of contact/discovery of intelligent extraterrestrials, envisioning contact analogous to that of the Columbian Exchange. He writes, "I don't think the discovery of life on other planets would pose a qualitatively different challenge than the discovery of the New World . . . we survived that, and in the end it deepened our understanding of Christ as a truly universal saviour."[131] O'Collins's stated definitive position here is untenable, as inhabitants of the antipodes after contact were concluded to be humans and hence belonging to the same family of man descended from the first parents. Therefore, the church determined they shared in the same guilt of Adam and hence were included within the same Christian divine plan of salvation on Earth. Intelligent non-human species would not necessarily inherit original sin, nor necessarily participate in the same economy of salvation as humans. In the same article, theologian and Bishop Joseph Augustine Di Noia stated in reference to intelligent extraterrestrials and an Earth-based economy of salvation for non-human intelligences, "If there are other persons in the universe, we can at least say that they too are involved in the same divine plan and are intended to share in the Trinitarian communion of life."[132] Both theologians conceded a possible future need for 'revising' the doctrine of original sin upon the discovery of intelligent extraterrestrials; and like Davis, to include a Christocentric economy of salvation to include all possible beings, however provide no further details. Di Noia makes the same claim as O'Collins in the flat assumption of a human economy of salvation for a non-human being.

Physicist Alex Mok, in his 2005 article "Humanity, Extraterrestrial Life and the Cosmic Christ in Evolutionary Perspective" also cites Colossians and argues in a similar vein to Moltmann, that Christ has restored the cosmic order and transformed the entire creation by means of his death and resurrection. Similar to others that espouse the inclusive view, Mok does not consider the historical cosmological context of a Colossians text, written in a pre-scientific Aristotelian cosmology and which did not include the notion of extrasolar planets contained in interstellar space but a rather

130. Balducci, UFOs and Extraterrestrials."
131. Allen, "This Time, the Catholic Church is Ready."
132. Allen, "This Time, the Catholic Church is Ready."

a "universe" composed of the Earth and an encapsulating heavenly sphere. Mok ties the text to de Chardin's idea of the Cosmic Christ:

> The cosmic character of the Logos is prominent in Colossians 1:15–20. The salvation of Jesus Christ is a once-for-all incident and its efficacy extends not only in time but also in space. This cosmic Christology of Paul is consistent with the conception of evolution that we have so far developed . . . The universal redemption of Christ essentially applies to all created beings, including any extraterrestrial intelligent life that might exist elsewhere in the universe.[133]

Mok dismisses the concept of multiple incarnations of the Logos in beings on other worlds as unnecessary because of an earthly unrepeatable and all-inclusive Incarnation of the Logos, and like Milne, forecasts space evangelization in the remote future, differing only in his space evangelization radio method, whereas Mok proposes future human interstellar missionaries:

> We now take this mission to bring the good news to alien civilizations should they exist. This is scientifically feasible on account of the colonization of the galaxy by our own species in less than 300 million years. Applying the space-travel argument to ourselves, we would have colonized the entire galaxy well before other intelligent beings could successfully evolve on their home planets.[134]

Mok falls in line with the thought of Milne, Moltmann, and Pannenberg in envisioning humanity as the sole possessor of absolute universal truth, with the singular duty to spread the Christian message to all intelligent beings regardless of the likelihood, if intelligences exists, how many entire worlds remain without access to evangelization while humans work to achieve space flight capable of reaching them.

A duplicate argument for inclusivism consistent with "Cosmic Christ" advocates was made by Sjoerd Bonting in his 2005 book *Creation and Double Chaos*, acknowledging "few contemporary theologians show much interest in the matter of possible extraterrestrial life."[135] He asserted that while worlds containing life may be rare, such life like Earth will have carbon-based chemistries, would not be radically different from Homo sapiens in physiology and biochemistry, have brains and neuronal systems resembling ours, and possess similar thought processes. He proposed that

---

133. Mok, "Humanity."
134. Mok, "Humanity."
135. Bonting, "Are We Alone?"

extraterrestrials are sinful and separated from God, but also seek and are offered redemption; however their sin is incomparable to that of the first parents as its source is universal, as argued by Teilhard de Chardin. According to Bonting, by the universality of Christ's sacrifice, the creative work of the Father, the saving work of Christ, and the communicative action of the Holy Spirit will apply just as much to any creature on another planet as they do humans:

> Sinful extraterrestrials will participate in the events in Palestine two thousand years ago, without needless repetition on their home planet, and as God has enabled the Christian message available to all on Earth for all times, he will likewise in his own way make the same message available to all intelligent creatures in the universe.[136]

Bonting's phrasing of the "uniqueness of Christ's one Incarnation" and its "cosmic significance and lasting validity" for all beings in creation which "has been groaning in labor pains until now"[137] encapsulates the inclusive argument among several of its formulators. The modus of communication of the Christ event on Earth to other planets according to their locales and timescales is not provided by Bonting; instead he claims salvation and reconciliation will come to extraterrestrials concurrent with humans upon Christ's return to Earth, and like other inclusive redemptive geocentrists, argues for a Christocentric salvation of all creatures. He rejects the doctrine of original sin as scientifically unprovable, and that sin is a universal phenomenon not originating with Earth inhabitants. Oliver Crisp also affirms a single, earthly Incarnation and the possibility of multiple Incarnations:

> Although such a divine act is metaphysically possible—there is no metaphysical obstacle to God becoming incarnate on more than one occasion—there is good reason to think that the Incarnation is in fact a unique event in the divine life . . . God could have become incarnate more than once, but he has not done so.[138]

Crisp explains that the primary purpose for the earthly Incarnation was the reconciliation and redemption of humanity,[139] and if humanity had not sinned the Incarnation would be rendered unnecessary; therefore, the Incarnation is a unique and one-time event sufficient for the purposes of human salvation. Further, considering the question of extraterrestrial

---

136. Bonting, "Are We Alone?", 587–602.
137. Rom 8:19, 22.
138. Crisp, *God Incarnate*, 155–75.
139. Crisp, *God Incarnate*, 155–75.

incarnations or divine manifestations, he argues that 'the emergence of intelligent life elsewhere in the cosmos is slim,'[140] and therefore, does not consider it further.

Multiple

The twentieth century demonstrated further evolution of the relation between Christological thought and the possibility of intelligent extraterrestrials, with consideration of possible multiple incarnations of the Logos becoming more prominent, dominated by Roman Catholic theologians who began to implicitly invoke Thomistic incarnational theology on the non-heretical position of possible incarnational multiplicities.[141] The traditional doctrine of the Incarnation of God's son on Earth was a unique and unrepeatable divine act, whose purpose was twofold: to communicate a divine revelation for the salvation for humans, and restore lost sanctifying grace in order to fulfill God's desire for reconciliation and redemption. Arguing from a position of a weak or strong Christocentricity, theologians E. L. Mascall, Zubek, Pittenger, Congar, and Delio considered repetitions of this motif and mode of a Logos incarnated according to individual species, which raised questions regarding the ultimate purpose for incarnation in other species. Multiple incarnations assumes extraterrestrials have fallen from grace as did humans, with incarnation considered in primarily two modes taken from the human example—as a requirement for atonement and salvific action on the part of the deity, and a medium of revelation. The *multiple* position contains several varying possibilities concerning which of the three divine persons incarnates; which *nature* is assumed, *when* and *how many* incarnations are possible within our understanding of the space-time continuum; if one rational nature (in our case extraterrestrial) can be assumed by more than one divine person *simultaneously*; and if more than one rational nature can be assumed simultaneously by one or more persons. For the majority of multiple theologians of this period espousing this view, only the incarnation of the Word is considered.

Bishop Frank Weston was an early proponent of the multiple view and in 1920 wrote a treatise titled *The Revelation of Eternal Love: Christianity*

---

140. Crisp, *God Incarnate*, 156.

141. Aquinas taught it possible for each of the three divine persons to assume simultaneously the same concrete nature, and for each divine person to assume more than one concrete human nature at the same time. See Pawl, "Thomistic Multiple Incarnations," 359–70. Also, see Brazier, *C. S. Lewis*.

Stated in Terms of Love, where he explored the possibility of multiple Incarnations on other planets:

> If other planets supported rational life . . . I am quite certain that Christianity is revealed to them in some way corresponding with its revelation to us. Our Christianity is the self-unveiling of eternal Love in terms and forms intelligible to us . . . their Christianity will be self-unveiling of eternal Love in terms and forms intelligible to them . . . It is only those who erect a false barrier between the universality of the Word and his incarnate life as a man who will boggle at the possibility of his self-revelation in a created form on another planet.[142]

Weston envisioned the Logos incarnating as a Christ-type figure, revealing an identical Christian message and mode of salvation as on Earth. His language of a 'false barrier' speaks to Paine's central criticism of the unintelligibility of a Christianity that claims a single saviour moving among civilizations or multiple, identical Christs on inhabited planets. The "terms and forms" cited by Weston represent that another "Christ" need not necessarily suffer and die in a manner equivalent to the terrestrial human Jesus; rather an extraterrestrial incarnation can take the form God sees fitting and the mode of salvation of each individual species or planet be according to their circumstances. A few decades later Anglican theologian E. L. Mascall in his 1956 book *Christian Theology and Natural Science* provided a fuller detailing of the idea of multiple incarnations by stating, "The arguments of both Ephesians and Hebrews[143] rest upon the unquestioned, but also unformulated, assumption that there are no corporeal rational beings in the universe other than man."[144] He noted the possibility of other varieties of hypostatic unions on other planets if deemed necessary by God's universal will of salvation, and if rational corporeal beings sinned and required redemption, the Son of God had united (or will one day unite) to his divine person their nature, as he has united to it ours. Further, Mascall considered that if the Incarnation took place not by the conversion of the Godhead into flesh but by the taking up of manhood into God, there seems to be no fundamental reason why, in addition to human nature being hypostatically united to the Person of the divine Word, other finite rational natures should not be

---

142. Weston, *The Revelation of Eternal Love*, 128–29.

143. Eph 1:20–23; Heb 2:7–9. These are references to what later was considered to describe the "Cosmic Christ."

144. Mascall, *Christian Theology and Natural Science*, 45.

united to that person too."¹⁴⁵ Mascall provided a basic critique against the inclusive or Christocentric maximalist position:

> For the latter, the essence of redemption lies in the fact that the Son of God has hypostatically united to himself the nature of the species that he has come to redeem . . . It would be difficult to hold that the assumption by the Son of the nature of one rational corporeal species involved the restoration of other rational corporeal species (if any such exist) . . . Christ, the Son of God made man, is indeed, by the fact that he has been made man, the Saviour of the world, if 'world' is taken to mean the world of man and man's relationships; but does the fact that he has been made man make him the Saviour of the world of non-human corporeal rational beings as well? This seems to be doubtful.¹⁴⁶

Mascall argued against the kenotic view that the Second Person's divine attributes were scaled down, as he described it, to the limitations of a human being, whereby another incarnation would be impossible,¹⁴⁷ leaving open the possibility of a multiplicity of divine incarnations among intelligent species. A few years after Milne, Anglican priest Norman Pittenger in his 1959 book *The Word Incarnate: A Study of the Doctrine of the Person of Christ*, considered the challenge posed by "Jesucentrism," that is, that the Incarnation of the Word in Jesus Christ provided a complete knowledge of God. Although Jesus is central to our understanding of God in his works, as teacher, as mediator of our salvation, and restorer of our true human nature, Pittenger argued these earthly acts did not disclose the entirety of an infinite God.¹⁴⁸ Further, he questioned whether the extent of the salvation won by Christ had application outside Earth:

> How can the Christian gospel, concerned with the salvation of men in this world, have any universal significance when we know that there may well be intelligent life on other planets?¹⁴⁹

Pittenger was open to the idea of extraterrestrial intelligent life and expected that what has been revealed in Christ would be commensurate with what God reveals elsewhere. A year later Catholic priest Daniel Raible took a similar position in his short 1960 article "Rational Life in Outer Space?" in support of multiple incarnations of the Second Person:

145. Mascall, *Christian Theology and Natural Science*, 39–40.
146. Mascall, *Christian Theology and Natural Science*, 37–39.
147. Mascall, *Christian Theology and Natural Science*, 40.
148. Pittenger, *The Word Incarnate*, 148.
149. Pittenger, *The Word Incarnate*, 248.

> Suppose that God intended to demand adequate satisfaction from a fallen (extraterrestrial) race. That would necessitate that God become a member of a fallen race in order to redeem it. Could it be the same Second Person of the Blessed Trinity who became incarnate for our salvation? Yes, it would be possible for the Second Person of the Blessed Trinity to become a member of more than one human race. There is nothing at all repugnant in the idea of the same Divine Person taking on the nature of many human races. Conceivably, we may learn in heaven that there has been not one Incarnation of God's son but many.[150]

It is worth noting that Raible mentions the Second Person taking on the nature of other *human* races, rather than necessarily non-human intelligences, avoiding questions regarding divine salvific action of alien races, cultures, and biologies. Fr. Theodore Zubek, in reference to the negative authority of scripture on the matter of extraterrestrials, was not reluctant to include an incarnation beyond the human, pointing out the existence of such beings is not opposed to any truth of the natural or supernatural order in his 1961 article "Theological Questions on Space Creatures." In his view, extraterrestrials would not inherit earthly original sin. "In any hypothesis, space creatures, not being offspring of Adam, would not belong to the human race and would not have Adam's original sin . . . they may be in one of several possible states."[151] He briefly speculated that extraterrestrials may be free of sin, or if sinful, receive an alternative means of salvation "Space creatures could have been punished by God individually and forever, like fallen angels . . . or God could have applied His infinite mercy by simply forgiving the sins of such creatures."[152] Zubek cited Aquinas, that the redemption could be possible with the incarnation of one of the other two divine personages, and conjectured that Earth religions may have much to gain in an extraterrestrial contact event with extraterrestrials.

> If we can understand that our way of encountering the universe and our views of spirituality only begin to express the range of ways that intelligent beings deal with ultimate reality; we are guaranteed to gain something very powerful: a more humble,

---

150. Raible, "Rational Life in Outer Space?", 352 (article condensed in Daniel C. Raible, "Men from Other Planets?" *Catholic Digest* 25 (December 1960) 104–8 and summarized in George Dugan, "Priest Suggests Rational Beings Could Well Exist in Outer Space." Also see McHugh on fallen and unfallen extraterrestrials in McHugh, "Life in Outer Space."

151. Zubek, "Theological Questions on Space Creatures," 393–99.

152. Zubek, "Theological Questions on Space Creatures," 393.

more realistic, and yet paradoxically more complete and more extensive understanding of our own place in the universe.[153]

French Dominican Cardinal Yves Congar in the 1960s asserted that revelation said nothing about astronomy; rather it illustrates God's generosity and autonomous creative act in relationship with humanity, which implies the real possibility of other, intelligent beings in the cosmos and that their intelligence, knowledge, and free will renders them fitting as images of God. He remarked regarding the absence of extraterrestrials in scripture, "Revelation being silent on the matter, Christian doctrine leaves us quite free to think that there are, or are not, other inhabited worlds."[154] Reasoning from the context of the new age of globalism and ecumenism in the second half of the twentieth century, he viewed the Incarnation of the eternal Word in Christ to the world as one plan of many given the endless love and grace of God.[155] Accordingly, he viewed Jesus as not necessarily superior to other incarnations, and saw the question about salvation not merely in relation to the individual but rather seeking an understanding of how God achieves salvation in differing cultures and religions. Further, he warned not to place conditions on an omnipotent Creator, arguing, "Earth should not limit divine power. There may well be other incarnations of the divine persons or Trinity of infinite persons,"[156] and "It is not contradictory, and therefore, not impossible, that the Word of God, or one of the Persons of the Blessed Trinity, should unite himself to any creature."[157]

Another Catholic priest, Fr. Francis Connell further affirmed a compatibility between extraterrestrials intelligence and Roman Catholicism in an essay entitled "Flying Saucers and Theology" in 1967, stressing that it is good for Catholics to know that the principles of their faith are entirely compatible with possibilities of life in other planets. Further, he briefly outlined potential extraterrestrial powers of mind and will, and whether, by means of free will, their ultimate relationship to God: "They might be beings like fallen angels, creatures with keen intellects, but with

---

153. Quoted in Wilkinson, *Science*, 129.

154. Congar, "Theologian of Grace," 371–400. John Polkinghorne accepted that another Incarnation of the Second Person to be revelatory and, therefore, takes the form of any given intelligent lifeform. "God's creative purposes may well include 'little green men' as well as humans, and if they need redemption we may well think that the Word would take little green flesh just as we believe the Word took our flesh." Polkinghorne, *Science and the Trinity*, 176.

155. Congar, "Non-Christian Religions and Christianity," 144. See Potvin, "Congar's Thought on Salvation," 139–63.

156. Congar, "Theologian of Grace," 188. See Congar, "Preface," 8–11.

157. Congar, *The Wide World My Parish*, 188.

wills strongly inclined to evil. But also may have innocence and benevolence greater than ours."[158] Connell, like most other theologians who have considered the extraterrestrial subject, did not pursue the subject further. However Fr. Kenneth Delano in his 1977 book *Many Worlds, One God* gave the subject of multiple incarnations chapter-length treatment, and echoing Aquinas and Congar, concluded that any of the three divine persons could become incarnate. He stressed humility in regards the transcendence of divine plans, the avoidance of any geocentric or anthropocentric attitudes, and respect for the silence of scripture. He considered the notion of multiple incarnations of one of the divine persons in extraterrestrial societies preferable to that of the inclusive view of a "Cosmic Adam," where the redemption of Christ on Earth would have equal effect for all beings in the universe, and that a multiplicity of incarnations does not impede humans from evangelizing and spreading word of our salvation history.[159] Franciscan Sister Ilia Delio in her article "Christ and Extraterrestrial Life" has given the subject of the Incarnation and extraterrestrials more extensive consideration than previous modern theologians. She writes regarding the dearth of research in this area,

> Speculation on the meaning of Christ for extraterrestrial (ET) life has received little attention in the science and religion dialogue, despite advances in astronomy, astrobiology, and space exploration. Perhaps the hesitation in undertaking this pursuit is the fear of disrupting the core doctrine of Christian faith, namely, the work of Jesus Christ.[160]

Delio notes that according to St. Anselm's satisfaction theory, human sin was the principal reason for the Incarnation, and as human sin was an affront to God's honor, divine justice demanded recompense either by satisfaction or by punishment. Therefore, the infinite magnitude of the offense of sin requires a like satisfaction, which can be achieved only by both a divine and human person. In reference to Anselm, Delio introduces the incarnational theology of Bonaventure and Scotus in providing other reasons for incarnation apart from redemption of humans in making the case of other intelligent beings who have not necessarily fallen from grace: First, she contends that by means of incarnating in a physical being, God's infinite power, wisdom, and goodness are manifested in a perfect manner. Second, incarnation brings perfection to the created order; since the first cause should be joined with the last (humans), it is fitting that the divine

158. Connell, "Flying Saucers and Theology" In *The Truth about Flying Saucers*, 258.
159. Delano, *Many Worlds, One God*.
160. Delio, "Christ and Extraterrestrial Life," 253–55.

Word be united to human nature. Since the entire created order is related to humanity, it finds its fulfillment in the perfect glorification of humanity. Therefore, in bringing the world to completion by the Incarnation, God instituted the perfect object for human contemplation. Third, the Incarnation was necessary to overcome sin, and in this way incarnation completes creation, as Christ and the world are not accidentally connected but rather intrinsically connected. This, she argues, is a redemptive completion and healing of humanity of its woundedness and restoration within divine creation. In regard to extraterrestrials, God's creative action through the Word means that any created reality, wherever it exists, would possess an inner constitution in relation to the divine Word. Delio writes,

> God creates not out of any need but out of desire to manifest something of the mystery of the divine truth, goodness and beauty outwardly and to bring forth creatures capable of participating in the splendor of the divine life . . . the created order reflects at some level the relation of the Son to the Father bound together in the love of the Spirit, for this relation is the ontological condition both of creation and of Incarnation.[161]

According to Bonaventure, the completion of creation by the Incarnation is not contingent on the fallen condition of humanity (contrary to the argument set forth by Crisp) but on the mystery of God as love. Delio, therefore, concludes that given this relationship between creation and incarnation, "a world without Christ is an incomplete world, [because] creation is structured Christologically."[162] For extraterrestrials incarnation must assume a form that includes the material reality of that creation, in whatever way that creation is constituted. She emphasizes an integral link between creation and incarnation, which form two sides of the mystery of God's self-communicative love. God's love is ordered, free, and holy—God loves his own self in others, his love is unselfish. Thus all creatures are predestined for salvation, not condemnation; all beings created with intelligence require spiritual transformation, whether or not sin is present, manifest through some form of incarnation. Similar to Teilhard de Chardin, she holds that evil is universal, and that for God "creation and incarnation go together." Christ is the divinization of created reality in whatever way the divine Word can fully enter into that reality. In short, Christ enters into a created order through an incarnation or *Word-embodiment* and completes that order through a self-giving act of love. Delio maintains that while extraterrestrial

---

161. Delio, "Christ and Extraterrestrial Life," 256–57.
162. Hayes, "Christ, Word of God, and Exemplar of Humanity," 3.

beings may receive multiple incarnations, there remains only one Spirit and one Christ to the glory of God the Father:

> I suggest that every created life-bearing order is Christologically structured so that, following Rahner's lead, there may be multiple Incarnations but only one Christ. The reality of Christ, therefore, is the union of God and creation and, as a symbol, mediates the divinization of every created order in its relation to God.[163]

> The conditions for the possibility of the Incarnation are the two terms of the relation must be capable of entering into such a unique and intense union; and there must be a unity of person; for if this were not the case, then the history of Jesus would not be the history of the Word but a history only extrinsically related to the Word, and granted the possibility from the side of God and from the side of man, it is yet required that there be a power adequate to effect the union.[164]

Delio has identified the terms *Christ/Word* for humans as identical with an incarnation for extraterrestrials, whereas other *multiple* and *inclusivist* thinkers differentiate between *Christ* and *Word/Logos* incarnate in non-human bodies. Where multiple theologians require a separation of the God-man Jesus Christ and the *Word/Logos* to accomplish multiple incarnations throughout the cosmos in order to maintain the Jesus Christ for humanity, Delio uses the term Christ, the Second Person of the Trinity, and Word synonymously (without referring to the human Jesus) which incarnates in other intelligent beings under the appearance of their natures. As each divine person is an infinite being, inclusive and multiple thinkers consider the Jesus Christ known to humanity as one Incarnation of a God-man; while the *Word/Logos* is an infinite, eternal Person, that Person would not necessarily be a Christ to extraterrestrials as we understand but a divine personification with its own purposes similar to or different than our own.

Karl Rahner considered the possibilities of multiple incarnations but was reluctant to pursue any concrete formulations or solutions. In his article "Natural Science and Reasonable Faith" he writes "In view of the immutability of God in himself and the identity of the Logos with God, it cannot be proved that a multiple incarnation in different histories of salvation is absolutely unthinkable."[165] He describes incarnation as "the 'unsurpassable

---

163. Delio "Christ and Extraterrestrial Life," 261.
164. Delio, "Cosmic Christology," 107–20.
165. Rahner, "Natural Science and Reasonable Faith," 51. Logos here is used to

climax of revelation", when God's self-communication reaches its highest point.[166] For Rahner, the incarnation is the final word of God, because in it, God and the world have become one, forever without confusion, but forever undivided, and in it, all the plenitude of God is included for the world, and nothing of it is excluded.[167] He agrees with Delio of incarnation as a means of a completion and divinization of creation:

> Accordingly, the Incarnation can occur once and only once when the world begins to enter upon its final phase . . . In this phase it is to realize its definitive concentration, its definitive climax and radical closeness to the absolute mystery which we call God. From this perspective the Incarnation appears as the necessary and permanent beginning of the divinization of the world as a whole.[168]

Rahner also argued in the vein of inclusivist thinking by envisioning humanity as pivotal to the ultimate fate of the universe: "The human being 'is a personal subject from whose freedom as a subject the fate of the entire cosmos depends.'"[169] In this view, another incarnation need not be required to further an already accomplished cosmic effect. Therefore, Rahner's consideration of multiple incarnations of the Logos seems incompatible with his understanding of the work and extent of Christ's role in creation. This is his 'Christological maximalism' (I have termed inclusive) which works contrary to multiple divine incarnations. Rahner's reticence to speculate on the nature of other intelligence or divine actions outside Earth was apparent with the exception of a few generalities.[170] He acknowledged that extraterrestrial beings live in their own social and cultural reality, and exist in what is collectively known throughout the universe's space-time.[171] "One could say

---

describe the creative principle in God and second divine person of the Trinity. See also Fisher and Fergusson, *Karl Rahner and the Extra-Terrestrial Intelligence Question*, 275–90.

166. Rahner, *Foundations of Christian Faith*, 174.

167. Rahner, *Theological Investigations I*, 49.

168. Rahner, *Foundations of Christian Faith*, 181.

169. Rahner, *Theology and Anthropology*, 15.

170. Rahner, "Sternenbewohner," 1061–62. See Rahner, "Landung auf dem Mond," 233–34. "In our context it is especially worthy of note that the point at which God in a final self-communication irrevocably and definitively lays hold on the totality of the reality created by him is characterized not as spirit but as flesh. This authorizes the Christian to integrate the history of salvation into the history of the cosmos, even when a myriad of questions remain unanswered." Rahner, "Naturwissenschaft," 56.

171. Space-time: the mathematical model that combines the three dimensions of space with the fourth dimension of time to create a four-dimensional continuum.

that these other corporeal and intelligent creatures in a meaningful way also have a supernatural determination within an immediacy to God (despite the totally unmerited reality of grace)."[172] However he did express openness to the hypothesis of modes of divine presence in other societies, while refraining to provide definition: "That dynamic bringing together of intelligence, matter, and divine presence may find realization in multiple ways in the galaxies."[173] "A theologian can hardly say more about this issue than to indicate that Christian revelation has as its goal the salvation of the human race; it does not give answers to questions which do not in any important way touch the realization of this salvation in freedom."[174] Ultimately, Rahner did not offer any specific solutions to this Christological problem. He and Delio appear as inclusive thinkers while nominally multiple as well.

Andrew Davison, in his journal article "Christian Systematic Theology and Life Elsewhere in the Universe: A Study in Suitability," discusses briefly the history of theological development with regard to divine work throughout the universe; and notes the paucity of scholarship in this area as theologians have found it difficult to integrate the central teachings of Christology and the Incarnation with the notion of life elsewhere in the universe. Consistent themes of the possibility of multiple incarnations and their plausibility, likelihood, and necessity have been examined by Tillich and Polkinghorne (each arguing for the necessity of divine work elsewhere); and conversely, on the non-necessity of multiple incarnations according to Vorilong, Melanchton, Mascall, or its impossibility according to Hebblethwaite. Against this thinking, Davison references Paul and Linda Badham's claim that Christianity is irrational as it seems to require multiple incarnations given the possibilities of extraterrestrial life, while maintaining the Christological impossibility within that scheme. Davison notes in summary, "The Badhams combine soteriological necessity with Christological impossibility . . . [while] Polkinghorne combines necessity and possibility; Vorilong combined impossibility with non-necessity; [and] Eric Mascall combined possibility with non-necessity."[175] Davison states his agreement with the thought of Vorilong, Campanella, and Peters against the notion of a "planet hopping Christ" as being unfitting; in this case, God's assumption of another nature would not be a second Jesus Christ known to humans. This

172. Rahner, "Naturwissenschaft," 59.

173. Rahner, "Naturwissenschaft," 59. See Weissmahr, "Die von karl Rahner," 175–80; Geister, *Aufhebung zur Eigentlichkeit*, 119–21; Fritsch, "Vollendung des Cosmos," 508–11; Edwards, "Resurrection of the Body," 357–83; Petty, *A Faith That Loves the Earth*; O'Donovan, "Making Heaven and Earth," 269–99.

174. Quoted in O'Meara, "Christian Theology and Extraterrestrial Intelligent Life," 9.

175. Davison, "Christian Systematic Theology," 452.

would be unsuitable or nonsensical, as other incarnations would not be Jesus Christ but "God's assumption of some differently creaturely nature ... in a multiple and parallel hypostatic union." This reveals that these approaches are contrasting, and many of them Davison argues are not grounded in a distinct tradition of theology, and therefore, lack precision in conversation with astrobiology. Instead, he argues the scholastic categories of suitability, appropriateness, and fittingness are better tools for the task, and allow for gradations of divine work, in which God's actions are not bound by necessity. In doing so, God's freedom is respected and other possibilities are left to divine counsel. In a position shared by Aquinas and Bonaventure, Davison starts from the "position of saying that whatever God does, wherever God acts, God does suitably, or fittingly."[176] In this way, God's actions are free but congruent with divine nature and creatures; God is not constrained by necessity but only in a sense by the nature of things he has made; that is by God's own nature and actions. God becomes, as Aquinas argued, a debtor not to creatures but to himself and his own purpose. This was echoed by Scotus to the effect that, "where creatures are concerned, [God] is debtor ... to his generosity, in the sense that he gives creatures what their nature demands." Davison, therefore, leaves open the possibility of the fittingness of divine work in creation and saving acts of creatures. He does not discuss any of the specific theological concerns with maintaining the inclusive or exclusive views with regard to the earthly Incarnation and life in the universe, but rather outlines the central questions which have been raised historically with multiple incarnations and the restrictive methodologies utilized in its evaluations by theologians. He maintains the preeminent role of divine freedom, while maintaining that God's acts will be consistent with his nature, allowing for more flexibility for theologians in understanding divine work. His consideration of these more appropriate categories of theological arguments allows for the supreme suitability of an earthly incarnation, as argued by Aquinas, while maintaining that God's actions outside Earth will be equally suitable. However, this methodology could be utilized to support several other views with regard to life in the universe. In support of an exclusive view, that God found it most fitting to provide a single rational creation of humanity, to the exclusion of extraterrestrial life or intelligence (in accordance with the rare Earth argument); or in accordance with an inclusive view, that one incarnation being the most supreme form of divine love, was suitable not only for humans but for all creatures in the universe; or in defense of the multiple view, that incarnation is *always* the most fitting and suitable manner to redeem creatures. In these cases,

---

176. Davison, "Christian Systematic Theology," 461.

arguments would have to be brought to bear on the unfeasibility of those positions as argued in this thesis. Overall, his position is most supportive of the arguments for the *varied* view (to include multiple) with regard to God's work in creation and his dealing with creatures.

We now will consider incarnation as a divine initiative for salvation, divine self-communication, and the interaction of these views with the inclusivist and multiple views of extraterrestrial soteriology. Rahner explained incarnation as a form of divine action when "'the world has reached its final phase,' where it "realizes its definitive concentration, its definitive climax and radical closeness to the absolute mystery which we call God ... [when] the Incarnation appears as the necessary and permanent beginning of the divinization of the world as a whole."[177] However, when it is questioned how it is *necessary* to creatures for a divine being to incarnate in their natures, contemporary theologians have provided two central arguments. In Ted Peters's recent survey, *Astrotheology: Science and Theology Meet Extraterrestrial Life* (2018) these are presented as the categories of revelatory or redemptive, that is, whether incarnation is a fixed principle which follows according to the divine will after creation of rational creatures for the purposes of completing and perfecting creatures and achieving the ultimate possible union with them, or rather incarnation is a divine prerogative in God's methodology of dealing with deadly sin.

Peters discusses that the theology of incarnation as principally revelatory has foundation in the theology of Bonaventure and Scotus; incarnation has no relation to the Fall as evil need not be a central theme for incarnation[178]—rather it is a self-communicative, completing, unifying action innate to the divine-creaturely relationship which is manifested at the proper time. He instead supports a redemptive model, following Augustine and Aquinas. From an orthodox perspective Vladimir Lossky in Peter's volume describes this view as follows, "The Fall demands a change, not in God's goal, but in His means. For the atonement made necessary by our sins is not an end but a means, the means to the only real goal: deification."[179]

Both these models have been used to support further arguments of whether there is a single incarnation on Earth for the entire universe, the *inclusivist* view, or that of multiple incarnations among other civilizations according to the *multiple* view. As we shall see, views of incarnation as ontological change to deal with deadly sin will tend to be persuasive to those proposing that Christ's incarnation was effective 'once for all' for

---

177. Rahner, *Foundations of the Christian Faith*, 181.
178. Scotus, *Reportatio Parisiensis*, III, d.7, q. 4.
179. Lossky, *Orthodox Theology*, 110–11.

the whole cosmos (the inclusivist view), whereas views of incarnation as divine self-communication, requiring conscious awareness on the part of the 'receivers', will attract proponents of the multiple view—that God initiated incarnations in every planetary civilization as part of the economy of salvation for that planet. According to Peters, a single redemptive incarnation for the entire universe, incorporating a high Christology rather than a strictly revelatory view is most appropriate. He admits this position might lead to accusations of geocentrism as our planet is given special status. In response, he argues that on Earth we received a prolepsis of a cosmic-wide transformation which Jesus promises, and that this is the case for the cosmos, whether conscious beings realize it or not.[180] This question, on the importance of the knowledge or ignorance of an economy of salvation by creatures will be discussed below. J. Edgar Bruns in Peters's book also argues for a single universal incarnation and writes, "The significance of Jesus Christ extends beyond our global limits. He is the foundation stone and apex of the universe and not merely the Savior of Adam's progeny."[181] Interestingly, similar to the arguments of Milne and Mok as discussed in chapter 3, he calls for missionaries to evangelize extraterrestrials upon discovery to spread news of the earthly Incarnation, a notion that has been the subject of much criticism, (notably by Mascall). Neils Henrik Gregersen, also in Peter's survey supports the inclusive model in his argument for 'deep incarnation' that all flesh are included in human redemption, as flesh refers to all material creation. He writes, "The New Testament nowhere states that God became human . . . rather the Logos of God became 'flesh' (John 1:14) . . . God's incarnation also reaches into the depths of material existence."[182] However it should be noted that Gregersen does not make any cosmic claims regarding extraterrestrial life within this perspective. Further, Joshua Moritz, following Gregersen, discusses the meaning of the Incarnation for animal theology where, in agreement with Gregersen's 'deep incarnation,' he argues that all biological life is united in the one Word to humanity. "Incarnation is the most profound expression of a solidarity which encompasses the whole of life."[183] In this way, God becomes flesh with the entire world of flesh which is brought into communion with God through Christ through

---

180. Peters, *Astrotheology*, 297.

181. Peters, *Astrotheology*, 284.

182. Gregersen, "Deep Incarnation," 174.

183. Moritz, "Redeeming Animals and ET," 341. Moritz also argues that Irenaeus's cosmic salvation included non-human creatures: "God made a covenant with the whole world through Noah, pledging Godself to all animals and humans." See also Linzey, *Animal Gospel*.

the resurrection in the flesh.[184] Therefore, the Word becomes biological life as the purpose of incarnation is the redemption of all biological life in the universe. According to Mortz's view, one man is selected by God as the perfect *imago Dei*; and the entire cosmos is thereby brought into God's plan of redemption. Extraterrestrials, as the ontological equivalents of humans share in this redeeming act, which Moritz sees as a redemption that includes all levels of being in which God in Christ participated in.[185] Therefore, both are in agreement with Gregory of Nazianzus's formula, "That which was not assumed is not healed; but that which is united to God is saved."[186]

In considering arguments in support of the *multiple* incarnational view, Peters claims that a revelational or exemplarist emphasis on Christology would be more supportive of multiple incarnations in extraterrestrial civilizations, as the redemption unique in Christ on Earth can be *communicated* to other species. Robert John Russell in Peter's book also argues for multiple incarnations, as extraterrestrials being rational beings and composed like humans would be gifted with the *imago Dei* similar to ours, and would share our gifts as well as our proclivities for sin. As a result, God would be present to the moral struggle of intelligent life everywhere; his grace will redeem and sanctify every species in which reason and moral conscience exist.[187] Karl Barth echoes this idea, in that God wills fellowship with the world, and in this context all worlds which he has created through the power of the Spirit. His conviction is that we humans know of God principally through the revelation in Christ—God is revelational in his actions with creatures. This is supportive of the multiple view of a God and who would provide analogously for other extraterrestrial civilizations. In this view each intelligent being is God's elected form of expression according to their own natures where God becomes God for them.[188]

Russell notes a distinction between the revelational view of the incarnation and the ontological view, which emphasizes how God redeems the world and nature in a transformative effect into a new creation away from sin and death. In a revelational view, God will offer a normative revelation to each and every extraterrestrial species, and that each revelation will be radically species-appropriate. Peter Hess, another multiple thinker, claims soteriologies since the apostolic era have been anthropocentric, terracentric, and carbonocentric. He considers it problematic to rethink

---

184. Irenaeus, *Against Heresies*, 4. 34.1
185. Moritz, "Redeeming Animals and ET," 342.
186. Gregory of Nazianzus, *Epistle* 101.
187. Peters, *Astrotheology*, 305.
188. Barth, *Church Dogmatics*, IV/1, 45.

Christology and soteriology to include alien culture which lived perhaps a hundred million years ago, or one millions of years in the future, and asks if the notion of sacrifice is intrinsic to soteriology or extrinsic—is it a particularity of the terrestrial sacrificial economy of post-exilic Judaism, and could salvation be accomplished on other planets without a sacrificial death?[189] He concludes that as God is not logically or theologically bound to become incarnate only once in the universe, and with Bonaventure and Scotus, sees the reason for incarnation as serving to complete what God creates; therefore, the saving of extraterrestrials is an act of love which brings creatures into divine communion.

David Wilkinson, in his *Science, Religion, and the Search for Extraterrestrial Intelligence* raises several concerns with regard to the *multiple* view. He cautions that the multiple incarnational argument has the effect of "driving a wedge between "Cosmic Christ" and the human Jesus,"[190] which can reduce the force of the idea of the eternal Logos in the temporal Jesus. Secondly, he asks if we are to consider multiple incarnations in other species, why have these not been witnessed on Earth? Although other traditions are acknowledged to contain some divine truths, the Incarnation in Christ is considered the fullest manifestation to humans. For humans the Incarnation is about revelation and salvation, but as we are the single known example of incarnation, he states we cannot say with certainty if incarnation always includes revelation with redemption.[191] Peters sees Wilkinson's concerns as cautious support for the inclusive view of a soteriological work of a terrestrial Christ for the whole of reality.[192] In another critique of the *multiple* perspective, Mark Worthing takes a firm stance against multiple incarnations when he writes, "If there is other intelligent life in the universe then God relates to it through Christ-the same Christ through whom God reconciles us to Godself. I do not think Christian theology can posit a multiplicity of Christs and remain Christian theology."[193]

Bearing this criticism in mind, the discussion now returns once again to the inclusive view, in which human civilization in the one incarnation and sacrifice of Christ are central to the redemption of all beings. Since the Incarnation and atonement applied to all creation in the universe, this necessarily includes all sentient beings, including those unaware of the Christian message, which nonetheless are held under its dominion. However,

189. Peters, *Astrotheology*, 324–25.
190. Wilkinson, *Science*, 158–59.
191. Wilkinson, *Science*, 158.
192. Peters, *Astrotheology*, 282.
193. Peters, *Astrotheology*, 283.

how would the salvation gifted by God through Christ be communicated to other civilizations throughout a vast universe? This has been termed the *scandal of particularity*, the claim that one person saved all persons; and that any particular historical event determines the ontological nature of all things universally. Here Peters defends himself against criticisms of geocentrism in his inclusive soteriology, and instead considers this position theocentrist, in that God is the center of reverence, and not Earth.[194] He claims God would communicate to rational creatures capable of understanding on other planets the divine reality simply to share communion with creatures; and argues, following Tillich,[195] that the divine Logos or divine reason maintains the same structure everywhere in the cosmos, so rational creatures would be by nature attuned to the presence of God, whether incarnated in flesh or merely apprehensible through mind."[196] Further, Peters states that as the historical event on Earth might never be known elsewhere, this implies we have exclusive access to a cosmic truth, although he does later appear to affirm the possibility of other forms of revelation elsewhere which might support multiple revelational incarnations.[197]

Robert John Russell, in the same volume, however points out several important weaknesses to the inclusivist position with regard to its communication and application to distant creatures. For those supporting the inclusivist view there cannot be a genuine, "personal revelation of the Good News without a historical and ontological act by which God redeems the world . . . without this revelation being received by human beings in the context of their lived religious experience."[198] Hence this ontological basis must be manifest, appear, and be received by people reflecting the specificities and diversity of human history.[199] Therefore, he supports the revelational view of the Incarnation in his argument for multiple incarnations. His view requires an ontological incarnation, and for Russell our participation by faith requires that this revelation be based on an ontological act of redemption and be known to all species needing redemption. Accordingly, a single incarnation is insufficient for the redemption of the universe due to lack of participation by extraterrestrials; thus multiple incarnations

---

194. Peters, *Astrotheology*, 297–98.

195. For Tillich, incarnation in Jesus is the "concrete universal." Also see Peters, *Astrotheology*, 284.

196. Peters, *Astrotheology*, 285.

197. Peters, *Astrotheology*, 298.

198. Peters, *Astrotheology*, 306.

199. Peters, *Astrotheology*, 306–7.

are required.²⁰⁰ Russell lists additional arguments against the revelational prohibitions inherent to the inclusivist view. The problem of distance to extraterrestrials is an obvious concern given the size of the universe for both the revelational and ontological of dimensions of incarnation. How would the gospel be communicated, and how would the universal change across an expansive universe be made by the unique initiating event of the Incarnation on Earth? Russell also considers what he terms the "problem of difference": how can our human salvation history with its certain context, narratives, and languages relate to those of extraterrestrials, and to the uniqueness of their lives and natures? It seems there would be formidable dissonances between the terrestrial account of salvation and extraterrestrial histories and perceptions. Therefore, Russell claims extraterrestrials must have their own access to the divine revelation and dispensation in their own histories.²⁰¹ Another problem is his "concern about absence." If there is but one incarnation in the universe, then the implication is it is highly unlikely that this single incarnation happened on Earth. The more planets with civilizations which exist, the less likely it occurred here on Earth among perhaps billions or trillions of civilizations.²⁰² This makes for a strong charge of pre-Copernican "Earth chauvinism" or geocentrism among those who hold a single incarnation for all intelligences.

This issue of communication and knowledge of an economy of salvation can also be related to the inhabitants of the antipodes. When discovered it was concluded that the indigenous people belonged to the family of man (descended from the same parents) and were included under the one salvation in Christ, despite having no familiarity with Christianity; this also applied to the Gentiles receiving the teaching brought by Paul and other apostles. On Earth, it was held that certain humans did not need to know about the Christian economy of salvation in order to be affected by it due to Christ's union with human nature. At the same time, Christians were given by Christ the divine commission to spread the Gospel on Earth;²⁰³ in this way, humans cooperate with God to bring the message and grace of salvation to other people. However it would be impossible for a distant species to receive news of an atonement, and no way for humans to carry out that commission, in order for distant extraterrestrials to be affected by it in comparison to our salvation history.

200. Peters, *Astrotheology*, 308.
201. Peters, *Astrotheology*, 309.
202. Peters, *Astrotheology*, 310.
203. Heb 2:10–18.

Since Christ redeemed a particular species by uniting the divine nature to it, all who share that nature would seem subject to that particular redemptive mode. But according to the multiple view, if Christ redeemed other races, then, it would not have been through his human nature but by his divine nature as the Second Person (according to the multiple view). There would be a difference between those saved in ignorance on Earth, which is few, versus an entire universe of beings. Therefore, those who argue for a *multiple* or *varied* soteriology find it impossible for the Christian faith to be communicated to the entire universe by any human means if Earth is its sole point of origin.[204] Further discussion on these reservations of the inclusivist view is found in chapter 5. Still, the following theologians expanded their perspective on what could be considered divine acts to redeem extraterrestrials beyond our human understanding of incarnation with the *varied* view.

Varied

The *varied* position represents the final set of possibilities in consideration of God's potential acts with other intelligences; it maintains the Creator's absolute divine freedom to create, manifest, and redeem as he chooses according to particular extraterrestrial beings, societies, and worlds. It proposes that incarnation is not necessarily integral to creation as it may not be appropriate to another world order, since it means literally taking on flesh, the embodiment of the divine Word in a creature. It cannot be assumed other worlds composed of alien societies have received a revelation paralleling a Christ event centered on the Second Person of the Trinity. Any insistence on incarnation as the only possible modus of divine presence is a presumptuous oversimplification of the theological, religious, and social constructions of potential intelligent extraterrestrial beings, in that every created intelligent society must ultimately manifest divine incarnation as a created reality. The *multiple* position makes incarnation a *requirement* for intelligent extraterrestrials rather than a free choice for the divinity to engage as a necessary part of the completion of creation and God's desire to love and plan to deal with sin. The inclusivist and multiple arguments assume God will manifest only as an incarnate being on other worlds, not in other types of theophanies or divine actions, failing to consider the absolute primacy of divine freedom and self-revelation in the creative possibilities and myriad modes of interactions with intelligent beings. To this point, philosopher David Braine provided a distinction between incarnation (God taking bodily form) and

---

204. Puccetti, *Persons*, 135–36.

indwelling (the spirit of God inhabits another separate, non-divine bodily being). Incarnation would be unique and specific, whereas an indwelling may occur multiple times and places. C. S. Lewis gave considerable thought to the Christological implications of extraterrestrials in several of his fictional works and commentaries. He argued against what he termed "theological imperialism," to which he stated, "To different diseases, or different patient's sick with the same disease, the great Physician may have applied different remedies."[205] One of his most important statements on the subject provides his conception of the *varied* view:

> If other natural creatures than man have sinned we must believe that they are redeemed: but God's Incarnation as man will be one unique act in the drama of total redemption and other species will have witnessed wholly different acts, each equally unique, equally necessary and different necessary to the whole process, and each (from point of view) justifiably regarded as 'the great scene' of the play.[206]

Lewis believed that discovery of extraterrestrials would have no more effect on Christianity than Copernicanism, Darwinism, or Psychologism.[207] However he did predict difficulties with the doctrine of the Incarnation in the event rational and sinful extraterrestrials were discovered, putting into question Christ's salvific role beyond Earth. Spanish Jesuit theologian Joaquin Salaverri was one of the earliest churchmen to espouse the varied view in a 1953 article titled "La possibilidad de seres humanos extra-terrestres ante el dogma Católico," claiming that a theology of humans as the only race of intelligent beings in God's creation was implausible; he argued for a variety of divine relationships with the Creator who might have several plans for the salvation of other beings.[208] However, he did not provide details on these possible divine relationships. Similarly, Catholic priest Angelo Perego in his 1958 article, "Origin of rational extraterrestrial beings" likewise considered the Christian message inapplicable to extraterrestrials in the case

---

205. Montgomery, *Christ at Center and Circumference*, 258.

206. Lewis, *Miracles*, 149–50. See also *Chronicles of Narnia*; Aslan, the Lion could be understood as a Messiah-type figure in the world of Narnia, who has taken the form of a lion in that specific realm, while maintaining his identity as the Christ in the human world.

207. Lewis, "Other-Worldly Faith," 64; Lewis, "Faith and Outer Space," 37. This claim could be considered highly presumptuous given a non-human species will be completely foreign to the realm of human affairs, a false equivalency is made in comparing human paradigmatic shifts to the short and long-term effects of types of interactions with extraterrestrial beings.

208. Salaverri, "La possibilidad de seres humanos extra-terrstres," 23–43.

they be without sin comparable to humans. He wrote that God is free to create any being anywhere, in which his freedom, wisdom, power, mercy, providence, and justice would be manifest. Extraterrestrials as non-humans would not be the descendants of Adam, and therefore, would not be guilty of sin, in which case it would be unfeasible to bring the message of Christ to other worlds by means of space travel as later suggested by Mok.[209] Paul Tillich took a similar position with regard to revelation and redemption of extraterrestrials, and in his *Systematic Theology* questioned how should we "understand the meaning of the symbol 'Christ' in the light of the immensity of the universe, the heliocentric system of planets, and the infinitely small part of the universe which man and his history constitute, including the possibility of other worlds in which divine self-manifestations may appear and be received."[210] He claimed incarnation as unique for the special group, race, or planet in which it happens, but it should not be considered unique in the sense that other singular incarnations for other unique worlds are excluded, as the sphere of humanity cannot claim to occupy the only possible place for incarnation.[211] While holding the idea of multiple incarnations on other worlds, Tillich included the possibility of other types of divine self-manifestations by taking example of our remote Earth within a divine plan encompassing the vast scale of creation:

> The saving power of God is available to all creatures everywhere, and the totality of all things "includes a participation of nature in history and demands a participation of the universe in salvation The manifestation of saving power in one place implies that saving power is operating in all places.[212]

Tillich's view of a God whose saving power if present in one place, must be present in all places and can be termed a type of "divine homogeneity" of action and relationship among species. Other beings will not be Christians, as argued by inclusivists and some multiple thinkers, as they did not inherit human original sin nor receive the specific message of

---

209. Perego, "Origine degli esseri razionali estraterreni," 22. According to Perego (*"Possibilitá di una redenzione cosmica,"* 121–40), Pope Paul VI considered the reasonableness of a reality of intelligent extraterrestrials and saw how the universal church would include more than Earth, reported by Reginaldo Francisco in a conversation between Jean Guitton and Pope Paul VI. See "Possibilita di una Rendenzione Cosmica," in *Origini, l'Universo, la vita, l'intelligenza*, ed. F. Bertola, et al, (Padua: Il Poligrafo, 1994), 121–40.

210. Tillich, *Systematic Theology*; See also Bradnick, "Entropy, the Fall, and Tillich," 67–83.

211. Tillich, *Systematic Theology*, 2.96.

212. Tillich, *Systematic Theology,* 2:95.

Christ; divine revelation would take another form fitting to them. Similarly, process theologian Lewis Ford in his 1977 book *The Lure of God* asserted salvation is not just limited to humans but applies to all intelligent beings in the universe, which have access to the divine apart from the work of Christ. "This creative purpose is hardly invariant in its specific manifestations: what God says depends upon the particular situation confronting that individual in his own world . . . God's dynamic Word knows no single form."[213] Twenty years later, another process theologian, Ian Barbour expressed a related view in his book *Religion and Science* in reference to the homogeneity of the universe.

> [A universe] with identical physical laws everywhere can produce intelligent life as it has occurred on Earth, even more advanced than humans. The Word of God, [identifying the initiator of divine relationship as the Second Person of the Trinity] was creating throughout the cosmos . . . [and] will also have revealed itself as the power of redemption at other points in space and time, in ways appropriate to the forms existing there.[214]

Arthur Peacocke in a chapter titled, "The Challenge and Stimulus of the Epic of Evolution to Theology," postulated that the gradual emergence of humanity renders moot the notion of the fall from an "original righteousness," and that sin, rather than resulting from a singular primordial event, is a consequence of our inability to transmute our latent animalistic tendencies with our modern sense of morality. Therefore, he argues for the revocation of the classical Christian doctrine of redemption and advocates a form of panentheism.[215] He questions, "What can the cosmic significance possibly be of the localized, terrestrial event of the existence of the historical Jesus? . . . Would ET, Alpha-Arcturians, Martians, et al, need an incarnation and all it is supposed to accomplish? . . . Only a contemporary theology that can cope convincingly with such questions can hope to be credible today."[216] For non-human intelligences, Peacocke holds that sin and Christian redemption is unique to humans, and without providing further detail, states that God intervenes in the history of other beings in a species-appropriate way. This view gained greater recognition in recent times when Vatican astronomer Jesuit José Gabriel Funes asserted in an interview in 2008:

---

213. Ford, *The Lure of God*, 63; See also Ford, "Theological Reflections on Extraterrestrial Life," 2, as quoted in Peters, *Science, Theology, and Ethics*, 128.

214. Barbour, *Religion and Science*, 215.

215. Peacocke, "The Challenge," in Dick's *Many Worlds*, 108–15.

216. Peacocke, *Theology for a Scientific Age*, 65–66.

> [A] multiplicity of creatures exist on Earth, so there could be other beings, also intelligent, created by God . . . This does not contrast with our faith because we cannot put limits on the creative freedom of God. To say it with St. Francis, if we consider Earthly creatures as "brother" and "sister," why cannot we speak of an "extraterrestrial brother"?[217]

Regarding possible sinful natures of extraterrestrials, Funes explained in agreement with Tillich, Ford, and Peacocke that extraterrestrial beings may be without original sin,

> We that belong to the human race could be precisely the lost sheep, sinners who have need of a pastor. God was made man in Jesus to save us. In this way, if other intelligent beings existed, it is not said that they would have need of redemption. They could remain in full friendship with their Creator.[218]

Funes considers that the Incarnation is unique to Earth and to humans and does not extend to other creatures—if they are sinful, they will be shown mercy in other, unknown ways: "Jesus has been incarnated once, for everyone. The Incarnation is a unique and unrepeatable event. I am, therefore, sure that they, in some way, would have the possibility to enjoy God's mercy, as it has been for us men."[219] Thomas O'Meara, in a chapter of his small book *Vast Universe* noted that although the Word and Jesus are one, the life of Jesus on Earth does not curtail the divine Word's being and life. All three persons could become incarnate because incarnation is one aspect of divine power and one divine activity, involving one creature as the object of one special divine relationship. It hardly represents all that God can do and is doing. He articulated this form of the *varied* view, whereby revelation and grace may have a multiplicity of divinely bestowed forms, serving a range of purposes:

> The cross is not the only theology of redemption, nor is it doctrinally the necessary or full purpose of Incarnation as Jesus could have died in many ways. If Jesus visited another planet he would be a divine messenger, not Incarnation of that species . . . supraterrestrial roles cannot be ascribed to Christ, the human/God, at other planets. The history of sin and salvation recorded in the two testaments of the Bible is not a history of the Universe; it is a particular history on one planet . . . the central

---

217. "The Extraterrestrial is My Brother," *L'Obssservatore Romano*.
218. "The Extraterrestrial is My Brother," *L'Obssservatore Romano*.
219. "The Extraterrestrial is My Brother," *L'Obssservatore Romano*.

importance of Jesus for us does not necessarily imply anything about other races on other planets. Incarnation is a form of divine love, would there not be galactic forms of that love?[220]

Accordingly, the Christian religion is restricted to humans and Earth, and we cannot make definitive statements about Christ or the Second Person's activities elsewhere within creation. Aquinas wrote, "The power of a divine person is infinite and cannot be limited by anything created,"[221] and Augustine remarked, "Christian doctrine does not teach that God was so joined to human flesh as to lose or resign control over a universe as though constricted by a baby."[222] For the handful of thinkers holding the *varied* view, God is free to manifest in a variety of forms, incarnation among them; God cannot be restricted to specific modes of presence, mediation, communication, and salvation on one specific planet or specific species within Einsteinian space-time composed of immense distances with billions of galaxies in the observable universe.

Conclusion

Some early thinkers argued the potential demise of a terrestrial Christian faith with a discovery of outside intelligences (see the aforementioned *Brookings Report*). Notably, Arthur C. Clarke in 1951 expressed concern of a loss of religious faith with extraterrestrial contact, and brought attention to long-standing concerns inherent with a terrestrial faith within the new scientific cosmic context.[223] Ernan McMullin stated the serious nature of religion's failure to address the implications of intelligent extraterrestrials, as "a religion which is unable to find a place for extraterrestrial persons in its view of God and the universe might find it difficult to command terrestrial assent in the days to come."[224] Conversely, as more data of the known extent and composition of the universe became available decades later in the second half of the twentieth century, surveys of clergy and laity began to suggest attitudes of extreme resilience of the Christian religion in the event of discovery or contact

---

220. O'Meara, *Vast Universe*, 47.

221. Aquinas, *Summa Theologica*, III, Q. 7, Art. 3.

222. Aquinas, *Summa Theologica*, III, Q. 1, Art. 4.

223. In Clarke's book *The Exploration of Space*, he stated, "[Some people] are afraid that the crossing of space, and above all contact with intelligent but nonhuman races, may destroy the foundations of their religious faith. They may be right, but in any event their attitude is one which does not bear logical examination—for a faith with cannot survive collision with the truth is not worth many regrets." Dick, *Many Worlds*, 198.

224. Quoted in Davis, "Search for Extraterrestrial Intelligence," 22.

with extraterrestrials.²²⁵ Theologian Ted Peters has argued that "Although there are partial grounds for thinking the Christian faith is so Earth centrist that it could be severely upset by confirmation of the existence of ETI, an assessment of the overall historical and contemporary strength of Christian theology indicates no insurmountable weakness."²²⁶ This is a premature assessment, and will be discussed in chapter 5.

A few early theologians, such as Catholic clergy Daniel Raible and T. J. Zubek, and Anglican E. L. Mascall, undeterred by the non-confirmation of extraterrestrial life in our solar system, simple or otherwise, recognized the scientific advances in space as an indication of a possible future discovery of intelligent life beyond current space exploration, and contemplated the theological implications within this greater context. Philosopher Roland Puccetti however, was pointedly critical of the claims of Earth religions within a greater cosmos of other civilizations and echoing Paine, openly questioned their declarations of possessing universal truth. In his book *Persons: A Study of Possible Moral Agents in the Universe* he posited that no single religion can claim doctrinal universality with the existence of extraterrestrials, which will inevitably lead to the abandonment of all particularized terrestrial religions.²²⁷ He argued, "[Given] the prospect of extraterrestrial intelligence, concerning which the principal sacred writings of Christianity, Judaism, and Islam are absolutely silent, generates a profound suspicion that these terrestrial faiths are no more than that (belief systems solely limited to terrestrials)."²²⁸ Philosopher Ernan McMullin rejected Puccetti's claim of the destruction of Christian doctrine of the Incarnation within an inhabited universe. "There is an odd, ungenerous fundamentalism at work here, a refusal to allow for the expansion of concept, the development of doctrine that is after all characteristic of both science and theology."²²⁹ Astronomer Paul Davies was also reluctant to consider a development of doctrine with regard the potential for other divine action, and has argued from Paine's naturalistic position in citing the impossibility of other divine incarnations: "The difficulties are particularly about the Christian religion, which postulates that Jesus Christ was God incarnate whose mission was to provide salvation for man on Earth. The prospect of a host of 'alien Christs' systematically visiting every inhabited planet in the physical form of the local creatures has a rather absurd aspect."²³⁰ Given scien-

225. See earlier discussion on the Peters and Alexander surveys.
226. Peters, "Exo-theology," 187–206.
227. Puccetti, *Persons*, 121–45.
228. Puccetti, *Persons*, 135–37.
229. McMullin, "Persons in the Universe," 69–89.
230. Davies, *God and the New Physics*, 71.

tific advances in radio telescopes and spaceflight technologies, some inclusive theologians such as Edward Milne incorporated their utilization as first steps of a space evangelization in efforts to maintain a universal Christocentrism for all creatures. Exclusive and inclusive theologians would adopt this "once-for-all" hypothesis to refute the suggestion of a planet-hopping saviour visiting a multitude of planets, living and dying repeatedly as sardonically coined by Paine, while multiple and varied thinkers expanded or abandoned this concept, respectively.

As evidenced above, the latter half of the twentieth and early twenty-first century produced a thin patchwork of Christian theological treatments of outside intelligences beyond the quite limited hypotheses of preceding centuries. Since the Sputnik launch of 1957, the Apollo Moon landings, the development and deployment of space telescopes *Hubble* and *Kepler*, the launching of 2,666 satellites presently orbiting Earth,[231] the Mars rovers *Spirit* and *Opportunity*, exponential growth in the discipline of astrobiology, and the unprecedented discovery of over four thousand exoplanets within an infinitesimal region of space, no comprehensive foundational principles for a Christian exotheology by the Church have been developed.[232] Nor were the NASA funded *Brookings* 1960 report recommendations for further research into religious and theological implications of space activities taken seriously by religious scholars. Similarly, the absence of official church pronouncements, encyclicals, or pastoral documents in order to clarify fundamental issues with regard to the core teachings of Christian dogma on the Incarnation and Redemption of Christ, the centuries-long central concern of the theme of intelligent extraterrestrials and its relation to Christian doctrine remained.

The most notable advancement in consideration of intelligent extraterrestrials by theologians in the last century has been a movement away from the classical *exclusivist* position towards acceptance of the *inclusivist* or *multiple* types (see chapter 5 for a summary of these positions). Those holding out for a position of exclusivism such as George, Steidle, Puccetti, and Hebblethwaite achieved resolution of the question of Christocentrism in an expanding cosmology, in the final analysis, by simple denial of

---

231. UCS Satellite Database, updated April 1, 2020.

232. Steven Dick in his book *Many Worlds*, to which he contributed the last chapter, sets the essential elements for what he terms a *Cosmotheology*; accordingly, he argues as a foundational principle for a Paine-type of naturalistic God to which terrestrial, localized Abrahamic faiths will have to dramatically adjust. This hardly suffices as a Christian accommodation of the implications of a vast, populated universe of intelligent religious beings. Dick, *Many Worlds*, 199–206.

extraterrestrial existence based upon a literalist reading of Hebrews 9:25–26[233] and Colossians 1:15–20, scriptural silence on otherworldly beings, and present lack of scientific evidence for their existence; Davies reflexively adopted the Paine position without further exploration. As David Wilkinson has observed, other contemporary theological thought continued to focus, as in previous centuries, on the importance of the central doctrines of the Incarnation and Redemption, and the role, or lack thereof, of human original sin within a potentially populated universe; and that God's acts of creation and redemption reveal a special concern for humans, however this does not imply that God is geocentralized.[234] More theologians occupied the inclusive and multiple columns, coinciding with the ecumenical movement beginning after mid-century (and the belief that the divine can be found in divergent religious traditions), globalism, and the rise of multiculturalism in Europe and North America. The inclusivist position can be considered the "Cosmic Christ" or Christological maximalist model, which make reference to the same scriptural passages as exclusivists: Col 1:15–20, Eph 1:20–23; Heb 2:7–9; and Rom 6:10 in their argument for a Christocentric universe.[235] Pannenberg, Milne, Mok, Davis, and Bonting argued for the human Incarnation of the Second Person of the Trinity as a unique event taking place only on Earth within the immense cosmos; a human Jesus Christ whose sphere encompasses the entirety of creation, and a humanity with a central role for the salvation of all intelligences in the universe by means of interstellar communication of the gospel. While Consolmagno, Pittenger, and Balducci offered no explanation how redemption is accomplished for extraterrestrials through Christ on Earth, Milne and Mok offer the possibility of radio transmission and space exploration as a means of cosmic missiology. As mentioned, given the "scandal of limited access," already seen with regard to the position of non-Christian religions, it is unsurprising that multiple thinkers find inclusive arguments unsatisfying due the logistics of transmitting or traveling in space and time within a context of possible civilizations billions of years before or after the human era, not to mention

---

233. That Christ would not offer himself repeatedly for the salvation of others; at the end of the age, he appears to take away sin by his sacrifice.

234. Wilkinson, *Science*.

235. Col 1:15–20, an early poetic arrangement, was most likely an early Christian hymn known to the Colossians and taken up in the letter from liturgical use. It presents Jesus's pre-existence, creation of all things through and for him; his preeminence among creatures; and reconciliation of all creatures through him on Earth or in heaven. Eph 1:20–23 refers to Christ's headship over all creation from his place in the heavens, above every power in this age and the age to come. Rom. 6:10, "As to his death, he died to sin once and for all; as to his life, he lives for God." Heb 2:7–9, especially v. 8, speaks of "subjecting all things under his [Jesus's] feet."

lack of participation in the human fall. These render the human Incarnation of the God-man wholly unfeasible for the salvation of other beings. As the Christological verses were written within the context of pre-scientific Aristotelian cosmology, some inclusivist readers of Paul do not take into account that the New Testament scriptures were written in a historical setting with a limited cosmology rather than in our twenty-first scientific age, leaving themselves open to criticism of how the earthly Jesus and singular Christian message is known and applied to beings occupying a vast universe in time and distance. As a result, neglect of the actual historical, philosphical, theological, and metaphysical setting of these texts and casting them into the modern cosmological era has been a principal cause for the limited and dilatory exotheological thought on extraterrestrials to the present. This issue is taken up in much more detail in chapter 5.

Those affirming multiple incarnations, such as provided by Mascall in its basic form, moved beyond the rigidity of the inclusivist position, recognizing its limitations where Christians cannot claim a human religious imperialism in a cosmocentric age and homogenous universe. *Multiple* thinkers generally affirmed the singularity of Christian revelation and its redemption for Homo sapiens on Earth, without any bearing on the destinies of other intelligent beings in the universe; all agreed on the possibility or even necessity of multiple incarnations of the *Word/Logos* in other rational corporeal beings, with an identical or nearly identical mode of salvation and revelation. However, this position remains the projection of a Christological composition to the universe, in an effort to subtly affix an earthly and anthropomorphized soteriological structure of Christian theology on other creatures of all times and places possible. The multiple solution is appealing to those who wish to maintain the salvation of extraterrestrials in either a strong or weak Christocentric universe or within their own religious heritage, while agreeing with Aquinas and Rahner that incarnation is the only and best way to achieve salvation of a species or population of a planet. Rahner regards any incarnation as the climax of revelation, occurring only once when a world is in its end stages where it operates to divinize a world as a whole. It may be claimed that a multiplicity of incarnations compromises the singularity of the human Incarnation of Jesus, as multiple thinkers such as Congar argue Jesus is not necessarily superior to other incarnations of the Word.

As Jesus is particular to humans, the *Word/Logos* incarnates in natures according to the historical epoch which requires it—in a universe of multiple incarnations, salvation must take place in ways other than through a single action of cosmic healing significance on Earth. The multiple incarnational position represented by modern theologians continues on a narrow trajectory by 1) its supposition that incarnation in a particular species operates solely as medium of revelation and salvation; with the exception of the argument made by Delio in her reading of Bonaventure in view of other divine motivations; and 2) that incarnation is only considered for the divine *Word/Logos* rather than possibilities of an incarnation of other divine persons either individually, collectively, or simultaneously in one or more creatures. Apart from Zubek, little to no attention is given to possibilities with regard to varying soteriologies or theoanthropological states of beings in conjunction with other incarnations, divine messages, or theophanies. While theologians who proposed a plurality of divine personifications on other planets can be credited for their efforts to resolve the inherent Paine quandary, the majority continued to hypothesize highly anthropomorphized extraterrestrial soteriological mechanisms, natures, personages, and media.

The *varied* argument constitutes a complete abandonment of geocentrism, cosmic Christocentrism, anthropomorphism, and theological anthropocentrism; expressed by Lewis, Tillich, Perego, and Funes in embryonic form. It is the least developed as theologians advocating the other types have been reticent to consider a potentially competing and parallel, non-parallel, or non-linear economy of salvation outside Earth that does not include or historically culminate in Incarnation of the Second Person in creatures. Multiple thinkers in particular remain tied to soteriological anthropocentrism as incarnation being the greatest example of divine communication and unification humanity has known; the assumption is God *must* incarnate in intelligent creatures if he is to reveal and redeem another species. One form of the more comprehensive varied view, conveyed by O'Meara and in partial agreement with Rahner, Congar, and Zubek do not presuppose this exotheological soteriological pattern. " . . . The Logos, the Second Person of the divine Trinity, indeed has a universal domination, but Jesus, Messiah and Saviour, has a relationship to terrestrials existing within one history of sin and grace."[236] This can be considered a *weak* varied view, given the continued presumption of the Word/Logos' capital role in the salvation history of other beings without consideration of the activity of the First and Third persons of the Trinity. Tillich and Perego make a comparable argument, as extraterrestrials did not inherit human sin nor receive the message of Christ; therefore, salvation must take another form, as Perego concluded it unnecessary to evangelize extraterrestrials to the Christian religion (given their soteriological position outside our own). Funes and O'Meara hold a stronger varied position with regard to divine

---

236. O'Meara, "Christian Theology and Extraterrestrial Intelligent Life," 20.

plans and activities outside Earth and beyond human range of reason and faith, and as stated by Ford, "God's dynamic Word knows no single form." According to the *varied* perspective, God's "omni-properties" are realized in creation in innumerable forms and modes and cannot be contained by our own human example. Specifics as to the possibilities of God's acts within individual species, theological anthropology of intelligent beings, responses to invitation to divine relationship, economies of salvation, and ultimate destinies of extraterrestrials are not examined by any of these thinkers. No further systematic development of exotheology along these lines of inquiry has been made as the majority of theologians have not engaged these questions; those few that have remain very general in their conclusions with few taking an official position. It is these questions that will be explored in the next chapter.

Chapter 4

Extraterrestrial 'Anthropology,' Xenobiology, Morphology, and Theological Systems

This chapter will examine the subject of possible extraterrestrials themselves, important for the theological discussion. Although speculative, it represents the beginning of the consideration of possible xenobiological structures, extraterrestrial environments, culture, and psychological and social compositions. This data will be considered on the basis of types of planetary or stellar habitats, evolutionary theory, competition models, and behavioral analogs. The putative psychological and sociological compositions of extraterrestrials will necessarily reflect their biological and environmental conditions. This 'anthropology' of potential extraterrestrial beings will be extrapolated from the resources available through human anthropology, and the scientific disciplines encompassed by astrobiology. As past theological scholarship has considered the extraterrestrial in a wholly generic manner and with little regard for possible environments, social structures, forms, states, capacities, and abilities, the paucity of consideration of the these fundamental aspects has resulted in little insight into the Christological possibilities and hence the repetitious and limited nature of theological work thus far. As discussed in chapter 1, this information will later be utilized to extrapolate potential theological anthropologies of extraterrestrials in Section B below.

Section A: Extraterrestrial Exoanthropology

Xenobiological Structures

Historian Steven Dick outlined these scientific premises regarding the development of intelligent extraterrestrials: Results from the WMAP[1] date

---

1. Wilkinson Microwave Anisotropy Probe, a NASA explorer probe launched in

the universe at ≈13.8 billion years. The first stars were formed about 200 million years after Big Bang, and the oldest Sun-like stars (population I) are between zero and ten billion years old. Heavy element generation and interstellar breeding through supernovae resulted in the first rocky and gaseous planets with solid cores to form; it may have taken another four to five billion years before life evolved on favorable planets. The maximum age of extraterrestrial natural or artificialized intelligence could be in the range of billions of years given the age of the universe and planets hosted by second-generation stars that contain high metallicity; allowing for the development of rocky planets containing heavier elements capable of producing sufficient gravity to sustain an atmosphere, liquid water, and lifeforms. Given this framework, the lifetime of a technological civilization is >100 years and likely much longer; and in the long term on a planet bearing intelligent life, cultural evolution (termed by Dick) can supersede biological evolution, eventually producing an artificial intelligence surpassing biological intelligence.[2] As the oldest Sun-like stars formed within a billion years of the Big Bang, and interstellar breeding produced the necessary heavy elements necessary for life to develop on rocky host planets, intelligent life could have developed up to 7.5 billion years ago using Earth history as an example.[3] In our terrestrial case, cultural evolution (meaning technological and social) has proceeded at an expeditious pace compared to biological evolution.[4] The main efforts leading the emerging field of cultural evolution in relation to our concerns here (and several Darwinian models have been explored),[5] are biotechnology, genetic engineering, nanotechnology, artificial intelligence, and space exploration. Dick proposes what he terms the *Intelligence Principle* to define the central idea of cultural evolution: that the maintenance, improvement, and perpetuation of knowledge and intelligence is the central driving force of a civilization, and that to the extent intelligence can be improved, it will be improved.[6]

2001 commissioned to provide fundamental measurements of the architecture of the universe.

2. Dick, "Cosmotheology Revisited," 294.

3. Norris, "How Old is ET," 103–5.

4. Dennett, *Darwin's Dangerous Idea*.

5. See Aunger, *Darwinizing Culture*; Dyson, *Darwin Among the Machines*; Lalande, and Brown, *Sense and Nonsense*; Richerson and Boyd, "Build for Speed," 423–63.

6. Dick, "Cosmotheology Revisited," 295.

Several AI experts[7] have envisioned the eventual dominance of intelligent machines according to the *Strong AI argument*.[8]

In pre-scientific cosmological and theological conjecture, extraterrestrial beings in ancient and early modern literature were typically conceived as essentially anthropomorphized and animalistic beings with little to no physiological, behavioral, or intellectual deviations. Well before science-fiction and Darwinian theory became the benchmark for what we may infer humanoid life forms to be, Christiaan Huygens's monograph *The Celestial Worlds Dicover'd, Or, Conjectures Concerning the Inhabitants, Planets and Productions of the Worlds in the Planets*, published posthumously in 1698, described possible beings as similar to humans but in other ways quite dissimilar:

> Nor does it follow from hence that they must be of the same shape with us. For there is such an infinite possible variety of Figures to be imagined, that both the Oeconomy of the whole Bodies, and every part of them, may be quite distinct and different from ours.[9]

Many early-modern astronomers like Huygens considered intelligent extraterrestrials a natural consequence resulting from favorable environmental conditions on other planets. Later, Darwinian theory provided a theoretical model of natural selection, variation, and environmental adaptation for our modern consideration of the potential morphologies of extraterrestrials. Scientists and astronomers from the mid-twentieth century onward were hence more skeptical of the probabilities of intelligent extraterrestrial life given their acceptance of the evolutionary synthesis and the increased understanding of the unique conditions needed to produce intelligence.[10] Of the few scientists that have speculated on the nature and morphology of extraterrestrial life, American geneticist and evolutionary biologist Theodosius Dobzhansky affirmed and emphasized mutation and natural selection:

---

7. See Searle, "Minds, Brains, and Programs," 417–57.; Moravec, *Mind Children*; Kurzwell, *The Age of Spiritual Machines*; Tipler, "Extraterrestrial Intelligent Beings Do Not Exist," 133–50.

8. The most modern form argued by Kurzweil, *The Singularity is Near*, of an intelligent machine that is capable of general intelligent action that rivals or exceeds the thinking capacity of humans. Strong AI refers to a computer capable of consciousness, rather than merely the running of pre-programed instructions.

9. Huygens, *The Celestial Worlds Dicover'd*, 74.

10. Morris, *Life's Solution*, 344; Barrow and Tipler, *The Anthropic Cosmological Principle*.

Despite all the uncertainties inevitable in dealing with a topic so speculative as extraterrestrial life, two inferences can be made. First, the genetic materials will be subject to mutation. Accurate self-copying is the prime function of any genetic materials, but it is hardly conceivable that no copy errors will ever be made. If such errors do occur, the second inference can be drawn: the variants that arise will set the stage for natural selection. This much must be a common denominator of terrestrial and extraterrestrial life.[11]

More recently, extrapolating from Darwinian models, mathematician Carl DeVito[12] and geneticist Norman Horowitz[13] hypothesize based on biological, psychological, and sociological equivalencies that intelligent extraterrestrials would have an analogous mathematical system, function according to a modern understanding of physics, and would be composed of and interact with similar elements according to our periodic table. Dobzhansky also argued for convergence, noting how various forms of aquatic life with disparate ancestral lines have similar morphologies in their adaptation to an aquatic environment. However, he also argued for divergent evolution for Earth life forms in similar environments.[14] Oceanographer Robert Bieri's article, "Humanoids on Other Planets?" asserted limitations inherent in biological chemical elements, as well as the available forms of energy seen in the limited range of terrestrial morphological variability.[15] Bieri states that due to these restrictions on possible biological adaptations, extraterrestrial intelligent beings will conform to the patterns we are familiar with on Earth. Scientist Zoltán Galántai[16] outlines several possible types of biology which may be present in the universe: Biology 1: Earthly life as it is known; Biology 2: an extension of our understanding of biology 1 in considering and searching for extraterrestrial life forms, known as today's astrobiology;[17] Biology 3: xenolife having an alternate form of biochemistry; and Biology 4: which refers to at present hypothetical other universes having different physical constants or different physical forces. In this chapter I will argue for a Biology 2 model. It can be safely

11. Dobzhansky. "Darwinian Evolution," 170.
12. DeVito, *Science, Seti, and Mathematics*.
13. Horowitz, *To Utopia and Back*.
14. Dobzhansky, "Darwinian Evolution," 168–69.
15. Bieri, "Humanoids on Other Planets?", 425–58; see also Beadle, "The Place of Genetics in Modern Biology."
16. Galántai, *Life, Intelligence, and the Multiverse*.
17. Astrobiology is founded on the uniformity principle, which claims that the physical processes of nature known to Earth are the same throughout the universe.

assumed for the present that extraterrestrials are likely carbon-based lifeforms, due to the unique ability of carbon to form the core of a very diverse range of macromolecules, and be water-based given the unique properties of water in the formation and maintenance of biological life. However other biochemistries may be possible based on silicon (although these to our knowledge do not allow for the same extensive variety of molecular combinations as carbon). It is no longer believed that oxygen is an absolute requirement for life as oxygen was absent from the surface of the Earth in the first few billion years while simple and multicellular life forms existed. Whether the genesis of intelligent life would require an oxygen rich atmosphere similar to Earth remains an open question.

As regards the development of possible alien life, convergence evolutionary theory posits that species with similar capabilities in similar habitats may evolve to look alike, as common functional demands that channel the solution of selection and shared molecular and environmental constraints limit the range of likely solutions. Different species with the capacity for swimming look alike.[18] Alien life on planetary surfaces in gaseous atmospheres and using intelligence to manipulate their environment with tools could have bilateral symmetry, with legs for locomotion, appendages used as hands for manipulating objects, and a pair of eyes to provide stereo vision.[19] However, it is possible there could also be much evolutionary divergence. Physicist W. G. Pollard has provided an example of the likely independent evolution of species on different planets, and how life tends towards divergent forms. He notes that about 180 million years ago Australia broke off from Gondwanaland and can be thought of as Earthlike planet "A," where evolution continued independently from a primarily reptilian stock. Similarly, South America broke off from Africa 130 million years ago and can be viewed as independent planet "S." Independent evolution also continued on planet "E" (meaning the rest of the Earth, especially Africa and the adjoining land). During the last 130 to 180 million years, independent evolution has diverged towards different kinds of animals on these three "planets," rather than converged. Humans appeared only on planet "E," certain primates on "S" and marsupials on "A." Humans on "E" appeared only about 4 million years ago and have existed for only 0.1 percent of Earth's history.[20] Natural selection appears to produce many species capable of occupying any habitable environment, therefore, given favorable environmental conditions, we should not be surprised if life has evolved from more elementary

---

18. Morris, *Life's Solution*, 147–223.
19. Darling, "Variety of Extraterrestrial Life."
20. Pollard, "The Prevalence of Earth-Like Planets," 653.

forms on another planet. However, given the magnitude of disparate life forms present in the myriads of environments on Earth it is possible that many forms of life on other planets given the proper conditions, including highly intelligent life forms, may appear humanoid or occur in a variety of physical realizations. Biologist Allen Broms once stated, "Life elsewhere is likely to consist of odd combinations of familiar bits."[21]

As lifeforms tends to expand their habitat until meeting a limitation, such as a food source or lack of geography, competition for resources is created as Darwinian evolution assumes any number of offspring will typically exceed replacement level. Charles Cockell and Marco Lee have argued that intelligent extraterrestrial life are likely to evolve at the end of a series of trophic levels, and for energetic reasons predation is likely to be widely represented, and would be influential in determining the morphological and behavioral characteristics of extraterrestrials.[22] Predatory pressures also contribute to the nature of a diversity of behaviors including aggression, speed and maneuverability, vigilance, flight, territoriality, and flocking, all of which can have an important influence on sociobiology and thus the potential characteristics of intelligent societies. Darwinian theory states modern humans resulted from a long struggle for existence, by way of violence, suffering, and death, which impacted our social and psychological makeup. Geophysical conditions of prehistoric Earth were quite different than our current epoch, and was inhabited by terrestrial creatures foreign to moderns. Therefore, it is reasonable to infer that other planets may have their own assortment of creatures adapted to their own environment, evolved from more primitive forms. Other variables, which cannot be measured or necessarily predicted in any meaningful way according to evolutionary theory even on Earth, include predation pressures, foraging patterns, metabolic requirements, genetic mutations, and developmental interconnections of the phenotype.[23] The biological classification of life forms on Earth includes kingdom, phylum, class, order, family, genus, species; and the five kingdoms include animals, plants, fungi, protists, and monera. It may be discovered that life on other planets could extended upwards to include super-kingdoms.

Physicist Gerald Feinberg and biochemist Robert Shapiro,[24] have argued in favor of evolutionary convergence, as has Simon Conway Morris, namely that historical contingencies may make it possible to predict certain

---

21. Broms, *Our Emerging Universe*
22. Cockell, "Interstellar Predation," 1.
23. Powell, "From Humanoids to Heptapods."
24. Feinberg and Shapiro, *Life Beyond Earth*, 411.

properties of extraterrestrial life forms.[25] However each have rejected the view of space scientists Roger MacGowan and Frederick Ordway's claim that the majority of intelligent extrasolar land animals will be of the two legged and two armed variety.[26] By means of mutual action of natural selection and mutation, great divergences are possible, however they agreed that "we will undoubtedly encounter [convergent evolution] on other worlds."[27] Robert A. Freitas Jr. has stated that xenobiologists have formulated a simple rule known as the Assumption of Mediocrity, whereas the Earth is considered as "typically exotic."[28] With Earth life as an example, as a means of survival, evolution devised solutions where we could expect to find parallels, but not necessarily their duplicates in extraterrestrial species.[29] The most obvious instance of convergent evolution is the "camera eye," developed independently in five major terrestrial animal phyla (chordates, mollusks, annelids, coelenterates, and protists), each having diverse developmental histories. The camera eye is the most ubiquitous because it clearly is the best evolutionary solution to the general problem of vision on this or perhaps any other world,[30] with lens, retina, focusing muscles, and transparent cornea—placed high in the body so to view obstacles and predators at a distance.[31] Other abilities that may vary from those of humans are power of vision, means of locomotion, hearing, and communication, or the ability to see in other bands of the electromagnetic spectrum, such as infrared, visualize heat waves, or display a sensitivity to magnetic waves, electric fields, or radioactivity. Each alien sense would have developed as a means to maximize survival in its particular planetary, geographical, and local environment and in competition for available resources.

As mentioned, evolutionary paleobiologist Simon Conway Morris has emphasized the ubiquity of evolutionary convergence, and argues against those claiming the impossibility of predicting extraterrestrial morphologies:

> [W]hat we know of evolution suggests ... convergence is ubiquitous and the constraints of life make the emergence of the

25. Morris, *Life's Solution*, 283–84.
26. MacGowan and Ordway, *Intelligence in the Universe*, 240.
27. Fienberg and Shapiro, *Life Beyond Earth*, 411.
28. The idea that Earth is unusual among most planets known to us in possessing an abundance of life forms, whereas Earth is considered special, privileged, exceptional, or even superior.
29. Freitas, "Extraterrestrial Zoology," 53–67.
30. Freitas, "Extraterrestrial Zoology," 58.
31. Ernst Mayr has argued the evidence of convergence of the eye in at least forty unrelated lineages in "The Probability of Extraterrestrial Intelligent Life," 23–30.

various biological properties very probable, if not inevitable. Arguments that the equivalent of Homo sapiens cannot appear on some distant planet miss the point: what is at issue is not the precise pathway by which we evolved, but the various and successive likelihoods of the evolutionary steps that culminated in our humanness.[32]

Biologist Richard Dawkins in his argument for "Universal Darwinism"[33] along with anthropologists Kathryn Coe, Craig T. Palmer, and Christina Pomianek, asserts that the principle of convergence is the norm; and "evolutionary theory, theoretically, should apply anywhere to anything that is living."[34] Confirmation of convergent evolution is widely evidenced on this planet; therefore, it is not unreasonable to hypothesize a similar convergence on an equivalent planetary environment capable of supporting complex life forms. This is best explained by Robert Bieri,[35] most notably in his argument on bilateral symmetry. The importance of bilateral symmetry in evolution is essential to maximum speed of movement in hunting and escaping and reducing resistance and turbulence in an aquatic environment; whereas those with more stationary habits tend to have radial symmetry and a lower level of organization, without the accompanying complex nervous system. Therefore, having a more complex nervous system is contingent upon a more predation-influenced way of life. Bieri states that an anterior mouth and posterior anus are the most effective method for ingestion and secretion for a predatory being. Additionally, the most important sensing organs and grasping organs and appendages are located in close proximity to the mouth, with the brain located closest to these sensing organs so to protect the brain from attack or damage. This is seen almost universally and independently among Earth creatures regardless of their evolutionary antecedents. Anthropologist Loren Eiseley made similar arguments and supported this view regarding its morphological advantages.[36] Therefore, it is reasonable to suppose that an extraterrestrial predatory species will have bilateral symmetry with a brain and sensing organs at the anterior portion of its body, and that higher complexity of the nervous system and brain achieve further development on land. Bieri believes large-scale brain development

32. Morris, *Life's Solution*, 283–84.
33. Palmer et al., "ET Phone Darwin," 215; Dawkins, "Universal Darwinism," 403–5.
34. Palmer et al., "ET Phone Darwin," 214–25.
35. Bieri, "Humanoids on Other Planets?" 453.
36. Eiseley, "Is Man Alone in Space?" 80–86. Eiseley states regarding cytologist Cyril D. Darlinton's opinion of Homo sapiens: "Darlington . . . dwells enthusiastically on the advantages of two legs, a brain in one's head and the position of surveying the world from the splendid height of six feet."

and conceptualization occur more easily as a result of social existence, speech, and use of tools. Claws would not be advantageous to an intelligent being, nor would feathers or thick scales from a being evolved from a land predator. Convergence is also evident in binaural hearing, which is essential for discerning the source of sounds. Smell sensors are ideally located nearest the mouth, to test the edibility of foods. Tactile sensors are ubiquitous to all organisms and provide additional self-defense. Walter Sullivan[37] states extraterrestrial creatures must be able to move about and build things. That is, they must have something comparable to hands and feet, have senses, such as sight, touch, and hearing, although the senses that evolve on any given planet will be determined by the environment. Vision in the infrared part of the spectrum (may) be more useful than sight in the wavelengths visible to human eyes. The amount of food available would set a limit on overall mass of a being, and the fixed sizes of molecules must limit the extent to which the size of a complex brain can be compressed. Therefore, given these and the aforementioned arguments, it is likely that an intelligent extraterrestrial species would have originated in a predatory environment with a basic symmetrical structure, large brain and sensing organs, exist in social groups, and use tools.[38] Therefore, it seems, given these morphological elements, it is likely that intelligent extraterrestrials will be generally humanoid in appearance, with variations in secondary features due to dissimilarities in physical environments and particular evolutionary tracks.[39]

Diverse extraterrestrial species would have certain morphological differences due to variations resulting from star types, planetary gravity, environmental conditions, food sources and predation, and evolutionary and social histories. As a result, many would likely behave and process information differently, and may have great differences in their subsequent technological achievements. Planetary orientation in space has a particular influence on the development of what we can surmise to be intelligent extraterrestrials. Super-Earths, several which have been discovered in the past few years, would have correspondingly higher gravity, resulting in a heavier endoskeletal (or exoskeletal) structure and more powerful muscles and connective tissues. Beings evolved on such a planet would possess shorter, stockier bodies and denser bones than those evolving in low-g environments

---

37. Sullivan, *We Are Not Alone*, 300.

38. Other variations from the prototypical humanoid can be considered, such as Larry Niven's "Puppeteers," a fictional race of intelligent beings having two forelegs, a single hindleg with hooved feet, and two snake-like heads rather than a humanoid upper body. This being uses their mouths to manipulate objects which contain finger-like knobs, enabling the use of tools by which they develop a high technological society.

39. Puccetti, *Persons*, 96.

as proper structural support depends on a bone diameter proportional to the square root of gravity.[40] The size and shape of an extraterrestrial will be partially determined by its source of energy, planetary gravity, and ambient density. For example, Pandora, the planet featured in the film *Avatar* (2009) has a lower gravity, a thicker atmosphere, more powerful magnetic fields, and differing day-to-night cycles than Earth, creating a variety of unique ecological conditions. Giantism of vegetation resulted due to lower gravity, and plants which absorbed metals from the soil utilized the planet's magnetic field for movement which the Earth's biologists referred to as "magnetonasty." Plants on Earth are green due to the presence of chlorophyll in their cells which processes the chemical compound necessary for photosynthesis; on other planets there may be other ways to achieve photosynthesis where green plants are not a requirement.[41] If Earth had double its present mass, higher gravity would have resulted in a stronger endoskeleton, which may have precluded bipedalism; an Earth analog with half its mass would have possibility resulted in quite different looking humans. Similarly, if Earth's axial tilt of 23.5° were altered to 60°, seasons and climates would be dramatically altered and with that our evolutionary adaptation to them. Similarly, our circadian rhythms, developed over the long period of our ancestry, allow for the opportunity for cells to replicate at night while avoiding DNA damage from ultraviolet radiation in sunlight. If an Earth day consisted of one hundred hours rather than twenty-four, mutations would have occurred in skin pigmentation, eye development, and metabolism, among others; modern humans would appear substantially different.

Given the above arguments, it is not surprising that a number of highly respected physicists and astronomers and a small minority of biologists hold that we are not entirely unique in our general physical structure, and that extraterrestrials would in many ways appear humanoid. Physicists such as Steven Weinberg and Sheldon Glashow emphasize similarity in terms of mental capacity and ability to perceive the same universal physical laws, while others claim physical similarity.[42] Astronomer Frank Drake writes, "They won't be too much different from us ... [A] large fraction will have such an anatomy that if you saw them from a distance of a hundred yards in the twilight you might think they were human."[43] Biologist Robert Bieri agrees that "they will look an awful lot like us."[44] Astrophysicist Joel

40. Freitas, "Extraterrestrial Zoology," 57.
41. Baxter, *The Science of Avatar*.
42. Basalla, *Civilized Life in the Universes*, 198.
43. Basalla, *Civilized Life in the Universes*, 184.
44. Basalla, *Civilized Life in the Universes*, 18

Primack reports that intelligent aliens will approximate the general size of humans; optimal for complexity and fast thinking, and that they may possibly share our fractal circulatory system, rates of energy use, and even lifespans.[45] Many biologists, on the other hand argue for uniqueness on the grounds that the many unpredictable historical steps leading to intelligence could never be duplicated. We cannot assume a binary male-female gender that defines Homo sapiens, and should consider the possibility of xenomorphs, hermaphrodites, neutrois, or transgenders. Extraterrestrials could be ovoviviparous[46] or be monosexual, parthenogenetic, or variable-sex. Propagation through cloning and genetic engineering are also likely outcomes in a highly advanced technological civilization, to be discussed in the following section.

Philosopher Roland Puccetti has stated that the development of human intellectual capacities was rooted in early social learning, which allowed for the maturing of symbolic speech[47] for transmitting collective knowledge of the environment. An extraterrestrial species evolved from land predators, similar to humans, could exist in isolated culture-groups during its immediate pre-scientific, early historical period; and would likely compete amongst itself in exploiting environmental resources.[48] Eventually these societies could reach, by means of scientific and technological advancement, a level of possible self-extermination due their inherent behavior in tribal warfare, rooted in pre-conscious predatory instincts. However it cannot be assumed that other races have not found the means to live peaceably prior to the advent scientific achievement, technology, and political institutions. According to the jurisprudence and philosophy of law of H. L. A. Hart, humans are characterized by "limited altruism," being neither totally motivated by self-interest and aggression, nor entirely benevolent and considerate of others.[49]

Humans require food, clothing, and shelter; since the sources of these necessities are limited it is necessary to obtain these by labor; thus some form of property, whether individual or communal, needs to be instituted and acknowledged.[50] Due to the logical advantages of cooperative effort and division of labor, rules and contracts become necessary, and given humans

---

45. Primack and Abrams, *The View from the Center of the Universe*, 224–28.

46. Producing young by means of eggs that are hatched within the body of the parent, as in some snakes.

47. Terrence Deacon describes the emergence of symbolic thought and language as a concurrent, co-evolutionary process. See Deacon, *The Symbolic Species*.

48. Puccetti, *Persons*, 105.

49. Hart, *The Concept of Law*, 219.

50. There are exceptions to this, as aboriginal Australians were discovered to have no concept of private property. See West and Murphy, *A Brief History of Australia*, 20.

have a limited understanding of their long-term interest in forbearance and compromise, and a limited strength of will to resist manipulation or abuse of others for personal gain, these contingent realities require systems of coercion and legal sanction for those who will not voluntarily submit to a system of mutual forbearances. Therefore, according to Hart, voluntary cooperation within a coercive system is what reason requires of beings constituted similarly to Homo sapiens within an equivalent environment.[51] Since convergent evolution provides a plausible hypothesis for some uniformity between Homo sapiens and intelligent extraterrestrials, we can consider that they will be descended from a predatory environment; a "limited altruism" can be suggested based on a similar sociobiological ancestry, and their social existence based on their achieved conceptualizing intelligence. Extraterrestrial biological entities will require physical nourishment, shelter, and perhaps clothing gained through labor and thus should be characterized by the same "natural necessities."[52] Advantageous to Homo sapiens in dominating Earth life forms were a long gestational period and extensive life-span, a highly developed brain, the evolution of arms and dactyls through arboreal ascent allowing for the manipulation of tools and weapons; and an extended parental dependence resulting in a longer period for maturation of the brain for complex cognition.[53] As outlined by Hart, "In the first instance is assured some selfish aggressiveness, in the second social egalitarianism and benevolence, reinforced by a long period of parental dependence."[54] We can, then, cautiously hypothesize according to the argument laid out for intelligent extraterrestrials a similar bipedal locomotion, manual dexterity, control of differentiated muscles of facial expression, vocalization, intense social and parenting behavior, stereoscopic vision, and forms of sexual behavior.

The genetic engineering of species, if chosen, could be beneficial for a host of reasons to an extraterrestrial race: to extend life well beyond its natural limit, enhance native intelligence, and repress certain negative innate characteristics, such as violent tendencies and extreme competitiveness, or increase one's passivity and willingness to obedience. In practical use engineering of the genome could be utilized to alter the body's ability to withstand radiation and other conditions necessary for interstellar travel or long-term habitats in biologically hostile environments. There might be sexually reproducing engineered species, while others might decide on cloning or purely genetically engineered biological life or synthetic life forms. Those having engaged in

---

51. Puccetti, *Persons*, 108–9.
52. Hart, *The Concept of Law*, 222.
53. "A Long Childhood Is of Advantage," Max-Planck-Gesellschaft.
54. Given the example in Homo sapiens.

non-sexual means of reproduction for very long periods, perhaps thousands or tens of thousand of years, might judge sexual reproduction a baser and less-advanced method of perpetuating a species, and consider engineering-produced life, with artificially introduced beneficial genetic variants less prone to hereditary errors and undesirable characteristics. Those societies where sexual means of reproduction have been engineered out for extremely long periods might have no concept of gender and exist as a homogenous, gender-less or androgynous species.

Extraterrestrial Psychology

Steven Dick maintains that Darwinian models for determining alien evolutionary psychology are problematic; lacking a robust theory, we are reduced to the extrapolation of current trends supplemented by the most general evolutionary concepts. Accordingly, the most relevant fields for understanding what may be extraterrestrial psychology can be best extrapolated from genetic engineering, biotechnology, nanotechnology, and space travel. While the evolutionary development and ultimate morphology of intelligent extraterrestrials will fundamentally depend on the general and specific physical environment of their given planet, moon, or artificially constructed habitat, as well as specific chemistry and biology, these same conditions determine thought processes and behavior. Basic biological heritage and early social and environmental conditioning will determine many of the later, developed intellectual functioning and cultural characteristics. Fear responses, aggressiveness, competition, hierarchical behavior, mating rituals, and innate curiosity are preconditions of pre-intelligent predatory and social behavior. Intelligent extraterrestrials, having developed along a dissimilar evolutionary trajectory may not possess identical human behaviors, thought processes, motivations, and goals. If it is possible for a species to survive the red giant phase of their parent star, civilizations may have lifetimes that greatly exceed our own, providing them with a far different psychology and world view towards other species and civilizations.[55] In fact, some intelligences may have consciousnesses that are structured and perceive reality wholly different than humans, or exist as a single integrated intelligence rather than a civilization of self-determinate individuals. Physicist Guillermo Lemarchand and science journalist Jon Lomberg refer to this as the *incommensurability problem*:[56] that

---

55. Billingham, "Summary of Results."

56. A conceptualization taken from epistemology of science to mean paradigm or worldview, a process through which an intelligent mind creates conceptions of its environment and utilizes them for regular functioning. Hoyningen-Huene, "Kuhn's Development," 185–96.

possible differences between the cognitive maps of humans and humanoid beings, or postbiological beings may be so great that any meaningful communication would be rendered impossible, or at least very difficult. Of these, there are two types: *methodological incommensurability*—a relation between different theoretical frameworks such that, because of the deeply cognitive map-dependent character of observation and scientific method, no rational method for exchanging intelligible information can be found (that is, a variation of sense experience of reality); *and semantic incommensurability*—a relation between two instances of the same term as it occurs in the confrontations between two different cognitive maps (that is, a discontinuity of reference to the same idea or concept between two species).[57] Lemarchand proposes an outline in the search for what he terms *cognitive universals* when considering the thought of extraterrestrials and their possible religious beliefs and systems. He holds that there are four types of cognitive universals 1) physical-technological 2) aesthetic 3) ethical 4) spiritual; this he considers more helpful than focusing on the believed universality of natural laws. He asserts that alien morphologies which originate in differing physical and social environments may have different ways of thinking, perceiving, interpreting, and relating, and our human abstractions and representations of what we perceive in the physical, aesthetic, ethical, and spiritual dimensions may not be necessarily universal in scope. Minimally counterintuitive agents,[58] in this case, extraterrestrial beings which are substantially similar to humans, might yet possess behaviors that violate intuitive expectations, and behave or appear in ways incongruous to humans. However despite micro or macro deviations in the operation of mind, imperative to the survival of any intelligence species is a highly developed social interaction which involve cooperation in sustaining and securing survival of the community, protecting against threats, and furthering the higher needs and goals of the community. There could be a collective conscious, where interaction of mind operates as a clearinghouse in which every conscious thought can be accessed and explored by anyone at any time.[59] A hierarchical social system is common in evolution due to its survival advantages; we may expect similar structures in advanced species with histories of competition for limited resources as part of their heritage.

---

57. Lemarchand and Lomberg, "Communication among Interstellar Intelligent Species," 371–95.

58. McNamera, *The Neuroscience of Religious Experience*, 194.

59. At present scientifically unproven but claimed psychic abilities in humans, such as mental telepathy, telekinesis, remote viewing or remote sensing, may exist as innate abilities of some extraterrestrials and perhaps integral to functionality, in other cases such abilities may have been achieved or enhanced by technologicalization.

Intelligence denotes conscious awareness, which implies self-determination;[60] self-determination implies or leads to the possibility of consciousness of the good and its absence; self-consciousness awareness leads to freely willed thought and behavior. There is the possibility of the awareness of sin (whether understood in relation to a deity or merely as a wrongdoing against one's own) and evil action on the part of sentient beings. Theology contains the fundamental categories of grace, sin, redemption, relationship, free will, evil, knowledge, and suffering; few theologians have utilized these theological principles to extrapolate what fundamental extraterrestrial consciousness may be or have attempted to formulate their cognitive and spiritual functions comparative to terrestrials. However, it can be argued that an effort to formulate extraterrestrial capacities and function in this regard would be inaccurate due to obvious anthropomorphizing of extraterrestrials. While it is theoretically possible that an extraterrestrial society may have developed to parallel ours, it is entirely plausible that it may not have developed theologies along the terms and standards according to certain Earth religions. Some extremely ancient, technological civilizations may have culturally evolved to have abandoned such concepts, or have biologically engineered and designed social structures to the extent that individuality and the sense of free thought are strained out of awareness. Highly advanced biological or artificial 'life' could attain "superhuman" attributes, having inestimable abilities of knowledge, intelligence, physical capabilities in controlling any natural environment, and/or the capacity to completely control or manipulate the minds and/or bodies of lesser-evolved species. Given these possibilities, there is some naiveté in the expectation that our categories of philosophy, psychology, and theology will directly apply to extraterrestrial societies or even have indirect parallels. Certain extraterrestrials and other life or non-biological intelligent beings might be solely motivated by self-preservation and perpetuation, and lack quintessential human ideals of equity, justice, benevolence, or freedom. Intelligent extraterrestrials have often been portrayed or hypothesized utilizing an overly simplistic model, what Alfred Kracher describes as *heuristic anthropomorphism*. That is, consideration of extraterrestrials in a narrow anthropomorphic sense that their problems, conflicts, and relationships mimic human affairs.[61] This is nearly categorically in the case of science fiction novels, television programs, and cinema purporting to describe and dramatize extraterrestrials individually and collectively.

60. There may be instances where a species or collective have mitigated, canceled or neutralized innate individual self-determinitive ability due to genetic and/or psychological manipulation or conditioning.

61. Kracher, *Extraterrestrial Altruism*.

'ANTHROPOLOGY,' XENOBIOLOGY, MORPHOLOGY, AND THEOLOGICAL SYSTEMS   141

I have outlined some recent thinking about the development and psychological nature and overall cognitive functioning of intelligent extraterrestrial biological life. As humans in an encounter with an unknown other, the task will remain for humans (and/or possibly them) to "bridge the gap" in communication and meaningful understanding, in whichever manner information is provided, whether explicit or implicit, according to our own established knowledge base and conceptual framework, religious dogma, moral codes, and relational patterns.

Constitution of the Extraterrestrial: 'Exoanthropology' Morphology

There are several morphological possibilities and innate/acquired capacities of extraterrestrial intelligent beings. Our own progress within the last 150 years since industrialization has been dramatic; if our own history may be used as an example, civilizations that are hundred of thousands or millions of years old,[62] if having survived any epochs of tribal warfare capable of mass extinction may be extremely advanced. Technological, biological, intellectual, psychic, communication, and travel capabilities will be far beyond ours, perhaps beyond our comprehension in some cases. Various capacities and abilities of each type of being may be achieved naturally, through material technological augmentation of innate mental or physical abilities, and/or achieved by highly sophisticated artificial/synthetic means. Morphologies discussed earlier in this chapter could ultimately possess these characteristics over long periods of scientific and cultural evolution. These may be categorized as follows:

1) Purely biological entities, having intelligence and a developed or highly sophisticated civilization, possessing an independent thinking mind, physical senses which may parallel or differential from humans; capable of physical movement, thought, ideation, memory, imagination, creativity, and recognition. They are descended from an independent, natural evolutionary process native to their own planet, and travel/occupy a planetary, interplanetary or interstellar habitat.

2) Biologically enhanced entities, similar to purely biological beings described above, genetically enhanced through biotechnology, nanotechnology, gene-editing, brain implants, and/or neuro-pharmaceuticals. Directed evolution or rewriting of a genome utilizing reprogenetic technologies such as germline editing[63] may enable societies to produce highly ad-

---

62. To a maximum age from 1.7 to 8 billion years. Dick, "Bringing Culture to Cosmos," 468.

63. Altering of the genome of a sperm or ovum, enabling changes in future DNA

vanced, physically and intellectually superior beings. These known methods would result in a trans-speciesism, allowing for dramatic enhancements in intellectual, physical and mental capabilities. Possible inter-species hybridization might exist, by technologically overcoming any inherent biological incompatibilities between species, and may be utilized programmatically, via colonization or free interaction between inhabited planets or separate species sharing the same habitat. Other, unknown highly advanced technologies may enable the design and creation of an entire race of disease-free, superior beings with lifespans many times that of humans, some essentially immortal. These types of beings may use cloning as an effective measure to ensure continuation of the most desirable qualities for certain functions within a given society.

3) Biological hybridized entities, similar to purely biological beings in the first case, combined with the engineered capacities of the second type, in addition to technological enhancements such as cybernetics, prosthetics, and merged with artificial intelligence resulting in a dramatically altered/implanted hybridized biological/partially mechanistic being. Capacities could include enhanced communication, augmented reality systems[64], enhanced senses, multi-dimensional thinking, bodily modifications such as powered endoskeletons or exoskeletons allowing for greater strength and physical capabilities, enhanced problem solving, outsourced memory or mind uploading,[65] consciousness-assisted technology,[66] technology-assisted consciousness,[67] technologically created and imposed virtual reality, photographic memory, and artificial nutrition.[68]

---

of every cell in an embryo, allowing for genetic changes to embryos affecting all future generations within a family lineage.

64. Augmented reality is a technology in which computer generated images are superimposed on the users natural vision, providing the user with enhanced, additional information, designed to heighten sensory awareness of one's environment.

65. A hypothetical transferring or copying information from a conscious, biological mind to a non-biological substrate by a mapping or scanning technology to a computer system.

66. A computer system programed to interpret and implement a thought signature to carry out commands

67. Specialized computer systems interfaced with the brain to assist or enhance the function of mind, thought, consciousness, and states of consciousness. An example is the Monroe Institute's Hemisynch tones, designed to assist in achieving states of relaxation, and perhaps future heightened mental capabilities, including the ability for mental telepathy. This technology could be feasibly utilized by advanced extraterrestrials in teleportation, telekinesis, remote viewing, and higher states of consciousness.

68. Endogenous nutrition, i.e., by means of a radioisotope generator that resynthesizes glucose, amino acids, and vitamins from their degradation products, theoretically allowing for weeks without nutrients.

4) Artificial biological/synthetic life. Synthetic biology is the design and construction of artificial organisms with cells that can capture energy, maintain ion gradients, contain macromolecules as well as store information and have the ability to mutate. A synthetically designed biological being would be capable of reproducing many or all natural biological processes: possessing synthetic DNA, processing information, manipulating chemicals, fabricating materials, producing energy, removing waste, provide energy, and maintaining its own "homoeostasis." As mentioned, an artificially produced being might be able to exist in a variety of environments on planetary surfaces, planetary satellites, in zero-gravity, vacuum, high g-forces, and able to function in spaceflight for extremely lengthy interstellar journeys. This type of manufactured being may be fitting candidates for cloning; for military uses or to perform highly specialized duties, or work in hazardous environments. Artificially synthesized beings are by nature genetically optimized, and may in addition be combined with specialized technology from types 3, such as cybernetics, prosthetics, and artificial intelligence. Presently, our technology has not yet produced artificial cells but major progress is being made along these lines.[69]

5) Robotic/mechanistic entities, an artificially manufactured being endowed with artificial intelligence termed in future studies literature as postbiological. Robots, androids, or other type of artificial beings capable of performing tasks equivalent to biological beings will likely possess *superintelligence*, with the capacity to far exceed the intellectual (and perhaps physical) abilities of biological beings.[70] Nick Bostrom has outlined three aspects of superintelligence: Speed superintelligence—the ability to process information faster than any purely biological brain's neurons, neurotransmitters, and synapses; collective superintelligence—the ability to interface collectively with similar artificial minds, outperforming the collective abilities perhaps of an entire biological civilization; and quality superintelligence—the ability to attain an intelligence quotient far superior in any domain.[71] Such beings could be designed to perform a specialized set of repetitive tasks in specific environments, be commissioned to undertake journeys across great distances and interact with other species, to serve military purposes, or to act as a failsafe to ensure continuation of a species' memory and cultural and technological achievements in the eventuality of

---

69. Deamer, "A Giant Step Towards Artificial Life?" 336–38.

70. Schneider, associate professor of philosophy and cognitive science at the University of Connecticut has been at the forefront of research on artificial intelligence and consequences for human society. See Schneider, *Science Fiction and Philosophy*; Schneider, "Alien Minds"; Schneider, *Superintelligent AI*.

71. Bostrom, *Superintelligence*.

a catastrophic event. The structure of their consciousness may not necessarily mimic that of biological beings such as humans, and may have programming enabling awareness and processing of a singularity, or a single integrated intelligence shared on a mainframe. Beings of this kind may be capable of conceiving any concept possible and exploring each completely, simultaneously, and instantaneously.

6) Purely spiritual non-corporeal entities, with the closest comparison known to humans as angelic beings in the Judeo-Christian tradition, having superior intellectual capabilities and able to interact in a discrete and non-discrete manner with physical beings and inanimate objects. It cannot be concluded that earthly and heavenly angelic beings known to humans are the only species of incorporeal entities God has created throughout the cosmos; in fact there may be a great variety of such beings that serve purposes unknown as well as other intelligent biological extraterrestrials. Non-physical entities may have benevolent or malevolent motivations and be understood to possess preternatural or supernatural qualities and abilities.[72]

Sociological Compositions

Anthropologist John W. Traphagan wrote that throughout much of its history, anthropology did not have the ability to examine its subject directly; and early "armchair" anthropologists of the nineteenth century, such as James Frazer, E. B. Tylor, and Lewis Henry Morgan relied primarily or solely on their research on sources afar from their subject. In this way, they were not unlike today's astrobiologists and SETI researchers; there were gross limitations in technology, means of communication, and restricted means of interaction with their research subject. The interpretation of data was based primarily on theoretical frameworks within a set of Western socio-philosophical assumptions, some which later required correction with the advent of direct contact and participant observation.[73] There is a concern, as always when considering the possible nature, values, structures, and patterns of behavior of extraterrestrials, of operating within a restrictive anthropomorphized framework; however this bias should be viewed as a valid starting-point, as Earth and its inhabitants at present provides our only actual example of intelligent life among a potentiality of myriad others.

---

72. It is entirely possible that the beings humans term angels and demons are products of an entirely separate genesis or evolutionary process. There is no evidence these preternatural beings are native to Earth's 'celestial sphere,' or its temporal or spatial epoch, and could be equally described as supraterrestrial as 'extraterrestrial.'

73. Traphagan, "Anthropology at a Distance," 131–32.

As discussed in the previous sections, biological intelligent extraterrestrials would have undergone a historical process of natural evolution; it is likely this evolution took place via predator/prey relationships, social deception, and manipulation of the environment.[74] In the event of Etho-ethnological contact[75] astrobiologists and SETI astronomers generally believe that most extraterrestrials we contact will be intellectually and technologically superior. This follows from the view that since humans are a young species and exist in an early phase of industrialization and technological development, with a planetary economy based on fossil fuels and capable of rudimentary space exploration, much older civilizations would be more advanced. However, a technologically advanced society will not necessarily seek interstellar communication, and may not share our human inclinations of curiosity and exploration. Certain scientists and astronomers have made statements regarding the nature of extraterrestrial societies, often biased in favor of their personal religious or philosophical perspectives. Carl Sagan conjectured that intelligent extraterrestrials have experienced and solved Earth-type social and environmental problems, and established a communication network throughout the galaxy to spread their knowledge. His fictional alien in his book and later film, *Contact* spoke volumes about his view of extraterrestrials as masters of galactic travel: immortal and benevolent mentors with capabilities best expressed to lesser-advanced species as technological-spiritual powers - described by Sagan, an atheist, as many would term deities. Similarly, Frank Drake, a proponent of the alien saviour hypothesis, and sharing Sagan's benevolent view of benign species, also speculates that extraterrestrials may have achieved immortality in part by curing all disease. Likewise, physicist and astronomer Robert Jastrow speculated that alien scientists may have achieved immortality by figuring out "the secrets of the brain" and "uniting mind with machine."[76]

In general, we can consider intelligent biological life individually and socially, primarily concerned with self-preservation, reproduction, and acquisition of resources for improved quality of life and furtherance of self-preservation. If extraterrestrials are natural beings, with any similarities to human societies, they could be organized into moral communities in their search for knowledge and truth, with a willingness to subordinate individual interest to social aims for a common benefit. In 1964, as a means

---

74. Lestel, "Ethology, Ethnology, and Communication," 229–30. For a detailed discussion on nonhuman social behavior from an evolutionary perspective see Emery and Clayton, "Comparative Social Cognition," 87–113.

75. Contact between two heterogeneous advanced cultural societies, as that between human and extraterrestrial cultures.

76. Basalla, *Civilized Life in the Universe*, 160–61.

to systematically categorize and develop an understanding of what forms extraterrestrial civilizations might take, Soviet astronomer Nikolai Kardashev proposed a hypothetical scale designed to quantify a civilization's level of technological development, based on the amount of energy it is able to utilize.[77] The scale includes three designated categories of types I, II, and III, and later proposed extensions types O, IV, and V. A type O civilization, on this scale considered equivalent to the state of development of twenty-first century Earth, obtains its primary energy from crude, organic-based terrestrial resources in the forms of fossil fuels, coal, wood, plants, and animals, and in limited quantities, solar and wind power. It has an advanced medical technology, is capable of modest extension of lifespans, planet-wide social communication networks, orbital spaceflight, and satellite technology. A type I civilization makes use of renewable, high density power sources such as fusion power and hydrogen, and the ability to harness all available energy of its planet. It is capable of interplanetary spaceflight and communication, megascale engineering, and an artificial intelligence singularity.[78] A highly advanced medical technology would include the capability of genetic enhancement and manipulation, cybernetics and cryogenics; resulting in potential extension of lifespans to 150 to 200 years (based upon the human species). Socially, a type I civilization may be led by a world government (or small group of planets within its host star); however it may remain vulnerable to extinction due to highly advanced weaponry capable to planet-wide destruction. Many science fiction works are placed within this period, an estimated from one hundred to one thousand years in the future. A type I civilization, according to this scale, would have a technological level close to the level attained on Earth perhaps in the next 100–200 years, with energy consumption at $\approx 4 \times 10^{11}$ erg/sec.[79] Guillermo A. Lemarchand describes a type I civilation as "a level of terrestrial civilization with an energy

---

77. Prior to Kardashev, anthropologist Leslie White argued that social systems are determined by technological systems, echoing an earlier theory of Lewis Henry Morgan. In it he sought to quantify societal advancement by the measure of its energy consumption. See White, *The Evolution of Culture*.

78. The hypothesis of the invention of an artificial superintelligence, that would begin a rapid process of 'upgrading' so to increase its intelligence and capability on an exponential level, resulting in an unprecedented intelligence explosion far surpassing the collective intelligence of human civilization. Author Vernor Vinge in his essay "The Coming Technological Singularity," 11–22, states this event would indicate the end of the human era and the beginning of an artificially-controlled world. See also Kurzweil *The Singularity is Near*, 135–36, Kurzweil has predicted the singularity to occur around 2045, while Vinge predicts a time prior to 2030.

79. Kardashev, "Transmission of Information," 282–87.

capability equivalent to the solar insolation[80] on Earth."[81] In theory, this can be achieved through application of the fusion of the Earth's available water resources of approximately 280 kg of hydrogen into helium per second,[82] equivalent to 8.9×10^9 kg/year. A cubic km of water contains about 10^{11} atoms of hydrogen, and as Earth's oceans contain about 1.3×10^9 cubic km of water, providing $\approx 1.3 \times 10^{20}$ of hydrogen atoms human civilization could sustain this rate of consumption over geological time-scales. Another possible although theoretical source of energy would use matter-antimatter collisions, where the entire rest mass of particles is converted into radiant energy. The energy density (that is, the amount of energy released per unit mass) is approximately four orders of magnitude greater than that of nuclear fission, and about two orders of magnitude greater than that achieved through nuclear fusion.[83] Several potential future sources of antimatter may be available given technological advances.[84]

A type II civilization's primary energy source (according to known methods, albeit theoretical) is fusion energy, obtained from planets in multiple star systems. It is capable of interstellar travel and communication, stellar engineering and terraforming of planets. Its technology would enable evolutionary intervention and the possibility of extreme extension of lifespans to hundreds of years. The *Galactic Federation of Planets* of the fictional series *Star Trek* would fall under in this category. A type II civilization is capable of utilizing the entire energy produced by its host star by one of several methods, (for example, using a *Dyson's sphere*)[85] with an energy consumption at $\approx 4 \times 10^{33}$ erg/sec.[86] This amount of energy utilization would be comparable to that of our Sun. Type II civilizations may use similar means employed of a type I civilization, achieved through a large number of planets

---

80. The amount of electromagnetic energy produced by solar energy incident on the surface of the Earth.

81. Lemarchand, "Detectability of Extraterrestrial Technological Activities."

82. Souers, *Hydrogen Properties for Fusion Energy*, 4.

83. Borowski, "Comparison of Fusion/Anti-matter Propulsion Systems," 1–3.

84. Than, "Antimatter Found Orbiting Earth"; Adriani et al., "The Discovery of Geomagnetically Trapped Cosmic Ray Antiprotons."

85. The most oft-cited method for capturing and utilization of the entire energy produced by a star is one of the hypothetical megastructures conceived by Freeman Dyson, designed to partially or completely enclose a star to retain most or all its energy output. In his 1960 paper, "Search for Artificial Stellar Sources of Infrared Radiation," he theorized a megastructure in space designed to encapsulate a star in order to capture and harness its entire energy output. He speculated that a structure of this kind would be necessary in order to serve the energy needs of a large, technologically advanced mega-civilization, perhaps encompassing several planets.

86. Kardashev, "Transmission of Information."

in a solar system or systems. Other methods, although highly theoretical, include feeding a stellar mass into a black hole and collecting photons emitted by the accretion disc,[87] or achieving the same by reducing a black hole's angular momentum, known as the *Penrose process*.[88]

A type III civilization is capable of extracting fusion energy and any raw materials as energy sources from all possible star-clusters, and may be able to traverse large distances in space by the use of theoretical wormholes, have intergalactic communication, mega-engineering on a galactic scale, and exert galaxy-scale influence, similar to the *Galactic Empire* of *Star Wars*. A type III civilization would be capable of harnessing and utilizing the entire energy output produced by its own galaxy, with energy consumption at $\approx 4 \times 10^{44}$ erg/sec.,[89] comparable to that produced by our Milky Way. Type III civilizations may use some of the same methods employed by a Type II civilization, applied to all possible stars within a given galaxy. Other hypothetical possible energy sources could be to utilize energy supplied by a supermassive black hole, such as are known to exist at the center of most galaxies, capturing energy produced by stellar gamma-ray bursts, and emissions from quasars which offer a massive power source if technology is available to harness it. Astrophysicist Michio Kaku has suggested that human civilization may attain Type I status in one hundred to two hundred years, Type II in a few thousand years, and Type III status in one hundred thousand to one-million years.[90] In accord with these civilizational models, engineer Csaba Kecskes identified several possible evolutionary and social levels of technological civilizations: planet dwellers, asteroid dwellers, interstellar travelers, interstellar space dwellers, intergalactic travelers, faster-than-light travelers, and parallel universe travelers.[91]

87. Newman, "New Energy Source"; Schutz, *First Course in General Relativity*, 304–5.

88. Also called the *Penrose Mechanism*, a process theorized by Roger Penrose, wherein energy can be extracted from a black hole, possible due to the rotational energy in the region of the ergosphere. In this region all particles are dragged by rotating spacetime, and as they enter into the ergosphere, are split in two. Any escaping particles of matter can possibly have greater mass-energy than any original infalling matter, which has negative mass-energy. This process results in a overall decrease in the angular momentum of the black hole, whereby the reduction corresponds to a transference of energy. The maximum amount of energy gain for a single particle is 20.7 percent. See Chandrasekhar, *The Mathematical Theory of Black Holes*, 369. In this process, the black hole can in time lose all its angular momentum, becoming a non-rotating, i.e. Schwarzchild black hole. The maximum theoretical energy available for extraction in this process is 29 percent its original mass. See Carrol, *Spacetime and Geometry*.

89. Kardashev, "Transmission of Information."

90. Kaku, "The Physics of Interstellar Travel."

91. Kecskes, "Evolution and Delectability," 316–19.

Futurist Allen Tough has provided a list of possible capacities of an advanced alien technological civilizations: virtual unlimited energy, technology so advanced as to be interpreted as miraculous, enormously evolved individual brainpower linked with implantations or computers, the capacity to live and travel anywhere in space, elimination of individual and collective behavior that is violent, destructive, or harmful; altruism and compassion combined with sensible public decision-making; individual self-understanding, self-acceptance, and exceptional mental health, along with the collective ability of relating effectively and harmoniously with members of one's own species; excellent skill at interacting with a vastly different species and culture; knowledge and wisdom unimaginable to us; excellent control over biological reproduction and evolution, including very healthy long-lived bodies and super-capacity brains; the technological and/or psychic ability to send and receive information, or detect and observe objects and events across great distances; the technological or psychic ability to influence virtually any object, the ability to transfer one's body or consciousness from one place to another; organic or psychic connections to other members of the species, or to a central organism or brain; and the possession of extremely rapid, accurate, versatile, and powerful weapons.[92]

Transitional periods between each civilization type could entail tremendous social and cultural pressures, including local, national, and international conflicts as each liminal period entails surpassing any limit of energy resources available in a given civilization's territory. Physicist Freeman Dyson[93] repeats a common view of such transitional periods, with transition from Type 0 and Type I entailing a high risk of self-destruction due to physical limits for expansion of a civilization, as in a *Malthusian Catastrophe*.[94] Conversely, it may be the case that a type I civilization, having reached this level, has also achieved the capability for advanced space travel and massive off-world engineering projects, to include the colonization of other planets or *O'Neill-type* colonies.[95]

---

92. Tough, "What Role Will Extraterrestrials Play?", 492.

93. Dyson, "Search for Artificial Stellar Sources," 1667–68.

94. A prediction of a civilization's forced return to subsistence-level conditions once population growth has outpaced agricultural production.

95. A type of space settlement designed by American physicist Gerard K. O'Neill in his 1976 book, *The High Frontier: Human Colonies of Space*. Also termed an *O'Neill cylinder*, whereby each colony would consist of two counter-rotating cylinders, using materials harvested from the Moon and asteroids. The cylinders cancel out any gyroscopic effects to keep them aimed at the Sun or other stellar mass. Artificial gravity would be sustained by the centrifugal force exerted on the inner surfaces.

Planets hosted by first-generation stars would lack the heavy elements needed for life to develop; subsequent supernovae provided these elements resulting in a second generation of rocky planets. The possibilities for life to have developed on these planets, no more than six-billion years older than Earth, allow for extraterrestrial civilizations and societies of extreme age, thereby potentially possessing extremely advanced technology, and social and spiritual development perhaps incomprehensible to Homo sapiens. Civilizations on or from such planet could evolve into multi-planetary mega-populations, who in ages past multiplied and expanded their empire to encompass hundreds, if not thousands or hundreds of thousands of planetary systems throughout a galaxy or galaxies. Extraterrestrial societies could exist along a vast continuum of development, with some species super-advanced, being extremely long-living, with certain societies exercising some measure or total control of some less advanced. In fact, we should not be surprised to find in the event of an encounter with another civilization, a super-advanced species having created a genetically engineered underclass who exist without knowledge of their creators, or who are subserviently designated to fulfill certain assigned roles and functions. Given the possibility of a hyper-advanced technology, the achievement of extreme life-cycle longevity could enable certain extraterrestrials to personally witness many historical periods of their society, such as the birth, maturity, and either the self-destruction or eventual joining of a type of galactic community of many, or perhaps hundreds of intelligent races.

Socially, we may find some populations exert control and domination of their own or other races through fear and militarization, and may engage in aggressive colonization among a large number of planets. Others may have a more benevolent relationship with species. Those cultures which exist in extreme isolation from others may be more likely to be cooperative, while those having spent centuries or millennia in their encounter with other, opposing species could have developed a more competitive or militaristic attitude. Some civilizations might have emigrated from their original home planet either due to environmental hazards, their host sun going supernova, or because of population growth, and have found a suitable moon, planet or planets, or terraformed a new climate to suit their natural environmental conditions. Some species may be engaged with others solely out of conquest, absent certain moral codes and driven by an ethics of avarice and as a result become completely militarized; others could engage in exploration of planets with motivations similar to the conquistadors and colonizers of Earth. War between competing species might result in the acquisition of entire planets, energy sources and technology, resulting in further expansion for the surviving species. As a war-waging species would have difficulty achieving stability and survivability, theorists claim such aggressive intelligent life forms cannot colonize a galaxy. Therefore, destructive and war-prone forms of life are unlikely to develop interstellar travel on a significant scale, and might either self-destruct or revert to lower technological conditions. If such civilizations do manage to travel and disburse throughout the galaxy, they may be terminated by other, more advanced and peaceful civilizations. Some might trigger probes that sense advancement and aggression, and be terminated remotely and automatically.

It is possible that an extremely advanced species might utilize a lesser advanced race to interact with humans or beings similar to humans, having a similar biological makeup, social evolution, and method of communication in a contact event. Certain societies might be widely traveled and have interactions with a large variety of intelligences, while others may have restricted themselves from interactions with other alien races; certain others may be so ancient that they have witnessed the emergence of newly evolved races and facilitated their incorporation in the larger extraterrestrial community, as well as witnessed the self-destruction or deliberate destruction of an entire planet or species. There may be those who consider some species only transient given their proclivities for violence or greed, and choose not to engage with the less advanced; some ancient beings may be wholly incapable of relating to a species whose life-cycle lasts less than a century.

Advanced civilizations may intervene or interfere in the genetic and/or social development of a primitive species either for selfish or altruistic motives, possibly in violation of a non-interference directive from higher authority.[96] There may be fierce competition for resources among space-faring species, where some massive populations dominate large quadrants of a galaxy, an entire galaxy, or grouping of galaxies. Certain species may be engaged as ambassadors and helpers in assisting newly-arrived societies into a "galactic community" of sort, speeding their assimilation into the larger, more homogenous group of long-term members. As there is no doubt there could be a great diversity, not only in biological and psychological composition, the same could be expected in individual and collective governance among varying species. An advanced civilization might engage in colonialist behavior with the less advanced indigenous populations, or engage in spying and furtive infiltration of societies in efforts to influence or overthrow certain regimes, governments, or other social structures for a variety of motives. Conversely, some societies may have eliminated the desire to amass wealth, have rejected materiality, instead investing their resources in means of self improvement, discovery, and mutual cooperation

---

96. Similar to the "Prime Directive" known to the fictional *Star Trek* series.

with others. Others could have agreed to live a communitarian lifestyle, sharing equally amongst themselves and/or other species. A single civilization could be comprised of some or many of these same characteristics among varying populations and groups.

Extraterrestrial societies could exhibit some or many of the characteristics we find in our own human society, and engage in behaviors that have driven and guided humanity since the earliest times. These commonalities are due to what are innate to all biological creatures—the drive for survival, competition for resources, methods of communication, means of cooperation, protection of the young and weak, and contention with suffering, disease, and death. The desire (or curiosity) for understanding self, others, and the greater environment, emotional bonding with others, personal or social acknowledgement (or lack of acknowledgement), worship, and belief in a deity; birth and beauty, suffering and death, war and peace are all powerful motivators known to humans, and may provide impetus and context in contact scenarios with other civilizations. We cannot assume that an advanced civilization necessarily knows everything about the universe. They can be homologous[97] and have a shared ecology. Neither can we assume that all alien societies although capable, develop a technological civilization; in fact, some may have achieved a high technology but evolved beyond absolute reliance on technology. It would be an anthropocentric mistake to assume that intelligent beings on other planets would necessarily mirror our biology, psychology, or societies. Intelligent alien life may not be motivated by the emotions and ideals as terrestrials, nor can we necessarily expect that they live in ordered, democratic-like societies and belong to a peaceful galactic organization that seeks out new forms of intelligent life as assumed by SETI proponents.

Intelligent extraterrestrials can have a variety of motivations, some which we should consider fundamental to many species, while others depend on their specific social, industrial, biological, environmental, scientific, or even spiritual needs or goals. Among some fundamental priorities are biological and cultural survival, expansion of scientific and cultural knowledge, and exploration of bio-systems; due to a desire for self-protection and security, civilizations could be motivated by curiosity and scientific study; and perhaps engage in a field of comparative civilizations. In this case, we could expect intelligent beings to have little interest in an antiquated technology of an inferior society, (except for perhaps nuclear capabilities and those which can cause self-extermination); and be more likely to value art forms and biology, religion, philosophy, social development, and other cultural artifacts.

---

97. Having the same relation, structure, or relative position.

A highly advanced civilization could be tasked by a larger collective or association with the responsibility of monitoring, assisting, and promoting the survival and positive development of those less-developed civilizations they deem to have the capacity, in time, to contribute to a cooperative or alliance of advanced civilizations. Conversely, for other inhabited planets and perhaps in the case of Earth, we may expect extraterrestrial societies, if they inhabit our galaxy, to be well aware of our existence, however have maintained a rule of non-interference in underdeveloped societies. In other cases, leaders or managers may decide to follow an isolationist policy until a certain level of cultural evolution and/or technological achievement is reached before participating or collaborating with less-developed groups.

Hollywood science-fiction has well-worn the idea of opposing, violent, and predatory extraterrestrials driven solely by conquest of the planet Earth and subjugation and exploitation of its population (typically unsuspecting humans); taking advantage of natural resources, and colonizing entire planets while enslaving or supplanting indigenous races. Some species may have these tendencies indeed; however control, domination, or exploitation of a technologically and/or biologically inferior race could take more subtle forms: clandestine infiltration of societies in key hierarchical positions to provide direct or indirect control for a predetermined political, social/cultural, economic, or even religious goal(s); furtive parasitic beings who invisibly interact with a host species and surreptitiously harvest needed resources, artificial, natural, or biological; races that manufacture or genetically modify an indigenous primitive species to serve their own as a sub-species, with the possibility of subjects being uninformed or misinformed of their actual origin. Although higher intellectual ability and development does not necessarily indicate moral superiority, an advanced race, if having survived self-annihilation, could function according to an ethical universalism, motivated by technological, moral, and spiritual advances, a trend towards democratic governance, the end of wars, and the evolution of supranational systems that impose order, resulting in peaceful and benevolent exchanges with other civilizations. With consideration of the foregoing discussion on morphological, psychological, and social development of putative extraterrestrials, the following section will address what we can surmise of extraterrestrial religious and theological systems: their impetus, manifestations, theologies, and basic structures.

Section B: Extraterrestrial Religious and Theological Systems: Exotheology

A foundational principle of the *varied* view of exotheology is the *divine prerogative*—the absolute freedom of divinity possessing omni-attributes to create, reveal, communicate with, and redeem intelligent beings throughout the created universe. A starting point of this perspective is that given extraterrestrials are the product of a wholly separate genesis event, they would not belong to the family of humankind, and therefore, not be guilty of Adam's sin, nor have any connection with the history of human proclivities to sin. Extraterrestrials could be indeed be sinful, although the doctrine of original sin argues a sole transmission by human generation, and thus is only applicable to humans. Therefore, it cannot be asserted extraterrestrials share in its inception or effects. The divine prerogative is incompatible with an anthropomorphic projection of terrestrial religion and its context of proneness to evil and sensuality beyond Earth, thereby limiting God's free omnipotent creative vision for creatures. The existence of divine presence, revelation, and grace in other civilizations modifies our concept of the divinity and considers its presence and works as a vast plurality rather than a singularity. Divine plans, presence, revelation, personhood, freedom, sin, history, and evil might exist in a myriad of variations among other worlds. As creation, revelation, redemption, and eschatological fulfillment are foundational structural elements of the divine will for human beings, how would these acts correspond or contrast in an extraterrestrial race? All intelligences are created free, as evidenced by humans and angels, and are called to relationship with a benevolent divine Creator. As a result, sin becomes possible. Therefore, God may reveal and redeem by means according to different expressions from those with which we are familiar in the human Christian tradition, resulting in a faith and knowledge of the divine which may be quite diverse.

The role of religion according to our definitions in human societies, simply, is to provide answers to the perennial questions of existence and destiny, an interpretation of experience, an ethics and aesthetics, with immanent and transcendent elements; a moral code; and the provision of access for the monotheist to a greater reality—a superpowerful transcendent personal being who acts within and apart from creatures. We can consider extraterrestrial religions according to these categories, but should acknowledge some of their parochial limitedness. Certain extraterrestrial religions may be far advanced in their understanding of divinity, the supernatural, and personal and collective spiritual development; in others, primitive extraterrestrial religions could be comparable to our own historical animist, pantheist, or polytheistic traditions, or contain elements of non-linear Eastern religions. Not all religions will maintain belief in a supreme, personal deity, and may be more oriented to personal or collective wisdom, the ending of suffering, or a connectivity of mind with other intelligences. Civilizations, societies, or segments of societies having been formerly religiously-oriented, may have abandoned notions of religion or spirituality upon contact with more advanced societies. Human civilization has had the historical challenge of inconsistent and competing religious figures, truth claims, revelations, traditions, schisms, and divergence of interpretation of religious texts. Therefore, we cannot necessarily expect a monolithic manifestation of religion in extraterrestrial societies within particular planets, civilizations, societies, races, or epochs; but rather a potential plurality of religious organizations, belief systems, spiritualities, and practices existing on a continuum is some ways similar to Earth. A world-bound species would have an opportunity to "test" an indigenous, terrestrial religious belief within the larger context of other, external claims to divine truth and meaning in contact with an extraterrestrial species. It would be likely encounter a variety of religious incongruities which will require interpretation and theological resolution.

Highly advanced religions, generally within this context can be described as those which have evolved beyond certain geographic and parochial perspectives, and accepted an integrated knowledge of the sciences. These religions can be global or even trans-global in extent. A less advanced religion may be compared to those most familiar to humans akin in certain human societies,[98] which typically exhibit diverse mythologies, are oral without written texts, are limited linguistically, culturally, and geographically, maintain a focus on the present life rather than otherworldly, consider the motives and works of gods to be often mysterious and fearsome, and function according to a cyclical time-scale rather than an historical or progressive theological impetus and finality. There is often animal sacrifice and idolatry, morality is often viewed in terms of keeping or breaking of set rules with limited understanding of their relevance, rather than responsible participation in a divine relationship. A simpler description can be made between natural religion, that discovered by unaided reason, versus a supernaturally revealed religion. As evolution stresses the importance of higher forms of life, more advanced religions should provide fuller content to divine works and understanding.

---

98. Animist, totemic, shamanist, henotheist, pantheist, polytheist, monist, ancestral, or naturalistic.

'Anthropology' of Extraterrestrial Religions

The emergence of religion in extraterrestrial civilizations would be a consequence of the genesis and evolution of higher intelligence and self-consciousness. Although evolutionary psychologists view the emergence of religion as a result of natural selection or as an evolutionary by-product of other intellectual adaptations,[99] others have argued religious attitudes and spiritual states are innate to the human condition due to their ubiquity in human civilizations.[100] Intelligent alien biological beings, depending on their physiopsychological, cultural, and technological evolution, may or may not possess religious paradigms of consciousness. Extraterrestrial religious beliefs and praxes might be derivations from historical events and processes commensurate with those among human societies or receive their impetus directly from divine initiative.

Given their unique environmental, biological, and cultural particularities, each society would formulate their own (with or without influence from neighboring societies or races) interpretations of rational existence, which could include religious beliefs or attitudes. What created intelligent beings consider in their immediate experience—preeminent among them suffering and death, could ultimately give rise to thought of the infinite and transcendent, resulting in a religious dimension of questioning which reaches within and beyond individual and communal experience.[101] As in-

---

99. Steven Pinker, in his lecture, "The Evolutionary Psychology of Religion," hypothesizes the religious mindset as exaptations from ancestor worship, as a set of emotional predispositions that lend themselves to religious attitudes. Lewis Wolpert, in his *Six Possible Things Before Breakfast*, contends that causal beliefs emerged in the manufacture of rudimentary tools, which required mental conception prior to their creation. Lee Kirkpatrick's, "Toward an Evolutionary Psychology," argues for collective religious belief as a by-product of numerous, domain-specific psychological mechanisms that evolved to solve other (mundane) adaptive problems. These include mechanisms for reasoning about the natural world (naïve physics and biology), other people's minds (naïve psychology), and specific kinds of interpersonal relationships, or 'social solidarity theories' (attachment, kinship, social exchange, coalitions, and status hierarchies). Edward Burnett Taylor held religious beliefs resulted from a desire to explain natural phenomena.

100. Rudolf Otto asserted religious attitudes and experiences resulted from a non-rational function of the mind, whereas religions could arise from a particular culture or society; rather, these resulted from numinous experiences; See Kunin, *Religion*, 66; St. Anselm taught the concept of an omnipotent first mover as an a priori universal human notion with his famous phrase, "God is that which nothing greater can be conceived."

101. The subjectivity and consciousness of immediate limitedness of the creature can give rise to conceptualization or acknowledgement of the infinite and transcendent. Steven Gould states early human consciousness and experience led to the awareness of personal mortality, religion may have been one way of solving the individual and

telligent beings seek to know the unknown, to resolve inconsistencies, and formulate complex ideas which serve to explain and cope with experience, religious solutions could exist on a continuum of possible resolutions with or without conjunction with scientific understanding to explain profound questions of physical existence, death, and the afterlife. If such a society is advanced scientifically, at least in accord with that of human civilization or beyond (a type 0–1 civilization), it will possess a general knowledge of an expanding universe composed of matter, space, and time, the emergence of this particular universe as the beginning of our space-time matrix, and the general laws governing it; it would be expected advanced religions would be in accord with or at least acknowledge a scientific cosmology. Therefore, theistic civilizations likely would acknowledge a Creator (whether considered uncreated or itself a product of an evolutionary or other process of becoming), which either stands apart from the constructs of matter-space-time, or regard the divine or deity within a potentially incalculable diversity of physical, metaphysical, or theological arrangements.

Religious thought and spiritual states require a brain with a neocortex large enough for higher order cognitive functions such as self-consciousness, language, emotion, and an understanding of causality. Creatures as products of early predatory environments would typically possess an inborn fear of death and desire to continue in existence after death in a manner approximating or exceeding their experience of physical and intellectual life. Personal and social morality may precede or coincide with religious beliefs[102] necessary for the stabilization and successful perpetuation of any sizeable social group. Through the medium of familial and social interchange, a localized set of religious beliefs can evolve into an organized religion, having hierarchical and social dimensions, or conversely without hierarchal structure or systematic organization. Evolution ensures the survivability of a given species by means of gaining some measure of control over the environment in competition with others of their species and outsiders. These are vital for the development of civilizations, although their implementation and development may be quite different than that of Homo sapiens. A deity would be known and understood among a variety of extraterrestrial civilizations

---

communal experience of loss of life, and cites religion as a way the newly-conscious being dealt with the concept of personal mortality. See Gould, "Exaptation," 43–65.

102. Michael Shermer lists requisite features for moral behavior: attachment and bonding, cooperation and mutual aid, sympathy and empathy, direct and indirect reciprocity, altruism and reciprocal altruism, conflict resolution and peacemaking, deception and deception detection, community concern and caring about other's perceptions about self, and awareness of and response to the social rules of the group. See Shermer, *The Science of Good and Evil*, 24–65.

fundamentally and firstly through what is created: universal physical laws (gravitation, energy conservation, genetics, and thermodynamics) and subsequently, natural laws. These laws must be seen as the operation of a rational and comprehensible universe. Secondly, a deity can be conceived in relation to particular physical environments, as well as other life-forms, accounting for religious varieties analogous to primitive Earth forms.[103] The concept and experience of the transcendent can begin as a result of the mind's curiosity of reaching beyond the finite, its perceptions and ideals of perfection, beauty, and holiness according to linear and non-linear religious models. A religion, whether terrestrial or extraterrestrial, involves the belief in a deity of supernatural, creative, and controlling force which entails degrees of physical and spiritual separation between creature and Creator. It is expected religion would manifest in extraterrestrial societies in structurally organized hierarchical or non-hierarchical forms (e.g. church) and possibly non-structural forms (purely individualized mystical religions). Religious impulse can be defined as the relating of the created to the larger created reality, and in the longings and yearning for connectiveness with the divine who created it; it is the divinity's reaching out to creatures which create the dialectic of religion and religious life; as noted by Norman Lamm, "Religion is the human, social response to transcendence."[104] Religion can include historical narratives, symbologies, economies of salvation, and teachings on the afterlife. Within this framework, the fundamentals of extraterrestrial religion can be extrapolated in general terms.

Fundamentals of Extraterrestrial Religion

As mentioned, a space-faring civilization containing a highly advanced religion might well incorporate a scientific perspective in conjunction with cosmological principles[105] with a highly developed understanding of the transcendence and operating immanence of the Creator.[106]

---

103. As mentioned, these include animism, pantheism, panentheism, totemism, shamanism, monism, and polytheism, among others.

104. Weintraub, *Religions and Extraterrestrial Life*, 82.

105. If scientifically and technologically advanced far beyond our own, certain civilizations may develop a "metascience" whereby supernatural phenomena are established by a separate set of laws outside of natural science. Scientific rationalism, or irreligion may dominate an extraterrestrial mindset, where there is worship of technology, a certain rationalistic philosophy comparable to a 'religious Confucianism', or particular types of personality or political cults, or the worship of a neighboring and far advanced alien race.

106. Demonstrable supernatural events verified by use of advanced technologies

The cosmological perspective of a religion provides the spatial and conceptual framework and orientation to any given religion.[107] As less developed world-bound societies may contain limited or erroneous cosmological views, extremely ancient and advanced space-faring races could exhibit cosmological perspectives far more advanced than that known to twenty-first century scientists and theologians. An extraterrestrial society, if religious, might conclude that creation of the material universe was intentional and governed by a system of highly consistent and predictable laws which can culminate in life and eventually, intelligent beings capable of divine relationship. This could result in the belief that the creation of the universe, the ordering of local galactic groups, the conditions of individual planetary systems, and the appearance of intelligent beings manifest the omnipotent will of a Creator. If this is the case, an acknowledgement of the infinite and transcendent as a philosophical, metaphysical, or theological view of nature and/or supernature is likely to exist in an extraterrestrial religious view of ultimate reality. Technological and social evolution of surviving technological civilizations may inevitably result in space exploration and ultimately contact with other intelligent species; this eventuality should be viewed as an inevitable evolution of a religion's self-concept and doctrine within creation. Advanced, space-faring species with intercultural exchanges over very long periods in this case could abandon, modify, or recontextualize their particular or strictly "*Extrapomorphic*"[108] concepts of God to accommodate other divine revelations.

Historically Earth religions have expressed and symbolized religious and theological teachings and beliefs through use of the biological, geographical, atmospheric, and celestial particularities of this planet, as well as domestic and social aspects. The Old and New Testaments are quite rich in this imagery. Earth's location in the solar system which provides visual access to distant stars and the Milky Way; its distance from the Sun and Moon and their relative comparative sizes from Earth resulting in an unusual matching apparent size;[109] the Earth's particular revolutionary period

---

can lead to a greater knowledge of divine works in the material world for a given religion

107. See Matthews, *The Ecological Self*, 12: "Cosmology serves to orient a community to its world, in the sense that it defines, for the community in question, the place of humankind in the cosmic scheme of things. Such cosmic orientation tells the members of the community, in the broadest possible terms, who they are and where they stand in relation to the rest of creation."

108. The self-concept or self-characterization of extraterrestrial beings in relation to their particular biology, psychology, intellectual endowments, and environment.

109. Gen 1:16; 15:5; Deut 4:19; 17:3; Job 9:7–9; 38:31–33; Isa 13:10; 40:26; Pss 136:7–9; 148:3–6.

of 365 days, twenty-four hour rotational period, and 23.5° axial tilt which create four distinct seasons;[110] a multitude of climates and weather patterns, a steady and frequent pattern of day and night;[111] unique plants, flowers, and trees which provide examples of creative beneficence;[112] multitudes of animal species; the pattern of stars from our celestial location resulting in unique and permanent constellations;[113] atmospheric molecules scattering the shorter light waves of violet and blue light, resulting in our associating purity, divinity, and heaven with the color of sky blue[114] and beautiful clouds; the seasons of winter and spring provide an existential understanding of death and rebirth, of fertility and sexuality.[115] Solar and lunar eclipses provided occasion for special religious significance for many cultures.[116] The synchronous relationship between the human female menstrual cycle and lunar rhythm;[117] the ubiquitous cultural use of the color red to represent blood, life, sex, and death;[118] fire to represent purity; and the water molecule, essential to all life forms, to symbolize life.[119]

Terrestrial weather and seasonal patterns led to the creation of ancient fertility religions and cults, of animist religions, and nature worship among indigenous peoples; and particular star patterns resulted in the creation of the zodiac and Mesopotamian astral worship. The Aztec deity of the Sun, *Huitzilopochtli*, believed to lose blood, its life force when setting due to the reddish cast at sunset, provided the impetus for human

---

110. Gen 1:14; Deut 33:14; 1 Chr 23:31; 2 Chr 2:4; 8:13; Neh 10:33; Ps 104:19; Jer 8:7; Ezek 36:38. Thanksgiving is a harvest celebration near the autumnal equinox, and Easter in the West is typically celebrated on the first Sunday after the first full moon following the vernal equinox. Similar and varied religious celebrations on extraterrestrial planets may also coincide with other celestial arrangements, conjunctions, and seasons.

111. Gen 1:18, 8:22; Exod 13:21; Lev 8:35; Josh 1:8; Neh 1:16; Job 26:10; Pss 1:2; 74:16; Jer 33:25.

112. Gen 1:1–12; 2:9; 3:22–24; Lev 19:25; Num 17:8; Deut 8:8; 1 Kgs 6:18, 29, 32; 2 Chr 4:3, 21; Ps 103:15; Song 1:14; 2:12; Isa 17:10; Hos 14:5; Luke 12:27.

113. Job 9:9; 38:32; Isa 13:10; Amos 5:8.

114. Exod 14:10: "Then Moses went up with Aaron, Nadab, and Abihu, and seventy elders of Israel, and they saw the God of Israel. Under his feet was a work like a pavement made of sapphire, as clear as the sky itself." Exod 24:10: "and they saw the God of Israel, and under His feet is as the white work of the sapphire, and as the substance of the heavens for purity."

115. Best known among ancient Egyptian, pre-Columbian American, Asian, Greek, Roman, and Oceanic fertility religions.

116. See Dvorak, *Mask of the Sun*; Nordgren, *Sun Mood Earth*.

117. Law, "The Regulation, of Menstrual Cycle," 45–48.

118. Gen 9:4–5; Lev 17:11; 2 Sam 23:17; 1 Chr 11:19; Isa 63:3, 6; Jer 2:34; Phil 2:17.

119. Isa 12:3; John 4:13–14; 7:38–39; Rev 22:1–2.

'ANTHROPOLOGY,' XENOBIOLOGY, MORPHOLOGY, AND THEOLOGICAL SYSTEMS 161

sacrifice to replenish its energy for its work of rising each morning. Sacrificial death to the god or gods served as repayment for the sacrifices gods made in creating and maintaining the world.[120] The Sun was more often worshipped in cold climates and the Moon more often in hot, arid climates. In Islam, heaven is believed to provide shade, having originated in a desert environment.[121] The Sun was known as central to worship in the Egyptian religion as the god *Re*, and in Asia in the religion of *Mithra* of Persia. In Japan, as the imperial deity *Amaterasu* in Shinto, in Peru as *Inti*; as a subordinate deity to the God of heaven as *Helios* in Greece; *Sol* in Rome, *Shamash* in Babylonia; the Sun and Moon as a divine pair or "world parents" in India, *Singbonga*, and in Zambia as *Nyambe*, god and goddess.[122] The astronomical setting can be critical to certain religions, where the location of a planet within its immediate celestial environment of other planets, star patterns, comets, asteroids, meteors, nebulae, and proximity to other galaxies may have particular importance for the genesis of a variety of religious and cultic beliefs.[123] The Chinese, Sumerians, Babylonians, Assyrians, Egyptians, Mayans, and others were keenly aware of star patterns forming constellations, resulting in detailed astrologies in accord with each unique configuration of stars and motions of heavenly bodies. A planet's possession of a moon or moons, Saturn-like rings, unusual cloud formations, and other astronomical particularities could likewise have significant effect on religious perceptions. The planetary capture of a moon, meteorite impacts, or a nearby star going supernova could initiate or have impact on a set of religious beliefs. Gravitational effects, particular planetary axial tilt, and the distance/size/type of sun could in a variety of ways serve to delineate seasons, ceremonies, and identification with celestial deities. More primitive extraterrestrial religions may manifest these characteristics given their particular cosmic location and particular celestial arrangements. Conversely, celestial worship may be absent on worlds with suns permanently obscured by cloud cover or absent moons. Planets tidally locked or those in synchronous orbits with extreme light/day conditions or those hosted by multiple stars may have their own unique religions tied to orbital periods, gravitational effects, and associated geophysical phenomena.[124] Objects which are the subject of religious

120. Duverger, "La flor letal," 83–93.

121. Quran 36:56–57: "They and their associates will be in groves of (cool) shade, reclining on Thrones of dignity."

122. Dexter, "Proto-Indo-European," 137–44; Robelo, "Biblioteca Porrúa," 648–50.

123. The religion of the Dogon tribe of the central plateau area of Maui contain astronomical beliefs wherein they are descended from the Sirius B star system.

124. Red dwarfs are the most common type star in the Milky Way, comprising

worship are typically determined by geography and environment in more primitive religions. In places of infrequent rainfall, such as in India, rain is highly venerated; extrasolar planetary environments where water is scarce may develop water or cloud worship, or envision an afterlife with an abundance of water. On larger planets having greater mass than Earth and hence more tectonic activity, there may be specific religious beliefs associated with volcanoes and earthquakes. Therefore, we can consider the possible variations of extraterrestrial religion distinctive to a race's particular geophysical and cosmic location, where a planet's natural features and unique environmental setting contribute to the evolution and expression of certain religious ideas and ideals.

As with environmental conditions, people, places, and objects having direct or indirect contact with a perceived or accepted supernatural event or personage could become venerated and in this sense 'divinized,' segregating the holy or sacred from profane and ordinary. The dichotomy between sacred space and the profane is ubiquitous in many Earth religions;[125] sacred space can be physically constructed or exist as a natural feature connected to a religious historical narrative, and can become part of a local, national, ethnic, planetary, or planetary collective identity. Transitions from profane space to sacred space exist physically and situationally; a similar distinction can be made between profane time versus sacred time: early proponents of Judaism abandoned the pagan concept of cyclic time and advanced the notion of an irreversible historical progression of divine activity; later, Christian redemption sacralized all human history (as argued by Hegel). Extraterrestrial divine presence, activity, and sacredness in persons, places or things could be categorized according to these modes or combinations thereof, with varying demarcations and periods of profane versus sacred time on timescales incomprehensible to humans. The divine and holy exist in myriads of forms in human societies: holy sites, churches, scriptures, sacraments, clergy, scriptural books, graves of deceased holy people, relics, devotions, rites, sacraments and sacramentals, prayer, and apparition sites. Ordinary bread and wine becomes the physical reality of a divine person in transubstantiation and is consumed; therefore, it can be expected that certain extraterrestrial religions may present similar features of sacred/

---

roughly seventy-five percent of all stars. Planets in habitable zones orbiting such stars will retain very close orbital distances, resulting in tidally locked planets in synchronous orbit, with one side permanently facing the sun, experiencing continual day and other perpetually experiencing night. Unique religious patterns could develop as a result of such planetary and stellar arrangements if a species were able to survive in such environments.

125. Durkheim, *The Elementary Forms of Religious Life*, 35.

profane space/time, people, and matter, including forms unfathomable to us. Therefore, certain fundaments of extraterrestrial religions could have characteristic organizational structures familiar to humans, each preserving and encapsulating a set of conceptions about the reality of the created being, the deity, and the relationship between them. We may expect, in the event of contact/discovery, congruent or utterly contrasting categories particular to these religious aspects, beliefs, and praxis known to humans, as well as greatly varying sacred identifications.

The individual personal roles within the dynamic of the human family and society contain several important relational and hermeneutical means to metaphorically express divine relations. Ancient Greek and Roman polytheisms were rife with metaphorical allusions to human relations; many cultures contained elements of mother-goddess or ancestral worship. In the Old and New Testaments, the Godhead is principally revealed in terms of *human relation*; predominant among them fatherhood, spousal, and familial relationships. Israel is characterized as the bride of Yahweh, the firstfruits of his harvest.[126] Isaiah uses the metaphor of husband and wife, comparing a disobedient Israel to a deserted wife whom the Lord restores.[127] The divine covenant between the Lord and Israel mirrors a marriage contract between two devoted persons, symbolic of monogamy in contrast to Israel's polytheistic neighbors. The New Testament contains several metaphors to describe the divine-human relationship in terms of a bridegroom and bride.[128] John the Baptist describes his role as assisting the ministry of the bridegroom, Jesus;[129] Jesus portrayed his mission as the bridegroom and the disciples as wedding guests.[130] Paul expressed Christ as the divine bridegroom, who loves the Church just as a husband for his wife,[131] and the New Jerusalem as bride represented the people of God prepared for the coming of the Lord.[132] Jesus, primarily in John's Gospel, reveals and describes his relation to God as a loving relationship between Father and Son,[133] and most powerfully, the relation between the divine person of Christ and his disciples as one of deep

---

126. Isa 62: 4–5; Jer. 2:2–3; Josh 2:16–20.
127. Isa 54:5–7.
128. John 3:29; Mark 2:19; Luke 5:34; Matt 25:1–13; Eph 5:22–33; 2 Cor 11: 2–4.
129. John 3:27–30.
130. Matt 9:15–16; Mark 2:19–20.
131. Eph 5:25–27.
132. Rev 21:2, 9; 22:17.
133. Matt 26: 39; John 5:19; 6:44; 8:49; 10:15, 30; 14:28, 31; 16:28.

friendship.[134] Human beings are portrayed as children of God.[135] The mother of Jesus, in Roman Catholic and Eastern Orthodox tradition is the God-bearer *theotokos*, maintaining a capital role as spiritual mother of all humans. It is apparent that divine usage of these intimately human relational archetypes in scripture is the agency by which divinity reveals, communicates, and redeems the human species. The scriptural repetition of these archetypes of human relations and centuries of theological development serve as powerful and enduring metaphors for interpreting the nature of divine personhood, activity, and the divine-human relationship. In these ways through scripture God does not communicate directly to the human mind but rather makes reference to and operates through the medium of human relations: the divinity particularizes its presence, communication, and power, imbuing the creature with a divine aspect theomorphically.[136] In extraterrestrial societies relations of affinity and consanguinity may differ considerably; accordingly, archetypes by which divinity may engage creatures could take a variety of forms. This can result from diverse means of propagation, such as cloning, artificial wombs and mechanized, impersonalized raising of young; 'non-family' patterns of relation, institutionalized unions, diverse forms of the life-cycle, much longer lifespans; each perhaps confirming, modifying, or negating human affinitive archetypical concepts; in these cases divinity could be modeled,[137] characterized, and communicated through varying sets of metaphorical idealizations and relational patterns.

Exotheological Anthropology: Soteriology, Hamartiology, Pneumatology, Theosis

This section will propose a potential theological 'anthropology' of extraterrestrials—incorporating a hamartiology, pneumatological constitutions and abilities, and possibilities of theosis. As past theological scholarship has considered the extraterrestrial in a wholly generic manner and with little regard for possible environments, social structures, forms, states, capacities, and abilities; the paucity of consideration of the these fundamental aspects has resulted in little insight into the theological possibilities

---

134. John 15:12–17.

135. Gal 3:26; 2 Cor 6:18; Rom 8:14; Matt 19:14; John 1:12; Isa 8:18.

136. Humans are divinized in their contact with the holy and diverse forms of divine grace.

137. Considering the possibility where an extraterrestrial species consists of something akin to certain insect colonies, where one queen produces all the young; in such a case the queen may be worshipped as a deity.

and hence the repetitious and limited nature of theological work thus far on extraterrestrials.

A first consideration in the examination of the theological anthropologies of possibilities and potentialities of extraterrestrial intelligent beings is their ontological relationship to a creative God. All sentient beings are created for the Creator's glorification and a sharing in that glorification; all intelligent creatures are ontologically dependent on God, being creatures, and all a duty to glorify God in their own specific way. All creatures by their mere existence give unconscious glory to God,[138] and are created for the dual purpose of giving honor and glory to God and to live in some type of eternal happiness as their natural end. All intelligences, as purposefully and lovingly created creatures of the divinity, are called in their own mystical union in the joining of the divinity and the creature, between the immaterial and material, and the raising up the creature to share in the life of the deity to experience a unity of truth, beauty, and benevolence. Extraterrestrial races can have their own *Protoevangelium*; however the events and conditions that foretell of future events for a race of beings necessarily would be quite different. What action has God taken with each of these groups will aid in understanding the Creator's modes of salvation for rational creatures not of this Earth.

In the greater cosmic scheme, the creation of intelligences and ultimate calling into intimate relationship with God are fundamental and the self-determinative choice of ultimate destiny is a necessary consequence of rational beings. Extraterrestrials may have various created capacities, gifts, divine probations and testing, and general outcomes in use of their free will in comparison to the human. In consideration of putative intelligent extraterrestrials, we can consider a range of possible natural/supernatural/preternatural combinations and their natural and supernatural destinies given the unlimited creative freedom of the Creator. A fundamental starting point, argued by some theologians (made in reference to states of humans and angels) allows us to distinguish six possible states of nature in regard to biological intelligent creatures in the universe:

1. Pure nature, without preternatural or supernatural elevation.
2. Integral nature, with preternatural endowments.
3. Elevated nature (i.e., the prelapsarian state of humanity) with preternatural and supernatural gifts.
4. Fallen unredeemed nature, incapable of attaining its original intended end due to sin.

138. Fundamental glory.

5. Redeemed nature, superabundantly restored to its original intended state through divine redemption.

6. Supernature, possessing supernatural gifts.

1. Pure nature or natural state, without supernatural or preternatural endowments, possessing the discursive and imperfect knowledge acquired by the light of sense perception and reason.[139] A creature in a state of pure nature would be capable of knowing God as the first cause and ultimate end, and could by ordinary and natural means fulfill God's will. It would be aware of and informed of God's will solely by means of natural law proclaimed in creation, and their understanding divinely imprinted in their reason, understood according to natural law proclaimed through nature (as proclaimed by Paul in reference to the "first gospel" preached to the Gentiles before Christ).[140] All creatures who know God, including the angels, can only know him imperfectly as they remain creatures. Aquinas argued the closer a thing is to an end, the more it desires the end.[141] God is not forced to create a creature in grace by anything in creation, he could have easily allowed a creature to be satisfied with knowing him through his effects rather than in himself. He does this already in the natural world. We can also suppose intelligent creatures without knowledge of the first cause of things, and in this case, the natural desire would remain void.[142] A creature created without grace and thus a natural desire to know God, could, in fact either be recipients of his grace to know him either through his effects or through other, more direct means such as divine revelations or theophanies. A natural creature would not be the beneficiary of supernatural helps as scripture, prophecy, liturgy, church, or sacraments; its only revelation available through the medium of natural law in creation. Such creatures, comparatively speaking, would have difficulty in the acquisition

139. Theologians defending the gratuity of grace against the naturalism of Michael Baius (1513–1589) and Cornelius Jansen (1585–1638), invoked a hypothetical "state of pure nature" where human beings could have possibly been created with a goal proportioned to their natural powers, while not called to the beatific vision. This created a dualism between the innate natural gifts in comparison to supernatural gifts, added to the natural. By the twentieth century the theory had become so powerful that it was thought essential since its rejection, it was held, necessarily led to a denial of the gratuity of grace. See de Lubac, *Surnaturel*; de Lubac, *Augustinianism and Modern Theology*, 137–39. Nature related to the supernatural only by obediental potency, i.e., a nonrepugnance. Aquinas, for his part, continued the patristic tradition concerning the human spirit as the image of God, never considering a "pure nature" construct. See Duffy, *The Dynamics of Grace*, 296–97.

140. Rom 2:14

141. Aquinas, *Summa Contra Gentiles*, III, 50.6.

142. Aquinas, *Summa Theologica*, I, Q. 12, Art. 1.

and building of knowledge, and in the development of a civilized culture and society. Moreover, the powers of the soul (memory, understanding, and free will) would be in continual conflict with bodily desires, and the body subject to sickness, accidents, and aging. Such creatures, although in possession of free will and reason, would be subjected to their natural appetites and in certain ways opposed to God's will; therefore, they would have intrinsic weakness possibly similar to that of human nature, with none of the help known to the first humans. To exercise proper control and use of their natural appetites they would require a type of natural assistance from God, that is, endowed with some kind of extraordinary faculties of body and soul, allowing for obedience to the divine law without great difficulty. If such creatures sinned, God could pardon each individually or collectively according to natural contrition. After physical death those faithful could achieve a natural possession of God in eternal life of natural bliss similar to that of unbaptised infants, having never been the recipient of divine revelation. The *Na'vi* people of the film *Avatar* can be categorized according to a state of pure nature—as hunter-gatherers with an unsophisticated technology equivalent to an earthly Paleolithic epoch. Having a developed a non-technological yet sophisticated egalitarian social organization and culture, their primitive religion is monistic, with worship of an omnipotent entity *Eywa* in addition to certain animistic beliefs tied to two sacred sites known as the *Tree of Souls* and *Tree of Voices*.

A being created in a state of pure nature, as Karl Rahner described, in a state not ordered to grace, can be considered, as God is not required to create a being so ordered. It would not have had the supernatural endowment of sanctifying grace. Descendants of such creatures would be born lacking sanctifying grace as their natural state, rather than being deprived of it, and they would have no union with the creator. Such a being would have a physical body, rational soul, free will, natural appetites, and may have enhanced faculties, however would be without any supernatural or preternatural gifts. As their nature is intimately bound to destiny, rational creatures created and existing in a state of pure nature would not desire God in the intellect as do humans. In this case, God does not owe what a creature does not have by nature, as gifts are just that, gifts. God does not have to create with a desire for God, nor would it necessarily exist in a condition of sin. Its end would be eternal life of natural bliss, a state of natural happiness without benefit of the Beatific Vision, and if guilty of sin, God could pardon an offense according to natural contrition. Beings in pure nature could desire to know God by the light of pure reason, and in this case God could provide, in the absence of supernatural grace in their natures, that is, absent a natural desire for knowing God according to their nature, other extraordinary means for knowing their creator could be made manifest given certain demonstrations, theophanies, or divine interventions or manifestations which provide empirical evidences and thus create an existential desire to know the creator, although whereby grace is not a constitutive element of their nature. There may be an opportunity for each individual to experience the creator in a personal way. Galileo's statement that "God reveals Himself to us no less excellently in the effects of nature than in the sacred words of scripture"[143] encapsulates the perspective of God's revelation and relationship to humans through the natural world.

Pure nature is a theorized state not maintained by traditional Catholic teaching; although as a formulated notion it has proved useful in defining a human nature apart from the gratuitous gift of grace and ultimate supernatural elevation. A state of pure nature can be postulated in the case of humans in which humanity contains all the gifts and abilities that is inherent in human nature, and where a human being can attain a natural and final end. That is, God could have created humans constituted of a physical body and rational soul, with an end proportionate to its created capacities. A supernatural end of a human being cannot be attained without supernatural help from God, and Aquinas taught that a state of pure nature is possible in the sense of a strict justice, that is, humanity could have been created without any supernatural gifts as man did not require anything more, since it was not owed humanity. Aquinas also taught the fittingness that supreme good self-communicates, and to the highest degree possible as in the Incarnation, therefore, it was also fitting for infinite good to elevate humanity to a supernatural state. As grace perfects nature, the human soul is naturally *capax Dei*. Humans are created by God and for God, and the dignity of man rests above all on the fact the he is called to communion with God."[144] Cardinal Cajetan, on his commentary on the powers of the human soul according to *Summa Theologica*: "All Christians in any state or walk in life are called to the fullness of Christian life and to the perfection of charity."[145] Humanity was created in God's grace which provides the reason for his desire and fulfillment in heaven, and that theoretically, human capacity did not provide a sufficient ordering of his powers to supernatural elevation, that is, humankind could have been satisfied knowing God through his creation and effects as the pagans of antiquity. This is termed the *hypothetical state of pure nature*, useful for our discussion of potential intelligent extraterrestrials who may not necessarily share our supernatural end. The Council of Trent defined mankind

---

143. Galilei, "Letter to the Grand Duchess Christina of Tuscany," 93.
144. *Catechism of the Catholic Church*, 27.
145. *Catechism of the Catholic Church*, 2013.

as being created in a "state of holiness and justice."[146] Therefore, there are two ends for humanity, one hypothetical and one actual. If human nature was pure and had under its own capacities a supernatural destiny, it would not be human nature, or human nature, being human, required the ordering of grace to raise it to another type of nature. According to Karl Rahner in an article titled "Concerning the relationship between Nature and Grace," he termed human nature without an ordering to grace *Restbegriff*, or "remainder concept." He argues, "there is no way of telling exactly how (man's) nature for itself alone would react, what precisely it would be for itself alone."[147] The desire for the will of God, according to Aquinas, originates from the powers of the soul, not as a desire of the will. Although man is naturally inclined to an ultimate end, yet he cannot attain that end by nature but only by grace and this is because of the exalted nature of that end.[148] It is possible that purely natural beings be content without a divine-creature relationship, without any psychological desire for future intimate union with God, perhaps due to having long, peaceful lifespans or their habitat allowing more for less struggle for survival and hence less personal and social conflict.

Pure nature

Original State—Gifts

Intellect	Natural knowledge
Will	Natural
Emotions	Spontaneous virtue
Body	Suffering and death

Sin—Consequences

Intellect	Ignorance
Will	Malice
Emotions	Concupiscence
Body	Suffering and death

Redemption—Effects

Intellect	Enlightened
Will	Strengthened
Emotions	Imperfect joy
Body	Sanctified

Afterlife

Intellect	Natural happiness
Will	Imperfect love
Emotions	Imperfect joy
Body	Limited glorification

146. *Council of Trent*: Denzinger, *Enchiridon,*, 1511.

147. The position of Karl Rahner regarding the supernatural. See Karl Rahner, "Concerning the Relationship between Nature and Grace." 87.

148. Aquinas, *Summa Contra Gentiles* III, 25–50; Aquinas, *Summa Theologiae* I, Q. 12, Art. 1.

2. Beings with an integral nature would have a common natural destiny and the natural means to achieve it without the gift of a supernatural destiny. Endowment with extraordinary faculties of body and soul would enable them to more readily observe God's will and have control over the passions. Msgr. Corrado Balducci and Fr. Domenico Grasso hint as such a being.[149] With natural law imprinted by God, they would follow the dictates of innate reason of the natural law in fulfilling the divine will. Such beings would have several preternatural gifts innate to angelic nature, i.e. infused knowledge at birth and easily acquired additional knowledge throughout life; freedom from sickness, accidents, and old age; free of excessive concupiscence, possessing harmony in the workings of their bodily and spiritual faculties; and may be spared a difficult or painful physical death, ending their lives in a peaceful simple transition after a test or period of testing. They may possess these preternatural gifts in various combinations and degrees as willed by God. A race of beings, if in possession of an amount of infused knowledge would have a far advanced understanding of science and the physical world, surpassing humans in their ability to explore their native world and the universe. A civilization retaining these gifts driven my mutual cooperation, may have been capable of avoiding tribal warfare, slavery, the oppression of others, crime, and sexual debauchery, and live in greater cooperation and harmony than could be imagined in a human society. However, such a race would exist without elevation to the supernatural order, and would not necessarily be imposed a supernatural destiny without given the means to achieve that end. Nevertheless, the life of beings would have much greater potentiality for greatness; such beings could create great social and cultural development in comparison to humans. In the event of sin, repentance and natural contrition could be sufficient for forgiveness, individually and/or collectively. It remains possible that a sin of the progenitors may result in a loss of original preternatural endowments, rendering beings to a less desirable natural state. After death, they

---

149. Balducci considered a rudimentary cosmic hierarchy and theological anthropology of intelligent beings: "When we refer to extraterrestrials, which would have a spiritual and material nature, a physical body, having both a spiritual and material nature, with a relationship between mind and matter different that than known to humans. The physical body, with its passions and sinfulness influences the soul so deeply that man becomes unstable and rather tends toward the bad than toward the good. Therefore, it is highly probable that in between humankind and the angels, another life form exists, namely beings which have a physical body, but one which is more perfect that humans and influences the soul less in its intelligent acts and intentions" (Second Ancona Ufological Congress, "Alien Civilization"). Jesuit Domenico Grasso believed in the universe there may be a great number of possible worlds in terms of the intellectual creature, and that such beings would be "far ahead of us in science." See Grasso, "Missionaries," 90; Grasso, "*La teologia e la pluralità*," 255–65.

would know a simple natural happiness without admission to the beatific vision. God could reward eternally these with natural happiness or eternal punishment according to their actions.

Integral nature

Original State—Gifts		Sin—Consequences	
Intellect	Infused knowledge	Intellect	Ignorance
Will	Loving obedience	Will	Malice
Emotions	Harmony with body	Emotions	Concupiscence
Body	No suffering, peaceful death	Body	Suffering and death

Redemption—Effects		Afterlife	
Intellect	Enlightened	Intellect	Natural happiness
Will	Strengthened	Will	Imperfect love
Emotions	Imperfect joy	Emotions	Imperfect joy
Body	Sanctified	Body	Limited glorification

3. Elevated nature/State of Innocence/Original Justice is that state having a combination of those of an integral nature (preternatural gifts) and those of supernature; the prelapsarian state of first parents, created with gifts of immortality, impassibility, freedom from concupiscence, ignorance, sin, and lordship over the created world. The preternatural gifts are reflected in the ordering of the natural gifts of the powers of the soul to God's grace without the condition of sin; including infused knowledge, loving obedience to the will, spontaneous enjoyment of the virtues in the emotions, and no suffering or death of the body. Adam and his descendents were originally destined to live a life of holiness and virtue on a terrestrial paradise for a certain period and then be taken up bodily to heaven without the experience of physical death. If such races exist in this state, passed their particular test and never fell from grace, they would have civilizations and cultures which far exceed our own in beauty, knowledge, and justice.[150] Such a culture may have produced

---

150. The stigmatized Capuchin, St. Padre Pio canonized June 16, 2002, in a dialog documented and officially published by the Capuchin Order mentions the possibility of sinless extraterrestrials. Question: "Padre, some claim that there are creatures of God on other planets, too." Answer: "What else? Do you think they don't exist and

generations of beautiful, highly cultured creatures in material bodies.[151] We may even consider a civilization perhaps as a race of saintly beings, living in total conformity to the divine will—however it is possible certain individuals or groups may have deviated and fallen into sin. In such case we can hypothesize a civilization containing a population of highly spiritually developed beings and an associated pristine cultural life, as well as a contingent of those fallen in sin and as a result lost their original gifts. In this eventuality it is not unreasonable to suppose the former group would serve as a means of salvation for the latter and developed a theology to accommodate and reintegrate sinful beings without their original gifts into the larger community; in such case there could be a "baptism" whereby the original preternatural and supernatural gifts are restored. As beings that remain in a state of Original Justice are not plagued by sin and retain their natural and preternatural gifts, we should expect, even if their basic physiology and intellect are comparative to humans, they would possess a superior intellectual ability and quite different psychological functioning and development. A civilization can also be considered that remained in a state of Original Justice for millennia, only to lose it after becoming a highly developed society. We should not project the early failure of early humans as necessary to other races.[152]

Elevated nature

Original State—Gifts		Sin—Consequences	
Intellect	Infused knowledge	Intellect	Ignorance
Will	Loving obedience	Will	Malice
Emotions	Harmony with body	Emotions	Concupiscence
Body	No suffering, peaceful death	Body	Suffering and death

---

that God's omnipotence is limited to this small planet Earth? What else? Do you think there are no other beings who love the Lord?" Question: "Padre, I think the Earth is nothing compared to other planets and stars." Answer: "Exactly! Yes, and we Earthlings are nothing too. The Lord certainly did not limit His glory to this small Earth. On other planets other beings exist who did not sin and fall as we did." Castello, *Cosi Parlo P. Pio*; Poe, *To Believe or Not to Believe*, 264.

151. See Harford, "Rational Beings on Other Worlds," 21.

152. Theologian Domenico Grasso wrote, "Knowledge of extraterrestrials would help us penetrate the wisdom of the plans of God and the evil of sin. If they live in a state of justice they would not have committed original sin, and we would see the immensity of all that was lost by our ancestors through sin. In the case of a redemption like ours we would see the special love of God for us in terms of a further experience of this love." Grasso, *La teologia et la pluralitá dei mondi abitati*, 255, 263.

'ANTHROPOLOGY,' XENOBIOLOGY, MORPHOLOGY, AND THEOLOGICAL SYSTEMS

Redemption—Effects		Afterlife	
Intellect	Enlightened	Intellect	Beatific vision
Will	Strengthened	Will	Love perfected
Emotions	Imperfect joy	Emotions	Perfect joy
Body	Sanctified	Body	Glorified

4. Fallen unredeemed nature. Theologians have considered this possibility unlikely given the merciful nature of God in historical relation to human beings, however such possibilities may exist in an extraterrestrial civilization. Natural beings without grace (tested, failed, and unredeemed)—we may use the example of humanity after Adam and those living outside the Abrahamic, Mosaic, Priestly, and Davidic covenants, living in opposition to God. One can conceive of a race of beings having individually or collectively refused the offer to share in God's life, and exist in a complete and permanent state of opposition to the divine will. Noting that while human sin is forgivable given our limited and obscured vision, experience, and knowledge of God (that is, not knowing fully who we are rejecting when we sin) the fallen angels, having greater perfection and an infused great knowledge of God, maintained their rejection of a shared life with heaven, where punishment was immediate and irreversible. As present-day demonic beings were originally created as angels, with powers equal to those of the heavenly angels, demons that exist in damnation retained their gifts after their fall from heaven. Their loss of divine grace and relationship with the Creator, their unforgivable and irrevocable choice to choose self over God resulted in an eternal and binding transformation of their mind, will, and relationship to creation to that of evil. Similar beings having a corporeal existence, making full use of their original gifts and capacities, could achieve a very high degree of technological capability and scientific knowledge, and serve as a menacing force among themselves and other civilizations, akin to the *Sith* of *Star Wars*. Being egocentric rather than universalist or egalitarian in motivation, malevolent beings could be engaged in a variety of hostile and nefarious activities, such as large-scale predation, colonialization programs, genocide, slavery of entire races or inhabitants of planets, psychotronics, forced religious conversions, and/or the infiltration, exploitation, and even destruction of interstellar habitats, planets, or entire planetary systems. It can be conceived such beings could be quarantined or consigned to a certain planet or planetary system where they are separated from other races.

Fallen nature

Original State—Gifts		Sin—Consequences	
Intellect	Infused knowledge	Intellect	Malevolence
Will	Loving obedience	Will	Malice
Emotions	Spontaneous virtue	Emotions	Concupiscence
Body	No suffering or death	Body	Suffering and death

5. Fallen redeemed nature—Modern humans. God chose on Earth to restore our supernatural destiny, and restored the supernatural gift of sanctifying grace, but without the gifts of infused knowledge, freedom of concupiscence, suffering, and death. In another instance God may provide a different measure of restoration by returning all or some of these preternatural and supernatural gifts. In the example of Homo sapiens, the final and complete state of nature is in heaven, where all graces are perfected: the intellect experiences the vision of God; in the will love perfected; in the emotions, the joy of contemplation and the body without suffering or further experience of death after the general resurrection. The elect experience the prelapsarian state, however with a greater vision and the full participation in the life of God. Restored and elevated extraterrestrial beings may have a life of grace, a participation in the divine life, and access to divine gifts at varying degrees and types in comparison to humans of the Christian era. A species having re-received a portion or all of their preternatural gifts would expect a similarity or approaching a state similar to that described as integral nature. They may have extensive lifetimes, resulting in religious and cultural achievements that may be unfathomable to us. Sin could be a rare event, religious divisions unknown, and indeed their personal and collective history may proceed according to a different trajectory having a greater measure of gifts restored to their nature. There may be vast differences between the mystical and the sacramental; divine presence and action may not exist only amidst signs and sacraments or through the graces of interior life, but outward theophanic manifestations may provide regular affirmation God's reality in the midst of a material world. Events of grace may be outwardly visible or made known to others in an obvious and powerful manner. A sharing in the divine life may indeed be quite different from what is known through centuries on Earth. Immanence may overshadow transcendence. A share of the populace may possess extraordinary abilities known only to earthly saints (e.g., bilocation, reading of souls, prophecy, levitation, miraculous healings, etc.).

Redeemed nature

Original State—Gifts

Intellect	Infused knowledge
Will	Loving obedience
Emotions	Spontaneous virtue
Body	No suffering or death

Sin—Consequences

Intellect	Ignorance
Will	Malice
Emotions	Concupiscence
Body	Suffering and death

Redemption—Effects

Intellect	Enlightened
Will	Strengthened
Emotions	Imperfect joy
Body	Sanctified

Afterlife

Intellect	Beatific vision
Will	Love perfected
Emotions	Perfect joy
Body	Glorified

6. Supernatural state. Physical beings in a state of supernature would have been elevated, either at conception or thereafter, to a state without the requirement of natural needs, and possess or have the latent capacity to possess certain powers known to humans as supernatural. These may be described according to our understanding as an entire race of holy saints, and have capacity for mystical union and transcendental thought with a more advanced brain and intellect. Some types may be effectively immortal and/or seemingly omnipotent, be capable of energy-matter transformation, teleportation, and time travel. Additionally, beings having supernature may be in full possession of the charismata (*gratiae gratis datae*) healing, miracles, prophecy, and mystical abilities. Other abilities could include: bi/multi-location, intuition, instantaneous transportation, extra-sensory perception, mind reading, mind control, ability to move/interact within different dimensions/realities, ability to interact with on soul-level with humans, or have psionic power.[153] They would have a supernatural destiny—seeing, knowing, and loving God in a manner exceeding its own natural perceptions, and possess supernatural grace to a degree enabling them capable of semi-divine acts. Such beings would receive special divine revelations in regards their destiny and role in life, as well as obtain supernatural help to sustain their quasi-divine life, and be recipients

---

153. The ability to mentally manipulate surroundings, i.e. telepathy, telekinesis, extrasensory perception, precognition.

of supernatural assistance as well as special revelation collectively and/or individually. However, these special gifts could be lost by sin on the part of an individual, leaders, group, or race, and such beings made subject to common evils of ignorance, illness, accident, old age, struggle between flesh and spirit, and death if God limited these gifts.[154] If supernatural physical beings failed their particular original probationary period, the creator could withdraw any preternatural or supernatural gifts, without recourse to fulfill their original end and forever unable to attain their destiny in eternity with God; or upon an acceptable act or acts of repentance, God could have re-elevated beings to their original dignity, and restored in whole or in part gifts as an act of mercy as means to redeem them. He could simply forgive their sin and re-elevate them to their former condition, or been satisfied with a partial satisfaction, accomplished by a representative (a being or angel, or other divine being) or representatives of their race. If he required an infinite atonement, he could choose a multitude of means to redeem a race of supernaturally endowed beings. It is possible for other combinations of supernatural and preternatural gifts, or a supernatural order without preternatural endowments. Creatures could have a combination of supernatural and preternatural gifts or a supernatural order without any preternatural endowments. God could create a class of those with supernatural attributes and other classes with those of the natural attributes or other combination thereof within the same species.

Supernature

Original State—Gifts		Sin—Consequences	
Intellect	Infused knowledge	Intellect	Ignorance
Will	Loving obedience	Will	Malice
Emotions	Spontaneous virtue	Emotions	Concupiscence
Body	No suffering or death	Body	Suffering and death

Redemption—Effects		Afterlife	
Intellect	Infused knowledge	Intellect	Beatific vision
Will	Loving obedience	Will	Love perfected
Emotions	Imperfect or perfect joy	Emotions	Perfect joy
Body	No suffering or death	Body	Glorified

154. Briefly discussed by Raible, "Rational Life in Outer Space?" 532–35; Zubek, "Theological Questions on Space Creatures," 393–98.

Extraterrestrial supernatural manifestations and communications may take a form where a divine being makes itself present on continual basis, without recourse to inspired writings, sacramental life as we know it, a priesthood, or a hierarchical or organizational structure containing a set of teachings designed to instruct the faithful amidst an array of competing messages and traditions. There may be no need for any supernatural miracles as these are necessary only for beings prone to ignorance, suffering, and death having fallen from original grace. Public and/or extraordinary supernatural communications may be wholly unnecessary given each individual's high capacity of an advanced state of holiness and communion with the divine. In essence, we may regard such beings similar to some of the highly gifted saints, capable of being in constant communication with God or one of his representatives. If existing on a planetary surface, such beings could be capable to creating great societies, a high degree of culture and justice, live in total peace, and perhaps serve as guardians, overseers, judges, or some similar leadership role among other types of beings in their location within the cosmos. We can consider creatures exist in a genus that surpasses human but does not equal the perfection of the angels. Extraterrestrials may have a more perfected physical body than human and which influences the soul less in its intelligent acts and intentions. This concept is confirmed by the ancient principle defined by Lucrezio Caro as *Natura non favit saltus* (The nature makes no jumps); that there can be no gaps in human evolution, therefore, the probability of intermediate beings between the human and angelic is likely given this gap is too large to be natural.

In particular, the creation of the angels, their nature, and our relationship with angels can have direct bearing on the consideration of the potential nature of intelligent extraterrestrials. Angels should be understood within the context of the probability of a vast array of beings inhabiting the universe, and not necessarily as the only non-human intelligences created by God; and for the Christian, are evidence of God's capacity to create a species vastly different from humans with varying sets of capacities, gifts, destiny, and a economy of salvation preceding and separate from humans. Being pure spirit and immortal, they surpass visible creatures in perfection, having a separate genus. Each unique and genderless, they are given a prominent role in the genesis and salvation history of human life on Earth. Angels have intelligence, will, and imagination, and the angelic mind understands free will, its consequences, and reality with such clarity and totality their individual choices are forever binding. According to scripture and tradition they can appear in human form and interact with humans. Angels were created in sanctifying grace and a share of the divine nature with infused knowledge and gifts of the Holy Spirit, including a right to the actual graces which were necessary to preserve and increase their supernatural life. Created in a realm outside of heaven, only those were granted entrance after having passed a test of loyalty. The fathers of the Church refer to the fall of the angels in Isaiah as the sin of pride, "I will not serve."[155] St. Augustine saw their sin as unforgivable due to their endowment of supernatural gifts.[156] The heavenly angels are perfectly pure beings and ontologically and intellectually superior to humans. The creation and personal assistance of what we understand as angelic beings may or may not be the case with every intelligent extraterrestrial species; this may be more likely if a race was particularly gifted equivalent to or beyond what was provided human beings. It may be theorized that God originally provided a race an angelic or other divine or supernatural being an important role in establishing and maintaining a civilization, however lost its favor and presence due to serious sin.

There is no reason to assume that angels were created immediately prior to the creation of the Earth for the sole purpose of assisting humans, or that prior to and after the age of humanity are not tasked with duties for benefit of other, distant intelligent civilizations or charged with tasks unfathomable to us. They may have been created at the beginning of the physical universe or well before it. As angels are pure spirit, we should not presuppose a uniquely human appearance as they can take on a variety of forms. Our prejudicial and narrow view of the role and personages of the angels has been anthropomorphized and geocentralized, and we cannot make conclusive statements regarding the entirety of their activities. We know from revelation and tradition that the God has commissioned a certain species to interact and work in coordination with another, lesser advanced race of beings for a predetermined timeframe for the purposes of union with God. We know the main work of the angels is to serve as a resource for countering the efforts of evil spirits in leading humanity to sin and ultimate perdition. It can be argued that the angels were tasked with assisting humanity only as a result of the fall of humanity, and that if the original sin had not been committed, mankind would have benefited from a very different type of relationship to the celestial creatures; one in which there may have been direct communication, as described in Genesis in the garden, where an angel of God kept a permanent presence. Further, angels may be an earlier result of universal evolutionary processes and possibly have emerged elsewhere in cosmic history; in either case, they are extraterrestrial beings.

155. Isa 14:12–15.

156. "Since we know that the Creator of all good sent no grace of atonement to the bad angels, how can we fail to conclude that their sin was judged all the more culpable because their nature was so sublime." *Commentaries on St. John, Migne, Augustine*, 35, 1924.

Salvation Histories

In considering the possibilities of a race of extraterrestrial beings having fallen from their original created state of nature, and with an opportunity for redemption, there are several outcomes inherent in this scenario: the divinity may postpone or postpone indefinitely a salvific act to reunify a race to the Creator as a result of the actions of the entire group or leaders of that group, leaving them in a fallen state; such a race may not be offered an opportunity for reconciliation with the divinity due to the nature of their sin(s); a redemption contingent upon the cooperation of a large segment of the populace or certain of its leaders.

Extraterrestrials, if in a fallen state, may be punished collectively (if they exist in a state where they procreate and in which no means of salvation or restoration was given to them or their offspring) or individually as in the case of challenging angels. God could choose to forgive an "original sin" type event, or require personal or collective repentance, or accept the offering, suffering, repentance, or request of a representative, mediator, or type of redeemer as satisfaction for an offense against the divinity. He may also accept the actions of such an individual or group of individuals, with or without the cooperation of repentant sinners. God could have forgiven sinful beings as a simple act of mercy, or require personal or collective repentance as a condition for reconciliation. He could accept the satisfaction won by a certain act or acts performed by a mediator or representative of a race or group of beings or the act of a representative/redeemer and the cooperation of repentant sinners. Such a representative or mediator need not be a person of the Holy Trinity, and could be a creature, natural or supernatural, acceptable by God. Such a creature would be unable to provide infinite satisfaction as a vicarious reparation, however this could be acceptable as payment and the specific act or acts necessary may be divinely revealed. In a case where God required an infinite satisfaction for sin a divine reparative act be needed. It is also not theologically necessary for a divine person to undergo an act of death in order to make full recompense for the sins of others, as any divine mediator, a specific act, thought, or intention, or if necessary, sacrifice would be fitting if God so willed serve a infinite satisfaction.

Extraterrestrials fallen from their original created state of nature, and with an opportunity for redemption have several outcomes: the divinity may postpone or postpone indefinitely a salvific act to reunify a race to the Creator as a result of the actions of the entire group or leaders of that group, leaving them in a fallen state; such a race may not be offered an opportunity for reconciliation with the divinity due to the nature of their sin(s); a redemption contingent upon the cooperation of a large segment of the populace or certain of its leaders; the divinity may wait for the natural death of certain individuals or generations before offering salvation.[157]

Divine Presence, Revelation, and Theophany

As in the Judaic and Christian traditions, a theophany or series of theophanies to key individuals or groups can provide the basis for a claim to a revealed religious narrative, an organizational structure, and authoritative teaching. A linear, historical progression of religion may not follow the same pattern as Earth-type trajectories, receiving its impetus according to person, geographic location, culture, tradition, and specific revelation. As linear, monotheistic Earth religions exist on a historical trajectory, we may expect instances of extraterrestrial religious beliefs founded on historical personages, chronicled by means of scriptures and/or revered objects or remembrance of divine acts, performed through various rites and traditions; along with varying degrees in type and intensity of possible continued divine manifestations within a faith community. A divine communication, inspiration, imparted knowledge or spiritual experience can be given collectively at one or several instances by specific mode; for example, as an individual divine revealing, an interior locution, certain states of ecstasy, or an individual or shared physical manifestation. An extraterrestrial theophany or revelation may occur diachronically or synchronically in an extraterrestrial species. These are all modes of divine initiative known to humans, there is no reason these and other means of theophanic or supernatural messages cannot manifest collectively and simultaneously to a given species or civilization. Conversely, the dynamic dialectic of the historically hidden, silent divinity interacting with humanity may contrast with a more direct, definitive, existential knowledge of divine presence, communication, and relationship in other societies, where "knowing" supplants "believing."[158] Conversely, rather than self-revealing extrovertively in the created order by means of a historical series of interactions with individual creatures, divinity could manifest in a purely introvertive mystical union and communicate without the use of conventional language or rational concepts. Without externalizations, a religion could be void of sacraments, scriptural documents, liturgy, or hierarchical authoritative religious structures to serve as

---

157. Num 14:30. Taking example of the ancient Hebrews' forty years of wandering.

158. The divine hiddenness known to Earth religion, where indirect experience of divinity remains the standard, other religions may have total concealment, relative transparency, or a full revealing to the individual or community, or some combination of the three.

a medium through which one encounters the divine.[159] Historical theological development may be unnecessary or irrelevant as a result of received ineffable mystical knowledge. The use of symbolism in religion is nearly a universal established phenomenon, representing the holy, deity, concepts, and teachings, as a means to communicate the supernatural through the means of the natural, the non-visible via the visible.[160] Symbolism may be absent in a purely mystical religion where it is deemed unnecessary due to the personal immanence of the divine action on the individual, or where symbols are forbidden as in certain Earth religions and denominations. Therefore, modalities of divine revelation can exist on a continuum with individual and/or collective manifestations encompassing those known to Earth histories and unknown others.

General revelation may be unnecessary with beings far more capable of highly advanced states of spiritual perfection and communion with divinity than humans, capable of intimate relationship with God and/or certain divine representatives. Such beings would be successful in creating civilizations with a high degree of culture and justice, live in total peace, and perhaps serve as guardians, overseers, judges, or some similar leadership role among other, lesser developed societies. Certain superior societies, by means of an advanced technology and social responsibility, might be free from natural disasters by engineering a controlled living environment, as well possessing a biological integrity resulting in complete freedom from suffering; thereby such a religion might not contain any particular theodicy to integrate an experience of evil. In other societies, the instinctive struggle for resources and innate territorial drives, striving for hierarchical dominance within a community, and competition for mate selection can produce the phenomenon interpreted religiously or practically as personal and collective evil. However, the internal and seemingly insurmountable dynamic of selfish versus altruistic motives known to humans is one among a number of possible inner states. As personal freedom and intelligence provide the conditions for evil but not its guarantee, varying degrees of receptivity to and givingness of material and spiritual gifts inherent in creatures may result in evil being a rare outcome in some civilizations.

159. Extraterrestrials possessing saint-like abilities, who commune with the divinity, receive either some form of divine communication such as divine apparitions, interior locutions, or inspired writings; these experiences can be made available to a collective without the filter of persuasive argument, personal witness, or teaching. In such a case, there may be little need for externalizing religious belief, since all imagery and intention is shared.

160. Geertz defined religion as a system of symbols which acts to establish powerful, persuasive and long-lasting moods and motivations by formulating conceptions of a general order of existence and clothing these conceptions with such an aura of factuality that the moods and motivations seem uniquely realistic. See Kunin, *Religion*, 153.

As the scriptural account describes the origin of terrestrial evil as rooted in a prehistoric angelic act which influenced the first humans, extraterrestrial malevolence and disobedience need not proceed correspondingly. Evil could receive its impetus and be manifested individually, collectively, and structurally in ways inconceivable to humans. As such, extreme distortions contained in now hundreds of extinct Earth religions which included murder, human sacrifice, cannibalism, religious prostitution, and others might be manifest in certain extraterrestrials societies. Evil could take greater, large-scale structural forms in civilizations. Extraterrestrial beings could engage in surreptitious manipulation of extant indigenous religious beliefs by means of technological superiority, to willfully create religious systems by means of artificially manufactured religious phenomena, 'theophanies' or epiphanic events to a lesser-advanced species or planetary inhabitants. In this manner, religious and theological foundations could have their genesis on a home planet, or have been initiated by a known person or group from an alien race in which lesser-advanced species worship or serve them as superiors or deities, or a species they have clandestinely genetically enhanced or manufactured.[161] In another possibility, a technologically produced artificialized religion could be utilized to institute an ethics and incorporate a moral code in an effort to create a stable civilization out of a barbaric one, to increase docility for unchallenged subjugation, or to transition a society to a greater level of egalitarian responsibility, mutual cooperation, or other peaceful motive. Conversely, it is possible advanced beings could unwittingly cause the development of a religion by their mere detection, presence, and interaction with primitive societies. Further, religions could receive their impetus purely from indigenous sources such as those well known among Earth cultures.

We may theorize a civilization with similarities to our own historical trajectory. Prior to the introduction of Judaism and Christianity, the Earth was witness to the flourishing of divergent religious traditions of shamanism, animism, ancestor worship, and fertility religions. Judaism in the ancient near-East and Christianity during its expansion in the Roman Empire, greater Europe, and later in the New World encountered belief systems interpreted as false, man-made religions or the actual and unwitting worship of demonic beings. There is the possibility divinity may choose to initiate a divine election and establish a "covenant" with a certain individual, family, or single community (or several communities) of beings in which he provides his grace, presence, teaching, and protection; we may find in a future contact event that there is a corollary historical relational pattern in the choosing of the individuals, families, or groups to be sole recipients

---

161. Alien to a particular world inhabited by another intelligent civilization.

'ANTHROPOLOGY,' XENOBIOLOGY, MORPHOLOGY, AND THEOLOGICAL SYSTEMS   183

of divine graces and messages. In the case of human beings, we have an extremely lengthy historical period from the account of the creation of Homo sapiens and subsequent fall until the election of the Hebrews, and similarly another long historical period until the arrival of Christ and his message of salvation for the entire human race.

Extraterrestrial Religions

There are several types of potential inestimable numbers of extraterrestrial religions, some of which can be extrapolated from terrestrial examples.[162] Human, terrestrial religions provide a starting point of informed speculation of other types and categories of what can be defined as extraterrestrial religion.[163] These types could include, apart from what we would term primitive religions: an individualized religion, without benefit of any hierarchy or structure, common among certain Christian believers who have dissented from an organized church and worship as small groups or individually, or an organized religion supportive of a militarized world mission, similar to the early conquest by the Islamic Empire and its expansionist spread as far as India and Spain. More highly organized, centrally controlled and managed, hierarchical religions of the Roman Catholic, Orthodox, and Anglican variety are also examples; or as mentioned mystical religions. Mystical religions may be exist as forms perhaps considered superior to organized religions given its individualized and authoritative nature of revelation. Extraterrestrial religions could survive and adapt to similar evolutionary changes with passing scientific, cultural, and social eras; they may take a variety of these

---

162. Tylor, *Primitive Cultures*. Tylor argued the essential element in all religion is the belief in spiritual beings, and that religion evolved by means of three stages: animism (belief in spiritual beings, originating from attempts to explain dreams and trances), polytheism, and finally monotheism.

163. Anthony Wallace identified four major categories of religion: *Shamanic*—A Shaman a religious figure (mediums, spiritualists, astrologers, tarot card readers, palm readers, diviners) who mediate between people and supernatural beings or forces, which were mostly characteristic of foraging societies. *Communal*—Polytheistic religions, belief in several deities which control aspects of nature; communal religions, with harvest ceremonies and collective rites of passage, which were more typical of farming communities. *Olympian*—First appeared in states, had professional priesthoods hierarchically and bureaucratically organized. These were polytheistic and characterized by pantheons of powerful anthropomorphic gods with specialized functions. *Monotheistic* or *Ecclesiastical*—Have hierarchical priesthoods, all supernatural phenomena are products or manifestations of, or are under the control of a single, eternal omniscient, omnipotent, and omnipresent being. A world-rejecting religion, such as Christianity and Islam reject the natural (mundane, ordinary, material, secular) and instead emphasize the supernatural, sacred and transcendent realms. See Wallace, *Religion*.

forms: monism, pantheism, animism, polytheism, monotheism, agnosticism, atheism, or certain combinations of these. There may exist, like Earth, planets with varying societies of competing religious beliefs, planets having two or more different races of indigenous beings with divergent religions, a civilization composed of one unified set of beliefs on a planetary scale; or a shared religion among different extraterrestrial races of the same or collection of planets, outposts, and colonies. There is a general assumption that each highly advanced extraterrestrial civilization that humans may encounter will have a racially or planetary unified religion, however this is a philosophical assumption in futurist theory, as scientific and technological advances sufficient to allow a race to visit or communicate with our planet does not imply a religious or spiritual superiority capable of unifying an entire population. In the case of a civilization similar to our own, where there exists competing belief systems, it would be erroneous to assume these differences would be resolved through social and cultural evolution; it may be the case that a major theophanic event occur, which could provide the necessary impetus to unite all under one banner of faith, knowledge, and experience of the divinity. Another scenario would be the case of a militaristic expansionist religion, where opposing religious or philosophical beliefs are defeated through military campaigns, and where conversion is forced under penalty of death, and recalcitrants, heretics, and apostates are eliminated in order to preserve one religious dogma. Although this type of religion is not in accord with our conceptions of a divinely inspired religion, it nonetheless has occurred on our own planet among claimants of divine inspiration for atrocities committed in hopes of achieving religious, political, and/or planetary conquest.

Societies may be occupied with a large variety of natural belief systems, ranging from animist and polytheist systems, from an extraterrestrial *"extrapolatry"*[164] and autolatry, to planet-wide religions, or the worship of a little understood super-technology among a deliberately under-educated majority. Some races of this kind and having encountered other, more advanced races may in turn be led to or allowed to worship technologically superior races as gods themselves. Socially, civilizations such as these may exhibit some or many of the behaviors of egotistic and sinful races described earlier, having large-scale corruption, injustices to individuals and communities, poverty, and other social ills. Some, without divine revelation to provide direction for individuals and communities, may have developed useful philosophies, either based in egocentric or utilitarian motivations.

---

164. The worship of an extraterrestrial being, in this case one's own species.

Common scientific conceptions of futurist human models or extraterrestrial religions are either comprised of universalist,[165] naturalistic deities according to Paine or Dick's views,[166] or are impersonal-scientific-atheistic, as argued by SETI astronomer Jill Tartar.[167] Dick in particular argues that the widespread acceptance of a universe containing abundant life and intelligence is ushering in a new era of cosmic consciousness in which "cosmotheology" must transform older theologies. A "natural god of cosmic evolution and the biological universe," holds the promise of harmonizing religion and science and becoming the "God of the next millennium."[168] Therefore, he suggests an abandonment of terrestrial religion in favor of a cosmic naturalized religion, "using our ever-growing knowledge of the universe to modify, expand, or change entirely our current theologies."[169] He is speaking perhaps more directly to some in the scientific community, certain of which may share his naturalistic views; however most Christians will find unacceptable the notion that Christianity is a man-made religion that should be abandoned in the face of greater knowledge of space and/or contact with extraterrestrials. These represent narrow, bluntly projected anthropomorphized conceptions of 'advanced' religion without consideration of the principle of divine prerogative and creaturely response. Indeed, a much older, technologically and/or socially advanced species would not necessarily possess a more developed or greater revelation forming the basis of a sophisticated religion or theology. Civilizations having long abandoned forms of religious belief in favor of scientific knowledge may be spiritually undeveloped compared to humans, or scientifically advanced civilizations may be proponents of evidentialist religions,[170] determined to be at odds with certain earthbound or extraterrestrial non-evidentialist theologies in which acknowledgment of a deity is fundamental to being, hence not requiring empirical evidence. Space-faring extraterrestrials may maintain substantial religious societies and could view the divine as a personal being or an impersonal force as in Eastern religions, or other non-theistic conceptions of the ultimate. As certain Eastern faiths teach salvation through individual enlightenment and conceive the supreme reality in strictly impersonal terms; and monotheistic religions acknowledge the Supreme Being

165. A relativistic ideology, where components of each religious, theological, and philosophical concepts have universal applicability.

166. Dick, "Cosmotheology Revisited," 202–4.

167. Tartar, "SETI and the Religions of the Universe," 145–46.

168. Dick, "Cosmotheology Revisited," 287–301.

169. Dick, "Cosmotheology Revisited," 200.

170. Religious belief considered rational only if provided sufficient evidence. The difference between subjective religious experience versus objective verifiable evidence.

as personal, surrounded by saints and other heavenly beings inhabiting a kingdom of super-moral beings, we may expect these constitutions might exist within an array of other divine conceptions.

Human categories of "religion," founded on European culture and languages, may not necessarily find exact corollaries in extraterrestrial societies. The terms and elements fundamental to what we understand as religion in Western societies such as "redemption," "saviour," "church," "scripture," need not exist in extraterrestrial civilizations, evidenced by their absence in certain Eastern religions. Therefore, we should not seek to categorically impose strict religious or theological paradigms on extraterrestrial religion, but allow divinity in its infinite wisdom to determine the terms of revelation, message, and means of achieving unity with intelligent creatures. Extraterrestrial religions may emphasize one of these aspects over the other, or maintain one, some, or all characteristics among human religions. Religion conceived in modern life need not exist as a separate sphere of activity common in European and American societies, nor exist within or amongst other spheres, but rather can maintain an omnipresent existence in a civilization. Therefore, conceptions of the Creator, divine being or beings could take a variety of forms,[171] ranging from a non-corporeal being to that consistent with portrayals of an extraterrestrial race. Ninian Smart's comparative, phenomenological approach is useful; religions exist as a multifaceted phenomenon, with varying emphases according to their particular world view, philosophy of life, and praxis.[172] Where Roman Catholicism places more importance on liturgy, ritual, and systematized moral and doctrinal teachings, in Islam and Orthodox Judaism the doctrinal and legal details are stressed. Certain Christian denominations place great prominence on emotion and personal spiritual experiences.[173] In certain extraterrestrial religions, doctrinal, experiential, and ethical teachings, therefore, could be combined with explicit scientific understandings, or where religious teachings, consciousness, and praxis are infused broad scale in social institutions. Many religions however should be understood as a "snapshot" within an historical context as the Christian religion demonstrates. Divine revelation, participation, grace, and unity manifest on a historical continuum; extraterrestrial religion must accommodate planetary and in real sense cosmic timescales and a particular universal location or locations. The Christian revelation is manifested

---

171. In other instances where societies whose sexual means of reproduction have been engineered out for extremely long periods and exist as asexual, non-sexual, or androgenous beings, a society may have no concept of gender, and therefore, have no conception of gendered divinity or divine persons.

172. Smart, *Dimensions of the Sacred*, 10–11.

173. Pentecostalism in particular exhibits these characteristics.

according to a certain timescale and teaches divine action throughout all human history within this particular epoch; it may be the case that ours and certain extraterrestrial religions have impact and import beyond their time scale in accordance with divine plans.

In the next chapter I reconsider the four classical positions on soteriology in the light of all the data gathered so far in the thesis, and after detailed evaluation conclude that the underexplored *varied* position is the strongest and most generative for the construction of a contemporary exotheology.

Chapter 5

Exotheology and Traditional Christological Formulations

This chapter will review the strengths and weaknesses of the four major historical soteriological positions with respect to multiple intelligently inhabited worlds. Working from the Roman Catholic, and specifically Thomistic, theological basis outlined in chapter 1, it will seek to show the *varied* view is the strongest and most viable theological solution to the new discoveries in the universe.

Section A. Review of the Four Major Historical Positions

The four major historical soteriological positions with regard to intelligent extraterrestrials in relation to the Christian doctrine of redemption have been demonstrated as a historical development of theology as an *exclusive* type—a single, exclusive divine Incarnation of the God-man Jesus Christ on Earth which alone provides salvation by one mediator between God and all human beings past, present, and future until the end of the age. Extraterrestrials are considered nonexistent, sinless and without need of redemption, or sinful and without access to redemption. An *inclusive* type, based on one divine Incarnation on Earth for all intelligent creatures in the cosmos, providing redemption for humans and intelligent extraterrestrials if they exist; a *multiple* type, where the second divine Person, or Logos is incarnated on other planets and within alien civilizations, taking their specific forms and joining with their natures to redeem creatures as on Earth; and a *varied* type, where the Creator manifests to his creatures in a variety of the most fitting ways according to his own designs to divinely reveal, redeem, perfect, and unify with creatures. However none of these solutions, as previously formulated, has been demonstrated to satisfactorily resolve

EXOTHEOLOGY AND TRADITIONAL CHRISTOLOGICAL FORMULATIONS

the Christological conflicts and general concerns of theology with regard to human and extraterrestrial soteriology.

Paine had argued the errors and contradictions of projecting a simplistic and universalist Christianity upon the framework of a cosmos composed of other potential intelligences. Proponents of the *exclusive* argument provided this critical response to Paine in their rejection of his solution of a cosmic natural religion: redemption from original sin embraces the human family on Earth and does not apply to extraterrestrial intelligences, as incarnation does not apply to the whole of the created universe. The exclusivist position maintains the classical geocentric and anthropomorphic perspective, without provision for the salvation of extraterrestrials, if in fact sinful. The exclusivist argument has two forms; the first is most often invoked by literalists and fundamentalists, who comprise the largest segment of Christians that tend to reject the scientific and theological probability of extraterrestrial intelligence due to the limited view of scriptural prohibition.[1] As the presence of extraterrestrial intelligences seemingly would be an important subject relative to the actual extent of God's creation and redemption of all creatures, the omission of such creatures in revelation is interpreted as confirmation of their non-existence. The exclusivist view is best termed the "classic" argument, which has dominated theology in the Augustinian and Thomastic traditions;[2] however their formulations were never intended to incorporate our modern notions of biological intelligent extraterrestrials. Modern advocates of exclusivism, Breig and Steidl, therefore, opt for the simpler solution of a simple denial of extraterrestrials. George and Hebblethwaite reasoned that since Christ's salvation is restricted to humans and Earth, sinful extraterrestrial cannot access Christ's redemption, hence they likely do not exist. The second form consists of those who accept the possibility of the existence of extraterrestrials, either sinless without need of redemption, or sinful without access to Christian redemption. Those sinful cannot receive the Christian message due to cosmic distance, and cannot participate in Christ's redemption due to their non-relationship with humanity. By themselves they are without access to divine redemption. This argument would seem fundamentally incompatible with a loving Creator who

---

1. Traditionally termed 'negative authority of scripture.' For these religious groups, acceptance of extraterrestrials can lead toward acceptance of evolution, a *sine qua non*, and a denial of humanity as a special 'crown' of God's creation, and therefore, to consideration of non-Christian means of achieving salvation. As a result, many fundamentalists deny extraterrestrial existence. See earlier *Brookings Institute* study conclusions; also see Peters, *Science, Theology, and Ethics*, 130–31.

2. St. Thomas Aquinas rejected the pluralist and embraced the exclusivist position as the former seemed to deny the orderly unity of the Creator. See *Summa Theologica*, I, Q. 47, Art. 3.

calls creatures to relationship as illustrated in the Old and New Testament. The narrow view of a Creator restricting redemption to one species and planet is significantly brought into question given our present knowledge of an inconceivably vast universe containing the possibility of millions or even billions of intelligent civilizations.[3] The image of salvation revealed in the Judeo-Christian tradition was geocentric and anthropocentric as well as universal and transcendental, and the scope of God's creation in the original Hebraic cosmological context was limited to an Earth with a fixed dome in the heavens. These 'heavens' were composed of the sun, moon, fixed stars, and wandering planets contained within the dome that contained the Earth; God and the heavenly host existed just outside this sphere. 'Heavens' thus did not refer to an expansive physical space such as envisaged in later, more scientific developments in cosmology. Rather, within this limited cosmological framework neither a modern cosmos nor one inhabited by intelligent biological beings was possible. As extraterrestrials were not included, either hypothetically or actually, in the original formation of scripture, and knowledge of the universe has advanced, modern theologians who maintain the exclusivist position advocate for a theology incompatible with God's omni-properties known to the Hebrew and Christian traditions, and current knowledge of a vast cosmos containing other planets and potential environments conducive to life and perhaps intelligent life.

The *inclusivist* argument presents the first effort to reconcile extraterrestrials within a single Christian redemption, in accord with the inference from Hebrews 9:25–26 that Christ could not die again on another world as he "suffered once and for all." It maintains the uniqueness of the hypostatic union in Christ in the universe, being the greatest self-communicative act of God to creation among the multitude of possible foreign intelligences throughout creation. The Second Person of the Trinity, having come to Earth and assuming a human nature, became the God-man Jesus Christ of one composite being possessing two natures. As the divine nature (of which there is only one in the cosmos) will remain forever joined to the human in the hypostatic union (the argument of Hebblethwaite), it is considered impossible in this view for the Second Person to be incarnated in another alien nature. Therefore, this view requires the Second Person to remain known by his outward appearance as a human being to make himself known and present (not necessarily incarnate) to other races of intelligent beings to proclaim an identical message of salvation given to humanity. As the redemptive act applies to all humans, present, future, and

---

3. "The Christian story is based on a Pre-Copernican cosmology. The Christian story makes sense until astronomical discovery makes the world no longer the center of the universe." Burgess, "Earth Chauvinism," 1098.

past on Earth, so the same salvation is offered to beings which exist outside the human family. God's redemptive plan is framed and structured within a Christocentralized universe; the Judeo-Christian paschal mystery is not merely a geocentric and anthropomorphic event but a cosmic, universal, spatially and temporally transcendent phenomenon which includes and encapsulates all creatures and creation within the cosmos. This is the argument of the "Cosmic Christ,"[4] based on the Pauline hymn of Colossians 1:15–20 which affirms the supremacy of Christ;[5] Romans 8:19–22 which presents all creation "liberated from its bondage" through Christ;[6] and Ephesians 1:10, 20–23 in which a salvific act of Christ on Earth is expanded to include any rational creature in need of salvation or divine unification throughout the universe.[7] Revelation 1:8 describes Christ as the Alpha and Omega, beginning and end of all creation; and Romans 6:10 proclaims that Jesus died for all. It argues for a human economy of salvation for non-human beings, as argued by Di Noia and O'Collins, and of a Jesus Christ as the only proper "image" of God.[8] John 10:16 also suggests this broad ambit to Christological soteriology has sometimes been understood as a possible allusion to extraterrestrials.[9] Therefore, Christianity has no need

---

4. Heb 1:3–14; 2:5–18, a liturgical hymn, also describes the cosmological role of the preexistent Son to the redemptive work of Jesus, made heir to all things through his death and exaltation to glory, and through whom God created the universe.

5. This was a Pauline editing of a preexisting hymn; the Colossians were questioning the supremacy of Jesus. Since Norden it has been accepted that the hymn spoke of the supremacy of Jesus in revelation, creation, and redemption. See Norden, *Die antike Kunstprosa*, vol. II; Christ is *prototokos*, "firstborn" of all creation, physical beings and angelic beings. He is the efficient and final cause of the created universe (cf. John 1:3; 1 Cor 8:6). He sustains the created order while remaining in transcendent authority over it and imminently incorporated in the reconciliation of all bodies, souls, and spirits into the one true God. He is existent and pre-existent. It is the Earth and 'heavens' that are within the scope of Christ's redemption and creation.

6. Rom 8:19–22.

7. Eph 1:15–23.

8. Col 1:15, "He is the image (*eikon*) of the invisible God, the first-born of all creation"; and Heb 1:3, "He reflects the glory of God and bears the very stamp of his nature, upholding the universe by his word of power."

9. "I have other sheep that do not belong to this fold. These also I must lead, and they will hear my voice, and there will be one flock, one shepherd." This is highly suggestive, and offers no real ground for an exegesis of extraterrestrials. In its proper context, Jesus refers to the fact that after the resurrection, he will send to apostles outside of Israel to all nations (cf. Matt 28:19) to preach the Gospel (cf. Matt 16:15), beginning in Jerusalem and extending to Judea, Samaria, and the ends of the Earth. This is to fulfill the ancient promise to Abraham as father of all nations, and Israel as its center for the salvation of all nations (cf. Gen 12:1–3; 17:1–8; 18:19; 22:1–18; 26:4; 28:13–14; Ps 2:7; Isa 2:2–6; 66:17–19; Rom 4; Gal 3:7–9, 26–29).

to modify or supplant its theology and image of the Triune God with an unknown religion or competing economy of salvation outside Earth. A single, unified redemption of all intelligences sinful and fallen, and/or sinless and unfallen (as argued by Moltmann) share in the recapitulation and reorientation of creation to the Creator, centered on events of a first century Earth. Therefore, original sin is conceived as either a uniquely human tragedy affecting all creation, or a primordial evil arising independently among other civilizations. In either case, human civilization as a matrix for the one incarnation and sacrifice of Christ are central to the redemption of all beings. Since the Incarnation and atonement applied to all creation in the universe, this necessarily includes all sentient beings, including those unaware of the Christian message, which nonetheless are held under its dominion. Given cosmic distances extraterrestrials may or may not be aware of their redemption wrought by the terrestrial Christ known to humanity, therefore, a divinely mandated human-led "space evangelization" endeavor, either by radio telescope or future human space exploration could be necessary, as argued by Milne[10] and Mok.[11]

The Second Person of the Trinity united himself hypostatically to the human species in order to redeem them, and in turn offered his sacrifice to the Father as the God-man. In doing so Christ incorporated all humans into his sacrifice, and by cooperation in grace humans inherit an elevated, supernatural life. Scriptural texts,[12] dogmatic teaching, and common belief for centuries held that humankind was the only *material* created intelligence, existing in a limited physical universe according to the Hebraic, Aristotelian, Ptolemaic, and Copernican cosmologies. All material and immaterial creation was subjugated under Christ and the infinite effects of the redemption at the Cross extended beyond the confines of Earth to include these "contained" cosmological models. Scripture makes reference to the term "cosmos" in reference to whom and what is being saved.[13] This term in the Scriptures would be centered on the domain of humanity and the material and spiritual world in direct relation to humans. As humanity was created on Earth and sin, according to the Testaments began on Earth with the human species, the argument of inclusivity contains an implication that human sin affected other, incomprehensibly remote beings or that the headship of a divine-human Christ includes extraterrestrials, a nonexistent

---

10. Milne, *Modern Cosmology*, 153.

11. Mok, "Humanity."

12. Principally Eph 1:20–23. Heb 2:7–9, argued by E. L. Mascall, is interpreted wherein there are no materialized beings extant in the universe other than humans.

13. Matt 11:25; Mark 13:27; Luke 10:20; John 1:9; Acts 4: 24, 17:24; Rom 8:20; 2 Pet 3:7, 10; Rev 10:6.

notion in the cosmological and theological perspective of the evangelists and apostles. The argument seems improbable for another factor: all evidences from geology, archeology, and anthropology demonstrate that intelligent life on Earth is, on a cosmic scale a very recent phenomenon. To argue the Christ-event has complete applicability to other intelligent races within our modern Einsteinian temporal and spatial scale of the universe renders this view nonsensical. A Christocentric soteriology entails an anthropocentric exotheology; the rigid and parochial claim that humans are central to the salvation of extraterrestrials within this incomprehensibly vast context, and that they alone possess the unrepeatable, unique, and solitary role in transmitting a message of supernatural redemption throughout the cosmos seems a gross anthropocentric distortion.[14] Similarly, the claim that Jesus Christ, an incarnated human being has the effect on extraterrestrials of a cosmic-scale salvation achieved in the Near East of first-century Earth places severe restrictions on God's ability to create, reveal, and redeem according to the biological, cultural, and historical particularities of species and localities inherent in the micro and macro scales of a created ≈ 13.8 billion year old universe.

The *multiple* incarnational position represents a modest evolution of the inclusivist position, utilizing similar scriptural references to support the notion of a duplication of the earthly incarnational mode of divine revelation and presence. It affirms a cosmic centrality of the second Person or Logos within the material universe and his kingship over all creatures, intelligent and non-intelligent. Its argument is from a Christocentrism in the strong (universal) sense, receiving its main thrust from scriptural passages emphasizing a "Cosmic Christ" John 1:16, 3:17; Colossians 1:15–20, Ephesians 1:20–23; Hebrews 2:7–9; Romans 6:10, and 2 Corinthians 5:19, all of which portray a Christocentric-type redemption on a cosmic scale. It is accepted in this view that original sin is not limited to nor necessarily originated from Earth, therefore, nor is the salvation won by the cross of Jesus Christ; sinful inhabitants of Earth analogs receive an Earth-type divine-creature incarnational mode of redemption. According to this view, the Second Person of the Trinity is not limited to his identity known to humanity; rather, the God-redeemer, often termed *Word or Logos* to deemphasize the earthly, human Jesus, is able to be incarnated in an extraterrestrial material body in an unlimited number of intelligent civilizations throughout the cosmos. This type of incarnation does not necessitate suffering and death as the earthly Jesus did, nor does it prohibit it. However it does view the act of incarnation of the Second Person as

---

14. A view argued by Karl Rahner: "[The human being] is a personal subject from whose freedom as a subject the fate of the entire cosmos depends." Rahner, "Theology and Anthropology," 15.

the primary and most noble act of disclosure and communion with a race of intelligent beings, and considers the hypostatic union with creatures the final, fullest, and most fitting means of accomplishing salvation, the "unsurpassable climax of revelation" according to Rahner. Incarnation is also considered as necessary in the case of sinless creatures for the divine purpose of revelation, perfection, and completion of creation, as well as provision of a perfect object for contemplation, as argued by Delio.

The main weakness to the multiple position is the insistence and reliance on the mode of physical incarnation alone, while ignoring other potential methods which God may choose at his will according to the *varied* position (below) betrays a Christological soteriological structure projected into the universe as an unconscious anthropological chauvinism. Although, according to Thomistic tradition,[15] and echoed by Rahner, the human Incarnation was the best, most fitting way to redeem the human race (given the particularities of human nature, history, and divine prerogative); it is not necessarily the most fitting mode in which to redeem other species of various creatures with varying theological 'anthropologies' and historical trajectories. Incarnation cannot, in my view, be held as a requirement, as the only means of divine action for creatures; and other potential combinations of divine person(s) and nature(s) remained unexplored by multiple thinkers.

The *varied* position represents the final phase in the evolution of soteriological possibilities. It proposes this definition of divine activity in redemptive interactions with creatures: in accordance with the Creator's absolute freedom and 'omni-properties', incarnation is one of innumerable possibilities of divinity to manifest, reveal, redeem, complete, and unify with creatures. This view provides the maximum reasonability, flexibility, and feasibility for the divinity to act in worlds and creatures without the necessity to modify orthodox formulations of Christian doctrine for humans. Sin, divine presence, revelation, redemption, and final union can manifest in myriads of forms. Another intelligent species would have an alternate, non-competing economy of salvation which might share essential commonalities with those known to Christianity, but might also diverge and include unknown categories of divine action and creaturely response.

---

15. "A thing is said to be assumable according to some fitness for such a union. Now this fitness in human nature may be taken from two things, viz. according to its dignity, and according to its need. According to its dignity, because human nature, as being rational and intellectual, was made for attaining to the Word to some extent by its operation, viz. by knowing and loving Him. According to its need-because it stood in need of restoration, having fallen under original sin. Now these two things belong to human nature alone. For in the irrational creature the fitness of dignity is wanting, and in the angelic nature the aforesaid fitness of need is wanting. Hence it follows that only human nature was assumable." Aquinas, *Summa Theologica*, III, Q. 1, Art. 1.

The incarnation of a divine person on Earth is in no way minimized or rendered obsolete by a reality of divine creation, revelation, redemption, and final unification with intelligent creatures outside Earth. Rather, the scope of human understanding of divine work is extended and exponentially dynamized in accordance with divine acts within an inestimable species of a possible infinite universe. It is important to illustrate some examples of these possible modalities of extraterrestrial restoration to divine grace. God could simply forgive extraterrestrial transgressions out of divine mercy, without need for any individual or collective sacrifice on the part of the creatures, or demand personal or collective repentance as a condition for forgiveness. He could accept the efforts of a chosen mediator, either a divine person or a creature, who performs a redemptive act or acts with or without the cooperation of repentant sinful members of the community. In the case of a finite creature serving as mediator for a race, only finite satisfaction for sin would be possible. Only to effect an infinite satisfaction would a sacrificial act be necessary by a divine person. A full and adequate reparation could be made by a multitude of conceivable ways, in another type of incarnation by one or more of the divine persons of the Holy Trinity acceptable to God and of infinite value. Also possible to envisage is the work of an angelic creature or messenger, mediator, or representative, a creature appointed by the community or the divinity, and acceptable to God, and/or manifested in a special revelation could act to fulfill the requirements for a finite satisfaction for a society of less powerful and sinful creatures in an unknown way. There does not necessarily have to be other incarnations of one of the persons of the Trinity for an alien race to be saved. Extraterrestrials could be redeemed from any number of sinful states individually, collectively, or through some other unknown process. These conjectures are only a small beginning of what could be considered of extraterrestrial incarnations, indwellings, appearances, apparitions, theophanies, communications, and economies and modalities of salvation.

Section B: The Varied Hypothesis

The *varied* view represents a theological reframing and developmental expansion of Christian doctrine, as a natural maturation in the evolution of theological understanding in the engagement with new knowledge of the scope of the universe. This section will discuss the biblical, scientific, and theological elements of exotheology, followed by a hermeneutic of the *varied* view which will provide scriptural basis for an alternative soteriology of extraterrestrials from humans.

Principles of Exotheology according to the Varied View

In general, exotheology is some respects an exercise in informed speculation, as a product of an interdisciplinary study of extraterrestrials, considering the resources of reason and revelation. Certain cosmic, environmental, and theological constants should be maintained in such speculation if we are to consider our own Christian tradition, as supernaturally legitimate, foundational for intelligent extraterrestrial creatures. Utilizing knowledge gained through astrobiology, scientific knowledge and natural theology would be the best tools to develop methodologies to carefully explore and extrapolate these potentialities. Given the probability of a biological diversity in nature and potentially within the cosmos, we can consider a plurality of religious expressions within our own civilization as evidence of possibly a multiplicity of extraterrestrial religions given the vast variety of cosmic, environmental, morphological, psychological, anthropological, and spiritual particularities of intelligent beings. Within the *varied* hypothesis is the importance to remain open to the range of possibilities by which divinity might create, reveal, provides grace, and redeem and unify creatures. The sciences can illuminate on extraterrestrial morphologies. As the universe and planetary systems capable of supporting lifeforms operate and exist within strictly defined physical constants,[16] divine revelation to humans demonstrates certain constants between creator and creature. Fundamentally, divine relation to creatures as love-gift, divine presence, revelation, sin, grace, history, nature, freedom, mind, and spirit constitute modes of the Christian religion which might manifest in extraterrestrial civilizations, as well as other forms of religion, as examples among our own species indicate. Intelligences could have analogous knowledge of science and religion, each utilized in determining their perspective on the nature of reality. Creatureliness/personhood entails divine relationship and supernatural orientation, as love is motivated to union; and mediated by revelation and grace to free beings providing a framework for consideration of extraterrestrial religion. Religions are fundamentally a relational matrix, in which free beings engage within an historical and situational continuum with divinity and/or divine representative(s). A theological openness, operating within these parameters is necessary when considering the potential actualizations of divine activity with creatures. God's plan of salvation for humans on Earth demonstrates an internal consistency and ultimate unity; we may safely consider that divine disclosure, relation, and salvific action will maintain or exemplify these fundamental aspects. As natural life, manifested within strict environmental constants exhibits a great diversity,

---

16. Strong nuclear force, electromagnetic force, weak nuclear force, and gravity, as well as the particular conditions necessary for supporting life, including sentient life.

one cannot conclude there might not be a similar great diversity of supernatural life. Astronomical science has discovered a universe that reveals laws, structure, and dynamics that are diverse and carefully and systematically organized. There will be divine immanence and transcendence, however not all religions will be necessarily fully realized according to their particular historical timeline pursuant to divine plan and creaturely actions.

There is fruitful ground amongst human religious traditions for considering possibilities of extraterrestrial beings and their religions. We do not consider blindly extraterrestrials or their religions, as rich and diverse human examples provide useful, albeit limited analogs of divine action and relation with creatures which can be subjected to historical and evolutionary processes. Earth religions provide a testimony of the potentialities of religious beings on a particular planet, containing a variety of forms and historical narratives. We should expect in religious beliefs examples of communal relationality and complementarity, as well as plurality within and without religious structures containing a dialectic between persons and persons with deity. According to the Christian model, the relational dynamic and actions of the persons of the Trinity provide key reference points as to divine action in extraterrestrial beings and religions. The divine unity, and consubstantiality of the Persons, each exercising a distinct and separate function by appropriation, but where essence, will, and action are unified can manifest and operate according to varying modes in extraterrestrial religions. As divine activity has only a single source in the cosmos according to Christian revelation, a supernatural extraterrestrial religion may be attributed to the entire Trinity, although the manifestations and operations of the divine in a religion can vary, probably considerably and in no necessary way be Trinitarian in shape.[17] There may be varying activities of a divine person manifested in a supernatural religion, but each are, therefore, legitimate within their own unique expressions. Divine action maintains a unity, while manifesting in diversity.

Although every action of a divine Person is attributed to the entire Trinity, the Spirit, an independent agent, is God's manifest and powerful activity in beings and worlds who provide grace and gift, and through whom supernatural grace is provided to intelligent creatures. An individual and communal journey of the spirit in unification with divinity is likely to be ubiquitous given creaturely nature and a biological life-cycle. Divinity creates and plans, reveals, communicates, and seeks unification with created intelligences. We understand the data of revelation only under human concepts, hence human statements, understanding, and expressions of God

---

17. As evidenced in Judaism.

are limited to analogy and symbol. Eternity is appropriated in Trinitarian formulations to the Father, source of all things; the Logos to the Son, who proceeds by way of intelligence; and fruition to the Spirit, who proceeds through love. Among divine attributes of action and operation, causality and omnipotence is known in the First Person, wisdom and its works throughout the universe to the Second Person, and charity and sanctification to the Third Person.[18] Therefore, as divine revelation of the nature of the Godhead, its internal relations, and divine mission, as revealed to humans cannot fundamentally differ from what supernatural religions are encountered in other civilizations; there may be provided further content, insight, and operational outcomes well beyond human understanding of the mystery of the Trinity. Theological understanding continues to develop, according to the principles outlined in chapter 1. Every action of divinity in creation proceeds from the Persons differently, by counsel, command, or origination. Therefore, as the Triune God was present on Earth during pre-Messianic age, the Incarnation enabled Trinity to be present in an unprecedented way and initiate new action in creatures. Therefore, the mode of origination, presence, and presentation can vary not only in human civilization but extraterrestrial societies as well.

The interpretation of scriptural cosmological perspectives can and should be in conversation with data available through historical criticism and science. The six-day creation myth has been invalidated by geology (although throughout history for many theologians a literal reading was not necessarily fully accepted); the instantaneous creation of the first parents has been replaced by an evolutionary model, widely accepted by scientists and modified by certain Christian denominations to maintain its supernatural impetus. Moses is no longer considered the author of the Pentateuch due to scriptural exegetics. Copernicanism challenged the model of scriptural inerrancy,[19] resulting in a process of a spatially de-centered humanity. The new cosmology and evidence of a single habitable exoplanet would dramatically impact understanding of Earth as a unique creation in a paradigm shift from one world to many, with Home sapiens' place in the wider context reoriented biologically and theologically. New data available from science and other methodologies impact understanding of scripture but does not necessarily negate certain doctrines, but rather modify them towards a more accurate faith statement. Galileo famously remarked, "The intention of the Holy Spirit is to teach us how to go to heaven, not how the

---

18. Denzinger, *Enchiridon*, nn. 2–3, ets., 17, 47.

19. Ps 19:4–6; Eccl 1:15 describe a rising and setting sun in an altered theological understanding of Earth; In Josh 10:10–15, Joshua commanded the Sun, not the Earth to stop during the battle between Israelites and Amorite kings.

heavens go."[20] Many Christians consider the Bible, being a book of faith intended for personal salvation free from errors in matters of faith and morals, but in scientific and historical matters it should be examined in light of extrabiblical sources and data available from the sciences. St. Augustine of Hippo in his *De Genesi ad litteram libri duodecim* argued that where scripture did not agree with the observations of nature these should be understood as allegorical or metaphorical:

> With the scriptures it is a matter of treating about the faith. For that reason, as I have noted repeatedly, if anyone, not understanding the mode of divine eloquence, should find something about these matters [about the physical universe] in our books, or hear of the same from those books, of such a kind that it seems to be at variance with the perceptions of his own rational faculties, let him believe that these other things are in no way necessary to the admonitions or accounts or predictions of the scriptures. In short, it must be said that our authors knew the truth about the nature of the skies, but it was not the intention of the Spirit of God, who spoke through them, to teach men anything that would not be of use to them for their salvation.[21]

Both Augustine and Aquinas held that scripture can have multiple meanings where certain content to texts required future information to illuminate their deeper or more accurate meaning. Aquinas in particular argued that the use of reason contributes to the better understanding of revelation.[22] Scriptural inerrancy, held by early Protestants was considered heretical by the Roman Church, as a literal meaning of the text was considered only the beginning of understanding deeper truths in scripture.[23] The historical-critical method, foundational in Catholic teaching in the twentieth century brought fuller understanding regarding the ancient world that produced the texts. In a book resulting from a bishops teaching conference of the Catholic Church of England and Wales, *The Gift of Scripture*, the bishops noted that, "We should not expect to find in scripture full scientific accuracy or complete historical precision . . . We should not expect total accuracy in the Bible in other, secular matters."[24] Additionally, the *Catechism of the Catholic Church* states, "Many scientific studies . . . have splendidly enriched our knowledge of the age and dimensions of the cosmos, the

20. Machamer, *The Cambridge Companion to Galileo*, 306.
21. Augustine, "The Literal Meaning of Genesis," 42–43.
22. Pope John Paul II, *Fides et Ratio*.
23. Pogge, "Some Notes."
24. Catholic Bishop's Conference, *The Gift of Scripture*; Gledhill, "Catholic Church."

development of life forms, and the appearance of man. These studies invite us to even greater admiration for the greatness of the Creator."[25] In regards the question of scriptural texts describing cosmological ideas Pope John Paul II, addressing the Pontifical Society of Sciences in 1981 stated regarding scriptural interpretation and cosmology:

> The Bible speaks to us of the origins of the universe and its make-up, not in order to provide us with a scientific treatise, but in order to state the correct relationships of Man with God and with the universe. Sacred Scripture wishes simply to declare that the world was created by God, and in order to teach this truth it expresses itself in terms of the cosmology in use at the time of the writer.[26]

Scientific advancement and historical criticism, therefore, provide better understanding of the ancient cosmological world view in which the texts were produced, contrasted with what we now understand as the physical and compositional extent of the created universe. Scripture is ordered to human salvation, and humans were created purposefully with an intellect to understand creation and how it operates.

Hermeneutic of the Varied View

The *varied* hypothesis of exotheology utilizes a historical/critical and scientific, modern cosmological hermeneutical lens as a 'revisionist' model that combines hermeneutical modes of recovery and resistance.[27] The historical-critical analysis of the text will engage a hermeneutic of recovery, while the scientific analysis will engage a hermeneutic of resistance. The combination of these will result in a hermeneutical methodology which will provide a reading of the text where it can be appropriated by the *varied* model set forth. The exotheological biblical hermeneutic for the *varied* view will make reference principally to those scriptural texts traditionally interpreted as evidence for a 'Cosmic Christ' by certain exclusivist, inclusivist, and multiple thinkers in their consideration of soteriologies which include extraterrestrials. As noted, texts of particular importance

---

25. *Catechism of the Catholic Church*, 283.
26. Pope John Paul II, *Cosmology and Fundamental Physics*.
27. In judging appropriateness and feasibility of this hermeneutic, Ian Barbour outlines four criteria for assessing theories in normal scientific research: agreement with data, coherence with other theories, scope (and comprehensiveness and generality), and fertility (is the theory fruitful for generating new hypothesis). These same criteria can be utilized within the framework of religion. See Horrell et al., *Greening Paul*, 44.

include Colossians 1:15–20; Romans 8:18–25; and Ephesians 1:8–10, 1:22, 2:2, 6–11, 3:9–12, 4:10. Typically these refer to a cosmology including principalities and powers, cosmic rulers of darkness, and evil spirits in the heavenly realms. The most important of these texts is the Colossian hymn (Col 1:15–20) with its apparently explicit proclamation of the creation of all things through Christ and the redemption of all things through the blood of his Cross. This text will be the main focus in this section as on the surface it is the most problematic New Testament text for any view that presumes multiple salvation events in the universe. The reading of recovery of this text will illuminate a tension between the cosmic and non-cosmic affirmations by the author, while the readings of resistance will inform the text from the perspective of modern science. Together these will reveal that not only is the 'Cosmic Christ' not the only viable reading to be taken from the text, but that this interpretation has in fact led to the constrained and limited historical resolutions of the difficulties inherent in the text, as theologians' readings of the text in relation to extraterrestrials have tended to result in affirmations of a cosmic natural religion,[28] or either the *exclusivist, inclusivist,* or *multiple* hypotheses.

The fundamental question of how certain scriptural texts, Colossians 1 chiefly among them, are understood in light of the possibility of intelligent extraterrestrials must be considered as a development of doctrinal understanding within the scientific context of a twenty-first century Einsteinian universe, the traditional understanding of the omni-properties of God, and the historical evolution of theology on this subject. Of primary importance is to establish which biblical cosmology was operational in the texts, in order to properly understand certain cosmological statements and their import in interpretation from which inclusive, exclusive, and multiple thinkers universally form the basis of their soteriologies. The cosmological model adopted by the Hebrew writers was a tripartite formula derived from Mesopotamian and Egyptian mythology, and tended to view the Sun, Moon, planets, and stars as celestial beings and a host of heaven that served Yahweh.[29] This cosmological narrative had significant theological impact on the worldview of early Christians who borrowed their cosmology from Jewish theological and cultural foundations and also Hellenic mythology and philosophy. In this way early Christianity could be termed as a development of Hellenistic Judaism with a combination of Jewish tradition and Platonic/Aristotelian/

28. According to Thomas Paine and Steven Dick; See Paine, *The Age of Reason*, 704; Dick, *Many Worlds*, 145, 199–206; Dick, "Toward a Naturalistic Cosmotheology."

29. As described in chapter 2. References to this three-level cosmos are found in Gen 1:7; 7:11; 11:4; Josh 10:13; 1 Sam 2:8; 1 Chr 16:30; Job 9:6; 28:24; Pss 28:19; 93:1; 104:5; Eccl 1:5; Isa 40:22. Also see Ps 148:3–4.

Stoic cosmology.[30] Among the views informing the New Testament was a sense of a savior of both Israel and Gentiles against a malevolent hierarchy of traditional anthropomorphic gods existing as celestial divinities which had corrupted humanity.[31] This Judeo-Christian cosmology existed within the sociopolitical context of ancient Mediterranean civilizations which provided the intellectual foundations of the later western perspective on the human place in the universe *in toto*. Inside this cosmological framework, there is an implicit perspective in Colossians 1 and other texts that there exist no other intelligent *biological* beings in the created order, rather only the existence of non-corporeal intelligences in celestial realms. In 1 Corinthians 15:24 Christ in the end destroys these elemental intelligences, known as the 'elements,' 'principles,' 'powers,' and 'forces,' which dominate the world of man in the present 'evil age,' as the old composition of power in the cosmos is ending with the advent of Christ. Further, Paul describes his vision of paradise as located in the "third heaven";[32] and his mention of angels, principalities, and powers, and language of 'height' and 'depth.'[33] He speaks of a time "when we were children, we were enslaved by the 'elements of the cosmos,'"[34] indicating beliefs in occult forces as custodians of the cosmos and human history. As Paul's astronomical vocabulary is a combination of Greek cosmology and Jewish apocalyptic tradition,[35] in his mind, it seems that the portion of the sublunar realm which controlled and enslaved humans would be demolished by the new reign of Christ.

Within this cosmic context described by Paul, central is his use of the phrase of "all things" (*ta panta*), which has often been interpreted to mean all created things on Earth and in the heavens. However early interpreters such

30. Plato described a cosmos, adopted by the Pythagoreans, in his *Timaeus*: a transcendent god created a spherical universe composed of four elements in which Earth was placed at center and orbited by seven planetary rings or spheres, each realm governed by heavenly gods and surrounded by a rotating sphere of fixed stars.

31. Davidson, "The Structure of Heaven and Earth." An early, likely first-century text known as *The Ascension of Isaiah* describes Isaiah taken to seven levels above the firmament, a clearly derived from Platonic models of the seven planets. Here Isaiah learns Christ will take the form of a man and be crucified by the god of that world, conquering the angel of death and ascending. In this scenario, the world is corrupt, and air below the heavenly spheres is filled with nefarious spirits, consistent with Platonic cosmologies of Aristotle and Plutarch. See Wright, *The Early History of Heaven*, 158.

32. 2 Cor 12:2–4.

33. Rom 8:39. In Jewish tradition, the 'first heaven' (the atmosphere of the Earth), a second (the heaven of the stars), and third (the dwelling-place of God) are delineated.

34. Gal 4:3, 8–10.

35. 1 Enoch 61, where the heavenly Son of Man judges the angels and Daniel 7 LXX where 'powers' of the world submit to the Son of Man. See van Kooten, *Cosmic Christology*, 93–94; Lewis, *Cosmology and Fate*, 59.

as St. Chrysostom explained that reconciliation of 'all things' in the hymn meant the reconciliation of humans with angels, as heaven was already at peace with God;[36] and Aquinas referred to the reconciliation of 'all things' only in terms of human beings.[37] Some modern commentators also limit the scope of the hymn by its context, for example, in its reference to a church (v. 18), and addresses to the letter's hearers who have been saved, (vv. 21–23) where the cosmic scope is qualified by this more limited context. The focus of the author of those phrases seems anthropological and ecclesial, rather than cosmological.[38] Thomas O'Meara also understands the church and not the cosmos as the central topic, and that the focal point of the Letter lies in the future and not the past, as Christ as the firstborn of a new creation in eternal life.[39] Jerome Murphy O'Connor also provides an interpretation with a focus on the incarnational and ecclesiological as an effort to redirect certain erroneous cosmological and angelological teachings, "He directs the reader's attention to the physical existence of him who is now the Risen Lord . . . Paul's insistence that Christ is present in him and in all members of the Church draws the cosmic dimension of the Christological reflections of Colossians down into ecclesiology."[40] Other commentators also see the focus of the Letter as on the salvation of human beings from the enslavement to personalistic outside forces. C. F. D. Moule commented that Paul had "readily resolved *ta panta* into personal beings" and this interpretation is supported by the hymn in 1:16, which also references disarming rulers in 2:15 and worship of angels in 2:18.[41] In this case, commentator Roy Yates questioned, "Since reconciliation properly relates only to persons, how can it be applied to the universe?"[42] Matthew Gordley argues the hymn utilizes Greco-Roman philosophical ideas with a worldview informed by Jewish concepts,[43] and Lars Hartmann reminds us that the author would have envisioned himself within a "cosmos that was alive, filled and swayed by all sorts of living powers" and that "the planets were living creatures, belonging to

36. Chrysostom, *Homilies on Colossians*, Homily 3.
37. Aquinas, *Summa Theologica* III, Q. 22, Art. 1.
38. See Lohse, *Colossians and Philemon*, 41, 61; Schweizer, *The Letter to the Colossians*, 55–88; Bruce, *The Epistles to the Colossians, to Philemon, and to the Ephesians*, 74–77; Schillebeeckx, *Christ*, 187, 194; van Kooten, *Cosmic Christology*, 127. Lohse suggests the theology of Cross "arrests all attempts to utilize the hymn for the purposes of a natural or cosmic theology." Lohse, *Colossians and Philemon*, 60 n. 211.
39. O'Meara, *Vast Universe*, 46.
40. Murphy-O'Connor, "Tradition and Redaction in Col 1:15–20," 237, 241.
41. Moule, *The Epistles to the Colossians and to Philemon*, 71.
42. Horrell et al., *Greening*, 90.
43. See Gordley, The *Colossian Hymn in Context*.

the same world as man"; he does not in Hartmann's view mention the rest of creation.[44] During this period, Stoic cosmology (which appeared influential as a heresy among the Colossian church) was also influential as a perspective on the operation of the universe. Based on a cyclical movement of the cosmos which eventually returns to its primordial state, it was a widely accepted worldview, and its cosmological framework lent itself to Christological reflection.[45] Within it, there is no movement of creation towards unity with the divine or final fulfillment, rather an unending repeating pattern of a universe in tension which never achieves finality. In the Christian context, the fall of man is the cause of cosmic disorder which Christ restores and resolves, providing a linear movement towards completion. The Stoic cyclic model is broken, and the idea of a cosmic Christ is invoked, which brings meaning and finalization to the workings of the cosmos in him who brought it into being. Therefore, the focus of the author as he addresses the Colossian heresy is of human beings subjected to personalistic spiritual entities, which created mankind's bondage to sin, and cosmic language is utilized to demonstrate these powers were now subjected to Christ. Consequently it makes more sense to understand *ta panta* as referring to humans and their world (the word 'nature' is also not included in the hymn). Accordingly it appears, considering the cosmic language in conjunction with its likely anthropocentric focus, together with references to elemental spirits, and the claim that his hearers have been reconciled, providing a much more limited environmental scope of the hymn.

If therefore, the phrase *ta panta* is used with a focus on the human relationship to creation and reconciliation; then understanding the Letter's actual focus, and given its sociohistorical setting within a contemporary Middle Platonist philosophy, the understanding of Christ's role in creation is better understandable in this context.[46] When the use of the phrase *ta panta* is understood within this framework of the Letter taken as a whole, there is raised the consequential possibility of a more limited emphasis. So the notion of "all things" as related to the Colossians' anthropological concerns becomes entirely plausible in light of our present scientific understanding. Therefore, there is an important interpretive tension between cosmic and non-cosmic readings of *ta panta*.

This cosmological picture therefore, implies a limited cosmological perspective on the physical and spiritual domain of the God-man Christ and human beings. We are faced, then, with two possible interpretations of

---

44. Hartmann, "Universal Reconciliation, 112, 120.
45. Balabanski, "Critiquing Anthropocentric Cosmology," 158.
46. Van Kooten, *Cosmic Christology*, 126.

'all things' in the Colossian hymn. It is notable that, even outside considerations of possible extraterrestrial life, the impact of the scientific world view and the known extent of the cosmos and creation caused a number of scholars to consider Paul's cosmic statements as a clear reference to and concern about human soteriology rather than cosmology. According to Rudolf Bultmann, although Paul utilizes cosmic language, his actual focus was anthropological and soteriological, rather than to make definitive statements about the cosmos. Bultmann has provided a classic representation of this approach and has been followed by many other theologians.[47] In doing so, he argues that the Greek "*kosmos*" was used to denote heaven and Earth, and was utilized by Hellenic Judaism, including Paul. Bultmann argues, "*Kosmos* is not a cosmological term here, but an historical one, so it also is in the numerous passages where it is used in the sense of "the world of men', 'mankind'—a usage, moreover, which Hellenistic Judaism shows."[48] *Kosmos* is not Paul thinking of a cosmic stage, but as the sphere of human relationships.[49] "Thus in 1 Cor. 4:9 when Paul explains that "we have become a spectacle to the '*kosmos*', to angels and to men" he refers to the persons within this context, but not the context . . . *kosmos* contains a definite theological judgment; as an antithesis to the sphere of God, denoting the totality of human possibilities and conditions of life or implies persons in their attitudes and judgments. It is the sphere of earthly life and cares of this world."[50] Therefore, it is an eschatological-historical concept; it denotes the world of men, as well as the sphere of anti-godly power who dominates humanity."[51] It is the sphere "of demonic powers, angels, principalities, and powers, the rulers of this age."[52] Accordingly, the cosmic language is *incidental* and *contextual*, enabling Paul to express his new Christian soteriological teaching of salvation from impersonal forces outside the individual. By reducing Paul's cosmic language to its anthropological referents a more

47. See Bultmann, *Theology of the New Testament*, 227–32, 254–59; Robinson, "A Formal Analysis of Colossians 1:15–20," 270–87; Hegermann, *Die Vorstellung vom Schopfungsmittler*; Kasemann, *Essays on New Testament Themes*, 149–68; Gabathuler, *Jesus Christus*; Pollard, "Colossians 1:12–20," 572–75; Lohse, *Colossians and Philemon*; Hübner, *An Philemon*, 59. Pollard states that "the cosmology, if it is cosmology, it totally subservient to soteriology and by making it thus Paul runs true to form" (Pollard, "Colossians 1:12–20," 573). Hübner points out that the intent of the passage is to demonstrate there are rivals to Jesus in the personalities of cosmic religion.

48. Bultmann, *Theology of the New Testament*, 254–55.

49. 2 Cor 1:12.

50. Bultmann, *Theology of the New Testament*, 254.

51. 1 Cor 3:22; 7:31; 1 Cor 1:20, 27. See Bultmann, *Theology of the New Testament*, 253–54.

52. 1 Cor 2:6, 8; Gal 4:3, 9.

accurate understanding of his teaching is, therefore, comprehended.[53] As such, Paul's language is merely the context for salvation from spiritual forces which have enslaved humanity which cause sin and death. Further, Paul's adoption of cosmic language is highly situational and an ad hoc response to those Colossians advocating erroneous astronomical and cosmological doctrines in the Gentile Greco-Roman world.[54] These were composed of heresies containing elements of astrophysical and zodiacal dogmas spread by false teachers; and if not for these, language of Christ's relation to the cosmos might have been absent from the Apostle's teaching.[55] False teachers, suspected as Judaizers were attempting to insert Christ into a metaphysical scheme among other elemental powers, (in an effort to understand where Christ should be located within the pantheon of powers) whereby Jesus was depicted as an intermediary creature between corporeal beings and spiritual created beings and lower than the angels, whose powers were superstitiously worshipped. These same teachers, therefore, claimed Christ was not enough for salvation and did not free humans from cosmic powers nor give access to the wisdom of God. Paul corrected this by placing Christ above the angels, not below them. He made recourse to the language of contemporary mythology to convey the significance of Christ within this domain; his cosmic Christology was thus a pointer to the salvific intentions of God which predated creation.[56] His placing Christ at the top of this hierarchy was deliberate, as his position of superiority over all earthly and celestial forces was critical for the Colossian church, and therefore, the most significant and central focus of the hymn. In accordance with this, van Kooten concludes that Paul's references to "Christ's cosmic rule, as the author of Ephesians makes plain, does not yet extend over the entire physical cosmos. It began to be implemented when Christ was resurrected and installed in heaven. The benefit of this rule, however, is still limited to the church because Christ has only been given as cosmic head to the church."[57]

Here we see the influence of wider consideration in the letter as a whole on the reading of the hymn; its focus on the lives of the Colossians and references to spiritual beings elsewhere in the letter provide a more limited context for exotheological appropriation. Therefore this hermeneutic would propose that Colossians 1 reached its final form in a way designed to refute

---

53. Helyer, "Cosmic Christology and Colossians 1:15–20," 235–40.

54. See Eltester, *Eikon im Deuen Testament*; Becker, *Paul*, 380.

55. See Guthrie, *New Testament Theology*, 353; MacArthur, *The MacArthur Study Bible*, 1782; Becker, *Paul*, 380.

56. Helyer, "Cosmic Christology and Colossians 1:15–20," 237.

57. Van Kooten, *Cosmic Christology*, 157.

an opposing philosophical system that did not acknowledge Christ's headship over the cosmic landscape. Accordingly, in the *varied* perspective, Paul's description is of a headship of Christ as Logos over all creation, but whose redemptive act in Jesus is limited to humanity contained within a geocentric cosmology, and including the terrestrial physical world and a proximate spiritual world.[58] The *varied* view argues for a hermeneutic which considers the biblical text in its proper historical context as well as in relation to modern science, which demonstrates a vast universe, while leaving open for the possibilities of God's action throughout a greater cosmos. The historical cosmological context of the Colossians text, written in a pre-scientific Hebraic/Platonic/Aristotelian/Stoic cosmology did not include the notion of extrasolar planets inhabited by putative intelligent beings within myriad galaxies, but rather a "universe" composed of the Earth and extremely limited encapsulating heavenly sphere. Given the ambiguity of reference of his terminology of 'all things' in Colossians, it is reasonable to propose that Paul's language is not a literal account to describe the physical extent or composition of the material universe but rather a literary device for emphasizing the power and glory of God within a pre-scientific description of what was in antiquity a cosmological model of all known creation. The message of a divine redeemer who has taken on human form to destroy demonic and occultic forces born of Judaic tradition and Hellenic mythology is inseparable from the cosmological worldview which preceded it and gave it form. Historically, this 'cosmic' Christology has been understood to reveal Christ as the preexistent agent and redeemer of all creation and all beings contained within it. A re-reading of the text in light of the historical and scientific data makes it exegetically plausible to read Colossians and similar passages within a limited scope of Earth and its environs, as the world in front of the text affects our appropriation of it to prefer this interpretation.

These considerations cast serious doubt on the plausibility of an inclusivist view in utilization of the Colossians text given a vast cosmos. Secondly, as the exclusivist view makes no attempt to incorporate intelligences outside Earth into divine earthly revelation, it reveals the difficulty of holding an exclusive argument for a singular divine election of humans, forcing the question as to whether it is more plausible (and responsible) for Christian

---

58. Col 1:15–20, a hymn of praise of Christ's dignity as God and man, and of Christ's pre-eminence over all natural creation and supernatural salvation. "In him all things were created" (v. 16) (cf. John 1:3). Christ has reconciled all things, the world and mankind, including Jews and Gentiles to God. All celestial powers are under his authority. He is the first born of all creation, the one mediator between God and man (1 Tim 2:5). "Born of the Father before time began . . . , begotten, not made, of one being [consubsantial] with the Father" (Nicean-Constantinopolitan Creed).

theologians to consider alternatives given the changed context of humanity's spatial and theological place within creation. Knowledge gained by science, as well as exegetical considerations of Paul's context lead us to consider this new reading, which can be appropriated within the arena of exotheological formulations which seek to explain Christ's role in a universe which possibly includes other intelligent beings wholly unlike human beings, with their own histories which could predate humanity by billions of years. The difficulty of a once-for-all redemption within all planetary contexts leads to a preference for a limited context for Colossians of the defeat of earthly and a proximate spiritual world. This is an exegetically plausible interpretation in light of the present scientific knowledge and where it inevitably leads theology to a more accurate perspective of the text.

In conclusion, the *varied* hypothesis, in reading scripture through the lens of our knowledge of the vastness of the universe, views Paul's language as chiefly concerned with a human soteriological message and does not include a conception of an expansive universe, but rather a physical domain contained within a geocentric and anthropomorphic cosmological model with celestial realms and a physical heaven in close proximity to Earth. The scientific perspective has led to an understanding of Colossians which provides for a substantially nuanced presentation of cosmic Christology. The Pauline hymns and related high Christological verses were written within a pre-scientific cosmology, and have in modernity, interpreted by *inclusivist* and *exclusivist* thinkers to reveal the critical weakness of these views: principally, how the earthly Jesus and singular Christian message is known and applied to beings occupying a vast universe in time and distance. A thorough evaluation of each soteriological formulation in chapter 5 concludes the *varied* hypothesis is preferred over these other historical interpretative methodologies, and can be sustained within orthodox Roman Catholic Christological and soteriological teaching.

Section C: Christology and Intelligent Extraterrestrials

This section will present an exotheology to accommodate the possibility of divine action in extraterrestrial civilizations, including incarnation. It will discuss means by which the divine relationship might be established, communicated, maintained, and finalized with intelligent extraterrestrials in relation to the example of the divine relationship as mediated by Christ to humanity. The discussion will develop the previous section, focusing in particular on the theology of incarnation, and will include themes of creation, original sin, and the redemption. There are several fundamental questions with regard to divine incarnation and extraterrestrials: What is the relation of incarnation to other forms of revelation? What is the relation between revelation and redemption? Is incarnation always linked with revelation and salvation among other intelligent beings? Is the primary work of incarnation that of revealer of divine truths or only to redeem creatures? Are multiple incarnations necessary for extraterrestrials? Further, which nature is assumed, when and how many incarnations are possible in the universe? Can one rational nature can be assumed by more than one divine person simultaneously, and conversely, can more than one rational nature be assumed simultaneously by one or more persons? As indicated earlier, the Catholic theological foundations for examining these questions will follow Thomistic incarnational theology, with contributions of theologians who have considered multiple incarnations. Further exploration on the plausibility of the *varied* view will be incorporated along with the examination of the *multiple* hypothesis.[59]

Aquinas taught God is most generous to the highest degree,[60] and established the divine motive for both creation and the Incarnation as unlimited goodness diffusing itself by bestowing goodness on others; "God is a living fountain, a fountain not diminished in spite of its continuous flow outwards."[61] Zachary Hayes described a God who created not out of need but a desire to manifest something of the mystery of the divine truth,

---

59. For other references on the question of multiple Incarnations, see Adams, "The Metaphysics of the Incarnation"; Adams, "Christ as God-Man, Metaphysically Construed," 239–63; Adams and Cross, "What's Metaphysically Special about Supposits?", 15–52; Arendzen, *Whom Do You Say-?*, 161; Baker, *Jesus Christ—True God and True Man*, 47; Bonting, "Theological Implications of Possible Extraterrestrial Life," 587–602; Brazier, "C. S. Lewis," 391–408; Craig, "Flint's Radical Molinist Christology Not Radical Enough," 63; Crisp, "Multiple Incarnations"; Crisp, *God Incarnate*, ch. 8; Cross, *The Metaphysics of the Incarnation*, 230–32; Davies, "ET and God," 112–18; Fisher and Fergusson, "Karl Rahner and the The Extra-Terrestrial Intelligence Question," 275–90; Flint, "The Possibilities of Incarnation," 307–20; Flint, "Molinism and Incarnation," 187–207; Freddoso, "Logic, Ontology and Ockham's Christology," 293–330; Freddoso, "Human Nature, Potency, and the Incarnation," 27–53; George, "Aquinas on Intelligent Extra-Terrestrial Life," 239–58; Hebblethwaite, "The Impossibility of Multiple Incarnations," 323–34; Hebblethwaite, *Philosophical Theology and Christian Doctrine*, 74; Kereszty, *Jesus Christ*, 382; Kevern, "Limping Principles," 342–47; Mascall, *Christian Theology and Natural Science*, 40–41; Morris, *The Logic of God Incarnate*, 183; O'Collins, "The Incarnation," 1–30; Pawl, "Thomistic Multiple Incarnations," 359–70; Pohle, *Christology*, 136; Poidevin, "Multiple Incarnations and Distributed Persons"; Schmaus, *Dogma 3*, 241–42; Sturch, *The Word and the Christ*, 43, 194–200; Ward, *God, Faith, and the New Millennium*, 162.

60. Aquinas, *Scripta super libros*, D. 3, Q. 4, A. 1, ad 3.

61. Aquinas, *Super Evangelium*, Ch. 1, Lect. 3, 20; *Summa Theologica* I, Q. 20, Art. 2; and III, Q. 1, Art. 1.

goodness, and beauty outwardly and to bring forth creatures capable of participating in the splendor of divine life.[62] Intelligent beings are is some sense the culmination of evolved creaturely properties in the created universe; all creatures bear divine traces of goodness in their existence and bear the image of God in their absolute freedom and ability to know and commune with the Creator. The creation of free creatures entails the possibility of sin, in any number of manifestations and degrees; Teilhard de Chardin viewed evil is a necessary consequence of God's creative activity in fashioning free beings, but offered no explanation for the universal origin of evil. He described evil as terrestrial as much as it is cosmic:

> If there is an original sin in the world, it can only be and have been everywhere in it and always, from the earliest of the nebulae to be formed as far as the most distant.[63]

Given a populated universe of free intelligent creatures, it is probable sin pre-existed humanity in other spiritual or creaturely forms. In response to sin, the patriarchs were subjects of covenants, the Hebrews Mosaic Law, and Christians had access to grace and redemption through Christ to contend with and overcome a human propensity for evil. In the New Testament, Paul locates the origin of sin in the Genesis account in order to reveal the motive of the redemptive act in Jesus Christ.[64] The everlasting covenant made possible through the sacrifice of an incarnated God-man was the Creator's decisive act on Earth for collective human salvation. In regards salvation of sinful or non-sinful intelligent extraterrestrials, we first examine the essential elements of incarnation in the human species in order to understand other possible types of redemptive divine action in creatures. The possibility of extraterrestrial incarnations/manifestations of the divinity outside the human sphere can be considered in relation to earlier examinations on the necessity and fittingness of an incarnation on Earth. The question of the necessity that a saviour/messiah be incarnated to share in human nature to save humanity; that is, salvation from the 'inside,' was examined by certain patristic authors who insisted the Incarnation was necessary to save humanity from sin. Irenaeus expressed this formulation of the necessity of the Incarnation in flesh:

> If a human being had not overcome the enemy of humanity, the enemy would not have been rightly overcome. On the other side, if it has not been God to give us salvation, we would not

---

62. Hayes, *Bonaventure*, 112; Hayes, "The Meaning of Convenientia," 78.
63. Teilhard de Chardin, *The Phenomenon of Man*, 286.
64. Rom 5:12.

have received it permanently. If the human being had not been united to God, it would not have been possible to share in incorruptibility. In fact, the Mediator between God and human beings, thanks to his relationship with both, had to bring both to friendship and concord, and bring it about that God should assume humanity and human beings offer themselves to God.[65]

Other early patristic authors insisted on the necessity of an incarnation for human redemption. Tertullian and Origen in the second and third centuries were early formulators of arguments for the Incarnation, as was Basil of Caesarea in the fourth.[66] The teaching can be best expressed in its classical form by Gregory of Nazianzus in the fourth century as 'the unassumed is the unhealed.'[67] Leo the Great stated the equivalent, that Christ had to share in our humanity (through Mary's flesh)[68] in order that all flesh be divinized; as a redemptive battle "fought outside [our] nature" would not have succeeded in deliverance from the power of evil:

> If the new man, made in the likeness of sinful flesh, had not taken our old nature; if he, one in substance with the Father, had not accepted to be one in substance with the mother; if he who is alone free from sin had not united our nature to himself, - then men would still have been held captive under the power of

---

65. Irenaeus, *Adversus haereses*, 3.18.7; see 3.19.1.

66. "If the Lord did not come in our flesh, then the ransom did not pay the fine due to the death on our behalf, nor did he destroy through himself the reign of death. For if the Lord did not assume that over which death reigned, death would not have been stopped from affecting his purpose, nor would the suffering of the God-bearing flesh have become our gain; he would not have slain sin in the flesh. We who were dead in Adam, would not have been restored in Christ." (*Epistola* 261.2 in Bettenson, *The Later Christian Fathers*, 70). *Homoousios* is the shared identity of the three divine persons as regards the substance of God, the Father and the Son are the same. Basil wrote, and was accepted by the First Council of Constantinople, of one *ousia* (identical essence) and three *hypostaseis* (individual personal substances) in one God. The union of Christ took place in the persons and not in the natures; two complete substances are united in one hypostatic union as defined in the Council of Chalcedon. Baxter, "Chalcedon, and the Subject of Christ," 9.

67 Gregory of Nazianzus, in his *Epistola* 101.32, affirmed the full humanity and full divinity of Christ in his arguments against Apollinarius. This formulation can be applied to support the argument for the necessity of the incarnation for human redemption.

68. On the double generation of the Son: in his divinity before all ages and his humanity born of the Virgin Mary. Prefigured in a kerygmatic fragment cited by Paul (Rom 1:3-4) and almost articulated as such by Ignatius of Antioch (*Epistola ad Ephesios*, 7.2), this theme of the double, eternal/temporal generation of the Son flowered with Irenaeus (*Adversus haereses*, 2.28 6; 3.10.2) and a century later even more clearly with Lactantius (*Divinae institutiones*, 4.8.1-2). See O'Collins, *Christology*, 166.

the devil. We would have been incapable of profiting by the victor's triumph if the battle had been fought outside our nature.[69]

There was also wide agreement that only a divine person would be a fitting instrument by which the depraved human race could be fully reconciled to God. According to Cyril in his Christological dialogs, "If Christ had only received his own divine filiation by gifts without possessing it by natural right, how could he bestow on others the power to become children of God?"[70] Anselm also argued for the necessity of the divine incarnation as all sins offend an infinite God—therefore, no reparation by a finite, imperfect creature could ever provide infinite satisfaction, as humans already owe everything to God. Only a God-man could provide reparation of infinite value as Christ in the hypostatic union provides the freely-offered sacrifice, accepted by God for the sinfulness of humanity. This is in accord with Aquinas, "the goodness of someone who is merely a man cannot be the cause of good for the entire race,"[71] and Irenaeus's "No other being had the power of revealing to us the things of the Father, except his own proper Word."[72] This argument is further extended by stating the revealer must be humanly visible in order to properly and fully reveal divinity to humans; as noted by Gregory of Nazianzus, "It was necessary for sinful humanity to be "fashioned afresh . . . by one who was wholly man and at the same time God."[73] Accordingly, Jesus provides the fulcrum between the divine and the earthly realm; he must be divine and human in order to fully exist on both planes; because Christ is divine, he can perfectly mirror divinity and reveal it to humans. Because he is human, humans can identify the divine materially and perceive him as a model of human perfection. Through the combination of divinity and humanity in Jesus Christ, humanity enters the life of God; in this way humans are divinized by the reception of grace in the divine-human relationship in their movement towards ultimate perfection in eternal glory. Therefore, "God by assuming flesh does not diminish his majesty; and in consequence did not lessen the reason for reverence toward him which is increased by this further knowledge of him. On the contrary, from the fact that he willed to approach us through the assumption of flesh he attracted us thereby to know more of him."[74]

69. Leo the Great, *Epistola* 31.2; Bettenson, *The Later Christian Fathers*, 70.
70. Cyril of Alexandria, *Quod unus sit Christus*, 738c, e; 762; 768c-769a; 771c; 773a.
71. Aquinas, *Summa Theologica*, III, Q. 2, Art. 11, resp.
72. Irenaeus, *Adversus haereses*, 5.1.1.
73. Gregory Nazianzus, *Epistola* 261.2; 70.
74. Aquinas, *Summa Theologica*, III, Q. 4, Art. 6.

Aquinas questioned the necessity of the Incarnation as a result of earlier scholastic debates relating to the nature of the Redemption. His question, 'Whether It Was Necessary for the Restoration of the Human Race that the Word of God Should Become Incarnate?'[75] is relevant to the question of extraterrestrial intelligence and the necessity of an incarnation or other divine activity outside Earth. Contrary to the early patristic authors, he concluded an earthly incarnation was not absolutely necessary for humans making it arguable whether incarnation is equally unnecessary for external intelligences, "For God in His omnipotent power could have restored human nature in *many other ways*."[76] However, Thomas did not describe other redemptive modalities. The Incarnation on Earth for human beings can be considered to be necessary respective to humans, however not in the absolute sense. In the case of humans, provision was made for their salvation after their fall from grace, whereas the fall of the angels was complete and irrevocable.[77] Although Aquinas did not discuss extraterrestrials according to our modern view, his thought did not discount the possibility of God creating other rational natures outside of humanity.[78] He detailed the reasons why it was fitting for human nature and not the angelic to be assumed by a divine person, noting the dignity and rationality of each, both resembling the divine nature and surpassing that of the lower non-rational animals; with each possessing natural capacities which grace does not

75. Aquinas, *Summa Theologica*, III, Q. 1.

76. Aquinas, *Summa Theologica*, I, Art. 2.

77. Aquinas taught the angelic economy of salvation, examined within the context of extraterrestrial intelligence demonstrates a mode of salvation in contrast to Homo sapiens. Angels were created in a graced state with free will, knowledge, and infallible reason; individually offered elevation beyond their created and gifted capacities within the beatific vision after passing probation *in via*. The angelic revolt and fall from grace was individually determined, in contrast to the collective guilt borne by humanity resulting from Adam's sin. The angelic journey consisted of a single step and one eternally binding choice, as more perfect beings in possession of superior knowledge, the decision to serve or defy the Creator was full and final without opportunity for repentance. Tradition holds the majority of angels accepted the offer to participate in the divine life, while the latter aspired to likeness of God; rather than meeting their proper end by means of God's grace sought divine likeness through their own power. Aquinas taught that after the fall from grace into sin, Lucifer and other fallen angels committed the sin of envy, demonstrated by their continued efforts to thwart humanity, to which they were intended to be intimately attached, from reaching their same intended end of sharing glorified existence. See Wawrykow, *The Westminster Handbook to Thomas Aquinas*, 1–4.

78. Aquinas believed it not incompatible with faith that heavenly bodies existed as rational souls; this belief, held by some was the result of a limited cosmological knowledge.

destroy, but rather are perfected as a gift of God.[79] However, he concluded that angels, although possessing a dignity of being assumed, were not fitting due to their lack of need.[80] He indicated the conditions for a nature (in discussing human and angelic natures) to be assumed: as rational and intellectual, possessing dignity and need—clearly within the realm of possibility for intelligent extraterrestrials. Also, he argued the Father and Son can assume different human natures. This is the justification for Aquinas's multiple incarnations by different divine persons:

> Whatever the Father can do, that also can the Son do. But after the Incarnation the Father can still assume a human nature distinct from that which the Son has assumed; for in nothing is the power of the Father or the Son lessened by the Incarnation of the Son. Therefore, it seems that after the Incarnation the Son can assume another human nature from the one He has assumed.[81]

Given that the Second Person possesses an infinite nature, and by the assumption of a finite human nature within the infinite nature of the divinity, it is possible that other, foreign finite natures would serve as equally compatible to be assumed. In each case, the infinite nature of the divine person is joined to a finite nature, in which the divine personality and uniqueness is bestowed upon the creature and divinity is limited by appearance but not in its nature. Aquinas asserts that we cannot restrict God to one incarnation; the infinity and incomprehensibility of God cannot be fully contained within the finite—other incarnations allow this possibility. Given the nature of the second divine Person (if this is the Person we are considering will be incarnate, given the example on Earth), there is no difficulty of an unlimited number of finite natures being joined to it, while not being confused or joined to each other within the divinity, just as individual humans are not joined or confused in the life of grace. St. Maximus provided impetus to the argument that the Logos can divide himself, and that *logoi* find their being coming from one Logos as one body is manifested in varying forms, and therefore, it could be argued that multiple *logoi* are capable of taking on of a multitude of natures:

> [T]he Logos provides all to all who are worthy proportionate with the quality and quantity of each one's virtue ... the Logos divides Himself indivisibly ... [and is] 'paradoxically' present to each of the participants according to worth." "The one Logos

---

79. *Gratia non tollit naturam sed perficit.*
80. Aquinas, *Summa Theologica*, III, Q. 4, Art. 1.
81. Aquinas, *Summa Theologica*, III, Q. 3, Art. 7.

divides Himself, neither by becoming actually divided, nor in the way a Proclean monad divides itself—i.e. as giving rise to several participated entities—but by directing His logoi as His acts of will towards the creation of a plurality of essences. In this way he divides Himself in His creative activity in relation to many things and remains Himself, as the personal subject of this creative will, an undivided unity.[82]

Therefore, we cannot place human limitations on God's powers and freedom throughout a universe containing potential intelligent beings; divine creation and divine revelation, and divine incarnation can exist among many rational creatures. This is in accordance with Aquinas[83] and Congar, "Earth should not limit divine power. There may well be other incarnations of the divine persons or Trinity of infinite persons."[84] Aquinas described this limitlessness of the ability of divinity to incarnate:

> What has power for one thing, and no more, has a power limited to no one. Now the power of a Divine Person is infinite, nor can it be limited by any created thing. Hence it may not be said that a Divine Person so assumed one human nature as to be unable to assume another. For it would seem to follow from this that the Personality of the Divine Nature was so comprehended by one human nature as to be unable to assume another to its Personality; and this is impossible, for the Uncreated cannot be comprehended by any creature. Hence it is plain that, whether we consider the Divine Person in regard to His power, which is the principle of the union, or in regard to His Personality, which is the term of the union, it has to be said that the Divine Person, over and beyond the human nature which He has assumed, can assume another distinct human nature.[85]

We understand that in referencing human nature, Aquinas is speaking of a rational creature, which has specific 'need' of redemption. This can, therefore, be applied to many rational creatures which may exist. In this case, each created intelligent species touched by incarnation would have a similar position in relation to the divine Person; however biologically, culturally, and religiously would subsist proper to its own world. As no change in the Godhead takes place with the Incarnation on Earth, if the Incarnation was accomplished not by the taking of human flesh but of the taking up

---

82. Tollefsten, *The Christocentric Cosmology of St. Maximus the Confessor*, 217.
83. Aquinas, *Summa Theologica*, III, Q. 3, Art. 7.
84. Congar, 188. See Congar, "Preface," 8–11.
85. Aquinas, *Summa Theologica*, III, Q. 3, Art. 7.

of humanity into God, then other non-human beings can be incorporated into the divinity and be united hypostatically to the Second Person or other divine Persons. God can manifest as himself or as the entire Trinity, or in other ways unimaginable to us. Aquinas says:

> The Incarnate Person subsists in two natures. But the three Persons can subsist in one Divine nature. Therefore, they can also subsist in one human nature in such a way that the human nature be assumed by the three Persons.[86]

> Whatever the Son can do, so can the Father and the Holy Ghost, otherwise the power of the three Persons would not be one. But the Son was able to become incarnate. Therefore, the Father and the Holy Ghost were able to become incarnate.[87]

> Consequently, in order to judge of a word's signification or co-signification, we must consider the things which are around us, in which a word derived from someform is never used in the plural unless there are several supposita. For a man who has two garments is not said to be "two persons clothed," but "one clothed with two garments"; and whoever is designated in the singular as "such by reason of the two qualities." Now the assumed nature is, as it were, a garment, although this similitude does not fit at all points, as has been said above (2, 6, 1). And hence, if the Divine Person were to assume two human natures, He would be called, on account of the unity of suppositum, one man having two human natures.[88]

Therefore, Aquinas argued for the possibility of multiple incarnations of the same divine person, as "one clothed with two garments." If it is possible for the Second Person of the Trinity to be a member of more than one race, any person of the Holy Trinity could become incarnated to redeem a race as each divine person shares the same divine nature.[89] He further explains "God could have chosen to assume a glorified body, but since man has three states—innocence, sin, and glory; Christ assumed from the state of glory the beatific vision; from the state of innocence, freedom from sin; from the state of sin, the necessity of being subject to

---

86. Aquinas, *Summa Theologica*, III, Q. 3, Art. 5.
87. Aquinas, *Summa Theologica*, III, Q. 3, Art. 5.
88. Aquinas, *Summa Theologica*, III, Q. 3, Art. 6.

89. Aquinas, *Summa Theologica*, III, Q. 3, Art. 7. The Word is not limited to a single hypostatic union.

the penalties of life."⁹⁰ However, as we saw above, incarnation in a rational animal must be considered only one of a myriad of possibilities of divine manifestation, communication, and presence. Therefore, the God-human, Jesus Christ can be understood as a 'type' considered within an unknown number of mediators and redeemers throughout civilizations. There could be partial and implicit manifestations of God in other times and places in extraterrestrial histories in a variety of ways, degrees, and intensities. The Incarnation is a final, culmination of actions among a historical trajectory of earthly divine acts culminating in Christ;⁹¹ however these acts did not restrict God from choosing other means of redeeming humanity. In addressing the fittingness of the Incarnation, Aquinas emphasized the repairing of humans themselves by means of supernatural elevation by grace, rather than specific focus on sinful acts; he also maintained God could have pardoned sin without a fully adequate satisfaction: "In satisfaction one attends more to the affection of the one who offers it than to the quantity of the offering."⁹² Incarnation accordingly was not necessary for salvation, but it was necessary and fitting for humans given their particular history and legacy of interactions with God according to the Hebrew scriptures.⁹³ Hence, recompense can be made by a creature without a divine nature, providing partial satisfaction for sins; in this case, a partial satisfaction may be limited to certain persons, locales, and epochs; a revealing of divine truth can be made by finite creatures, although incompletely. Therefore, while incarnation is the highest self-revelation and unification of the divinity with the human species, God could have provided a means of salvation in an inestimable variety of ways in extraterrestrial civilizations.

90. Aquinas, *Summa Theologica*, III, Q. 13, Art. 3, ad 2. Four preternatural gifts are integrity, immortality, impassibility, and infused knowledge. Christ received integrity (the absence of concupiscence) and infused knowledge. He did not receive the others as it was willed that he suffer and die for the redemption of human beings. Divine or supernatural beings may possess some of these aspects.

91. God provided a means of salvation by a series of covenants, a cultic-sacrificial religion, the Mosaic Law, temple worship, and lastly by divine grace through Christ.

92. Aquinas, *Summa Contra Gentiles*, 79.5. The sins of humanity are forgiven by the passion, the supreme act of love.

93. Aquinas, *Summa Theologica*, III, Q. 1, Art. 2. Also, "For if man had not sinned, he would have been endowed with the light of Divine wisdom, and would have been perfected by God with the righteousness of justice in order to know and carry out everything needful" (Aquinas, *Summa Theologica*, III, Q. 1, Art. 3, rep. 1). Other salvation histories may not require the necessity of an incarnated divine being. The Hebrew scriptures provide the best known example of a supernatural religion without an incarnation. Other species, such as those with varying theological anthropologies and historical trajectories may not require an incarnation as divinity may will to interact in other modes.

As discussed earlier, according to Bonaventure, there are purposes other than redemption for incarnation;[94] as creation is not whole and fulfilled without being redeemed by God, wherever sin and evil has made its presence, creatures will not fulfill their ultimate purpose without God's intervention and salvation. The created order remains imperfect if sin is present; thus incarnation completes creation.[95] Incarnation and the message of salvation offer an opportunity for the reconciliation of the creation and reordering to the divine will and life. Accordingly, he asserts the Son is the full expression of all that God can be in relation to the finite.[96] Further, Duns Scotus taught that given that God's first initiative is love, willing good for himself as the end of all things and second, that the created be good for Him, God wills according to the perfection of that love; thus it would have been necessary for divinity to come to an Earth without a record of sin and need for redemption.[97] Aquinas also held this position.[98] Therefore, divine incarnation for purposes other than redemption of an extraterrestrial nature can be considered, among them as a representative or agent, as mediator or intercessor, counselor, amanuensis, or deliverer of revelations.

A further area of inquiry is what can be determined of extraterrestrial economies of salvation by examining that of the angels. Aquinas wrote extensively on angels in his *Summa Theologica*. The creation of the angels, their nature, and the human relationship with angels bears on the consideration of the nature of intelligent extraterrestrials. Angels can be understood within the context of the possibility of a vast array of intelligent beings inhabiting the universe, and therefore, not necessarily the only non-human intelligences created by God. The angelic economy of salvation, examined within the context of extraterrestrial intelligence demonstrates a theological anthropology and mode of salvation which in important ways contrasts that of Homo sapiens. Anthropologically, angels surpass visible creatures in perfection, having a separate genus while each is unique and constitutes its own species. According to scripture they can appear in human form and interact with humans, however as angels are pure spirit, we should not presuppose a uniquely human form but rather the ability to take on any form which will service divine goals. According to Aquinas angels were created in a graced state[99] with free will, knowledge, and infallible rea-

---

94. See earlier discussion on Bonaventure and Scotus in chapter 2.
95. Hayes, "Incarnation and Creation," 320–29.
96. Hayes, "Disputed Questions," 47.
97. Delio, "Christ and Extraterrestrial Life," 253–55.
98. Aquinas, *Summa Theologica*, III, Q. 1, Art. 3.
99. Aquinas, *Summa Theologica*, I, Q. 63, Art. 4–5. Augustine said the gift of grace

son, ontologically and intellectually superior to humans, and individually offered elevation beyond their created and gifted capacities to achieve the beatific vision after passing probation *in via*.[100] The angelic revolt and fall from grace was individually determined, in contrast to the collective guilt borne by humanity. Angels who passed their period of probation were confirmed in grace and elevated to participation in divine life. Therefore, the angelic journey consisted of a single step and one eternally binding choice, as more perfect beings in possession of superior knowledge the decision to serve or defy the Creator was full, immediate, and final without opportunity for repentance.[101] Angels may be an earlier result of a universal creative process, or possibly the material universe and angels were created concurrently as the non-corporeal sphere is intimately integrated with the corporeal sphere, according to Aquinas;[102] as all creatures, material and immaterial are ordered to each other. Therefore, angels were possibly created within cosmic history and not from eternity, and may have been the first creatures of creation. In this sense, they are extracosmic, as an order of creation preceding material composite creatures, rather than what we term biological extraterrestrials inhabiting physical spaces.

Therefore, it is possible that angels were designed for countless intelligent civilizations in accordance with the temporal and spatial scales of the universe; hence angels known to our economy of salvation may have been, or could be participants in other economies of salvation prior, concurrent to, and after Homo sapiens.[103] The modern scientific perspective of a big bang universe suggests that angels could have been created billions of years ago. This deepens exponentially the possible role and activities of angels on a scale equivalent to the physical universe. Angels as

---

was bestowed together with the gift of nature, so their creation and sanctification were simultaneous. Their full possession of natural gifts from the beginning, and their instantaneous knowledge and decision may have resulted in a very short period of probation.

100 Aquinas, *Summa Theologica*, I, Q. 52, Art. 6.

101. The fathers of the Church refer to the fall of the angels in Isaiah as the sin of pride, "I will not serve" (Isa 14:12–15). St. Augustine saw their sin as unforgivable due to their endowment of supernatural gifts.

102. Aquinas, *Quaestiones Disputatae de Potentia*, Q. 3, Art. 18. "If, however, the angels would have been created separately, they would seem to be totally alien from the order of corporeal creatures, as if constituting of themselves another universe." The Fourth Lateran Council in 1215, (*De Filus*: Denzinger, *Enchiridon*, 3002), professed that "God from the beginning of time made at once out of nothing both orders of creatures, the spiritual and the corporeal, that is, the angelic and the earthly, and then the human creature, who as it were shares in both orders, beings composed of spirit and body." See *Catechism of the Catholic Church*, 327.

103. Angels can communicate and cooperate instantaneously towards shared ends despite cosmic distances. See Aquinas, *Super Epistolas S. Pauli, ad Hebraeos*, #85–87.

depicted biblically and traditionally are anthropomorphized and bound to the supernatural-natural-preternatural matrix of Earth; this is a limited view. Therefore, what information is available from scripture and tradition is possibly a microcosm of an immeasurably larger reality of angelic activity. As such, their anthropomorphic form is very likely a merely adopted appearance as they are immaterial, and can be expected to take on a variety of forms according to particular planetary, historical, and situational need. Angels described in scripture may only be so in an analogous and representative manner. Therefore, certain extraterrestrial economies of salvation may include the involvement of angels in a manner known to humanity or differentiated according to the particularities inherent in civilizations and divine action within them. Angelic beings can be compared to extraterrestrials in that both are persons, with individual wills, knowledge, capacities, powers; they are created in grace and share a certain supernatural destiny.[104] Angels differ from humans and, presumably extraterrestrials in that they are of a higher order by nature; what angels have by nature is greater than what humans had by gift; extraterrestrials differ from angels as they are not purely spiritual beings, but composites of matter and spirit forming a unity. Demons, existing apart from angels while maintaining their angelic powers manifest two separate kingdoms of spiritual entities with connections to humanity. Hence there may be many sites of evil in the universe, which could produce fallen extraterrestrials, or even evil extraterrestrial societies or civilizations. As a result, opposed spiritual worlds may confront each other on innumerable worlds; in fact the accounts in Genesis and Revelation can be interpreted with a view of humanity co-opted into a gargantuan war between two primordial and immensely powerful adversaries.

Intelligent extraterrestrials would presumably be created, again following Aquinas, in a state of original grace,[105] ordered to a greater end, with their lower powers subjected to the higher. There would be a necessary balance between individual merit and final reward according to the specific anthropological template of creatures, as it would be contrary to God's justice to create species determined to fall due to certain incapacities in their nature. In this case merit may be achieved by extraterrestrials through probationary

---

104. In some cases, extraterrestrials can be theorized as ordered to achieve a natural happiness rather than a supernatural end.

105. "[I]t pertains to divine freedom to infuse grace into all who are capable of grace, unless something resisting is found in them, much more than the give natural form to any disposed matter. But angels from the beginning of their creation had the motion of free will, and there was nothing in them impeding [grace]. Therefore, it seems that he immediately infused grace in them." See Aquinas, II *Sent*. Dis. 4, q. 1, Art. 3, sed contra 3.

periods (a lifetime for humans and perhaps instantaneously for angels), of an order and type in ways both familiar and unfamiliar to us. In neither case creatures do not lose their natural capacities but only those received above their natures. Another possibility is that given creative differences among the same species of extraterrestrials, these could be composed of both fallen and graced beings, having different eschatological outcomes.

Christ's sacrifice and merits were not produced through or for the angels.[106] The angels achieved glory without Christ's assumption of their nature, indicating that incarnation was not necessary, nor was the Incarnation of Christ in humans necessary to secure angels' entry into beatitude as they pre-date Homo sapiens, potentially by billions of years. Thus, it cannot be *necessary* for extraterrestrials to be saved by means of the earthly Incarnation, given the example of the angels and in view of the spatial and temporal extent of the universe. Rather, the angels (and those which later became demons), and it would follow, extraterrestrials, achieve merit and beatification (or condemnation) through individual free acts within a separate economy of salvation unrelated to the Incarnation of Christ on Earth.[107] According to this view, an angelic economy, preceding that of material creatures was the first produced by God, to be followed and possibly patterned in certain material creatures. Therefore, extraterrestrials according to the *varied* argument would have an opportunity for merit, justification, and reward according their own particular soteriological modes.

Extraterrestrial Contact and Christianity

A contact/disclosure/discovery event would have varying effects on organized Christian religion; there are a variety of scenarios, ranging from radiotelescope messaging to discovery of artifacts on a planet or satellite, to direct contact. Social implications of certain scenarios have been briefly

---

106. Heb 2:16–17.

107 Aquinas is supportive of this view. As Christ in his humanity did not exist yet, angels were saved only through his divine nature. In his discussion on the merits of Christ as inapplicable through the old law. "Nothing prevents that which is posterior in time to move an agent according as it is apprehended and desired by him. But that which does not exist in the nature of things, does not move according to the use of external things. Whence an efficient cause is not able to exist posterior in being in the order of duration like the final cause can. See *Summa Theologica*, III, Q. 70, Art. 4, ad. 4. Further, in his commentary on Eph 1:8–10, "The effect of this hidden plan was to restore all things. For insofar as all things are made for the sake of man, all things are said to be restored . . . All things he says which are in heaven, i.e., the angels-not that Christ died for the angels, but by redeeming man, the fall of the angels was repaired." See Aquinas, *Super Epistolas S. Pauli, ad Ephesios*, #29.

explored by Peters,[108] Tough,[109] and Vakoch.[110] Herein I discuss some of the essential considerations of extraterrestrial evidence and Christian theology and religion. As mentioned, the *Brookings Report* was the result of the U.S. government's early research into the religious impact of a discovery of intelligent extraterrestrials. The report did not speculate whether an alien civilization would assume a cooperative, ambivalent, hostile, or other position with Earth leaders and scientists; however it predicted generally negative reactions among fundamentalist Christian denominations and theologies (typically advocating strictly anthropocentric and geocentric views) upon confirmation of extraterrestrial existence. It also warned of an inferiority syndrome, the notion that contact with a highly superior civilization would produce a racial inferiority complex for Earth scientists and religious denominations. It concluded with recommendations for further research given an unknown mode/discovery type/contact, study of types of relevant historical analogs, and contemporary sociological studies. Two aforementioned sociological studies, the *Peters Extraterrestrial Intelligence Survey* and *Alexander Report*, contrary to popular opinion at the time, indicated a high level of confidence in the resiliency of individual belief and of Christian denominations in an extraterrestrial contact event.[111] However, such surveys did not consider specific discovery/contact scenarios, potential extraterrestrial behavior and objectives, or information content, highly determinative of any social outcome. Several such scenarios/intent/content may be devised,[112] many which cannot be assumed to be mutually

---

108. See Peters, *Astrotheology*; Peters, *UFOs*; Peters theorizes four models of extraterrestrial motivations. Celestial saviour model: those evolved and progressed further in science, technology, medicine, and morality, and wish to share their knowledge with humans. Hybridizers or alien enemy model: extraterrestrials will be our conquerors, cannibalize our human and natural resources for their own civilization. The interstellar diplomat: an exopolitical diplomat skilled in interspecies exchanges. And scientist: those interested in studying Earth and its life forms as unique phenomena.

109. Tough, *Alien Worlds*; Tough, "What Role Will Extraterrestrials Play?", 491–98.

110. Vakoch, "Roman Catholic Views of Extraterrestrial Intelligence."

111. These studies are often cited as settled science on the socioreligious impact of discovery/contact. They are, however highly speculative due to lack of even general situational parameters mentioned above. A more recent study performed by the National Institutes of Health measured reactions to a faux news story of the discovery of extraterrestrial microbial life, concluding general positive responses. See Kwon et al., "How Will We React to the Discovery of Extraterrestrial Life?"

112. Contact scenarios can include: remote detection by radio interferometer; discovery of alien artifacts on a planet or satellite within our solar system; probes encountered in space; direct contact in space; on planet or satellite; in Earth orbit; contact on Earth on one site or several. Extraterrestrial objectives include scientific interests, exploration, resource acquisition, conquerors, saviours, observers; they may be hostile,

exclusive. In the event of an actual engagement with an alien intelligence, rather than speculation of its hypothetical possibility without any known factors mentioned above, it would be highly presumptuous to assume specific individual and institutional religious reactions. Studies indicating the high resiliency of particular faiths in the event of extraterrestrial contact project a confidence which is not consonant with the lessons of history of advanced cultures encountering the lesser advanced. Such a prediction may be grossly premature given the unforeseeable content or mode of a future contact scenario. One need only think of the American Indian or the indigenous peoples of Mesoamerica, whose experience gave witness to the complete transformation or utter destruction of ancient and cherished religious beliefs as a result of imperialism and colonialism. In accommodating the actual fact of extraterrestrial intelligent life present within our society, rather than its mere hypothetical possibility, it is possible that subsequent generations after contact may abandon earthly religions, as evidenced in our own history. In fact, a peaceful exchange with a religious, benevolent species, rather than one hostile or ambivalent may paradoxically create greater religious turmoil in certain groups due to a perceived superiority to Christian praxis and doctrine. Study of historical analogs with regard to extraterrestrial contact, although arguably imprecise considering our subject, demonstrate short-and long-term total transformation, utter destruction, or marginalization of religious beliefs within the context of colonialist or imperialistic motivations and programs.[113] Most importantly, these surveys did not consider long-term effects of contact on religious organizations as well as individual believers. Subsequent generations may have a propensity to modify key doctrines of past generations, develop syncretic theologies, or abandon them altogether in the face of a much older, technologically advanced species possessing a compelling, competing belief system.[114] Our advancing cultural climate of religious and philosophical relativism throughout Western Europe and the Americas could serve to exponentially enhance that potentiality; as such, an otherworldly religion

---

benevolent, ambivalent, curious, helpful, or harmful; information may be purely scientific, historical, technological, religious, anti-religious, undecipherable, or non-rational according to human perception.

113. Carl Jung noted his concern with encountering intelligent extraterrestrials, similar to that of indigenous civilizations in the encounter with Western powers would suffer accordingly: "the reins would be torn from our hands and we would, as a tearful old medicine man once said to me, find ourselves 'without dreams,' that is, we would find our intellectual and spiritual aspirations so outmoded as to leave us completely paralyzed." Finney, "The Impact of Contact, 16.

114. Over time these may result in large and small scale schism among mainline Christian denominations.

beyond the parochialism, competition, and divergences in praxis of Earth religions may be attractive. Aloysius Pieris has argued:

> Mass conversions from one soteriology to another [e.g. Christianity to Islam] are rare, if not impossible, except under military pressure. But a changeover from a tribal religion to a metacosmic [world] soteriology is a spontaneous process in which the former, without sacrificing its own character, provides a popular base for the latter.[115]

A more primitive religion,[116] or religion held by less advanced, remote societies would likely pose little challenge to Christianity, but one more complex or belonging to a more ancient, culturally and/or technically advanced race could prove especially damaging for certain believers.[117] In an encounter with advanced extraterrestrials devoid of religious beliefs, their explicit or implicit acceptance of agnosticism or atheism may discourage religiosity among certain human populations. Some thinkers have portended the nullification of terrestrial religion by an extraterrestrial reality and consider Christianity falsifiable upon discovery/contact; and that it will renounce its claim as a truly universal religion while accepting the implications of excessive particularism.[118] Therefore, it is argued that discovery could verify the Christian religion purely as a terrestrial faith; and any world where there was an incarnation and resurrection would invalidate orthodox Christianity's claim as the only true universal religion. Angelo Perego states "[T]he universality of the Redemption . . . would be impeached by the discovery of intelligent extraterrestrials,[119] and philosopher Hans Blumenberg has claimed, "The realization of the hope for interstellar communication would necessarily result in the death of Christianity as well as of any religion."[120] According to SETI astronomer Jill Tarter, to be plentiful enough to afford us a chance of detecting extraterrestrials, they must be very long-lived and hence abandoned religion, a primary source of conflict on Earth.[121] The

---

115. Pieris, *An Asian Theology of Liberation*, 99.

116. For purposes of argument, "primitive religion" can refer to religions similar to aboriginal cultures, animist, pantheist, and shamanist belief systems.

117. Biblical literalists, fundamentalists, and those espousing views on the silence of scripture with regard to extraterrestrial nonexistence would be most challenged.

118. Roland Puccetti has argued if there are multiple independent religions on different planets each claiming absolute truth, this falsifies any claim to absolute religious truth. See Pucetti, *Persons*, ch. 5, "Divine Persons."

119. Perego, "Rational Life beyond the Earth?", 178.

120. Blumenberg, *Die Vollzähligkeit*, 145.

121. Tarter writes, "For one of the nearest 1,000 solar-type stars in our galaxy to host another technology, the average longevity L must be measured in tens of millions

implicit assumption of these arguments is that an extraterrestrial religion must either annihilate or supplant terrestrial religions; or that technological and cultural evolution directly correlates with decreased religiosity; it fails to acknowledge that the innate nature of the human (and presumably extraterrestrial) mind contains a religious dimension. Van Huyssteen has argued accordingly that an innate religious categorical structuring to intelligence has always been part of the mind's greater search for meaning, and essential to the development of higher cognition.[122] Extraterrestrials could positively expand their religious consciousness in their contacts with other races, including humans; or conversely become a source of conflict, or result in a trivialization or relativization of indigenous religious beliefs and values. Species with advanced religions may avoid direct contact with the less advanced so as not to disturb their societies and natural cultural evolution according to their particular revelation and history.[123]

We cannot assume extraterrestrials we contact will communicate a divinely revealed religion. A civilization that humans contact may possess a divinely revealed religion, an illegitimate, or manufactured religion, or syncretistic combination. Christianity would be tasked with distinguishing between what may be understood as an extraterrestrial natural religion versus revealed religion; certain cultural convergences of history demonstrate this in operation. Canaan's indigenous Middle Eastern pagan religions were considered by the Hebrew writers from who we learn in the Old Testament to be dominated by false deities. During the Christianization of the Roman Empire, the early church fathers (e.g. St. Clement of Alexandria) viewed Christianity's Hellenistic competitors as products of fallen angels resulting from man's confusion and wanderings as a result of the fall. The Catholic conquistadors interpreted the Mesoamerican Aztec's indigenous religions of

---

of years . . . to live so long, such societies must have greater wisdom, knowledge, and social stability than ours . . . they 'either never had, or have outgrown, organized religion.'" Dick, *Many Worlds*, 145.

122. "In this sense one could indeed say that, even though we may aspire critically to understand the cultural pressures that have been influencing metaphysical views and religious convictions in the course of past millennia, our deepest beliefs and firmest convictions reach back further than any cultural influence . . . the fact that religion, and religious intelligence, has always been the response to the holistic search for meaning in our experience." Van Huyssteen, *Alone in the World*, 107–8.

123. In this case, rather than public contact or disclosure, a culturally sensitive and responsible species may interact with humans furtively or clandestinely with certain groups or individuals in order to avoid public or large scale social destabilization. Long-term models of contact or disclosure would be preferable in acclimating society to extraterrestrial existence on a scale of decades to hundreds of years, to accommodate social consciousness and cultural evolution for an eventual acceptance and integration of an extraterrestrial reality with human civilization.

the New World as demonic creations and their extermination a fulfillment of the divine will. Modern Christians would be tasked with the substantiation or invalidation of extraterrestrial religion; an interpretation of religion would likely follow according to their distinctive theological predilections, texts, doctrines, and their corollaries in Christian tradition.[124] As such there may be areas of agreement, areas of small adjustment, areas of wholesale redefinition, and areas of complete incompatibility. We may encounter extraterrestrial religions wholly incompatible with Christian thought, as in the discovery of the New World or the failure of missionaries to convert any sizeable numbers of the native populations of China and Japan. There may be a repetition of terrestrial history in the encounter of those without any concept or willingness to accept the notion of a universal, personal Creator God. Conversely, the human, terrestrial Christian religion need not conflict with those recognized as supernaturally instituted religions, as absolute truth cannot contradict absolute truth; other revelations of the divinity cannot exist fundamentally juxtaposed with divinely revealed truth in another intelligent creation. Christian categories could naturally find their equivalencies in other cultures due to its insistence on the primacy of relation between creature and Creator. As Christians recognize the divine validity and veracity of the Old Testament theophanies, covenants, and teachings imparted to the ancient Hebrews as those of the Christ, a foreign belief system can remain foreign while supernaturally legitimate. Those harbouring, from our estimation, invalid religions may not accept the idea of a personal or terrestrial-bound God, finding such belief an indication of religious primitiveness. Steven Dick has argued human religions must adjust to disclosure of new religious information from extraterrestrials (in his pursuit of a naturalized, cosmic religion); however this is to deny any supernatural presence or intervention in the human species providing specific divine truths for humans.[125] Conversely, it should be considered that Christianity may be further validated on a cosmic scale by the disclosure of another, divinely inspired corresponding extraterrestrial religion.[126]

---

124. Comparative theologians practice "hermeneutical openness" in the reading of other religious texts where comprehension should precede judgment. This may influence the conceptual framework for understanding one's own religion. See Clooney, *Beyond Compare*, 208.

125. Dick, *Many Worlds*, 202–8.

126. Extraterrestrial contact could provide more information on origins of angels and demons, history of heaven, greater knowledge of the creation of the universe, galaxy, solar system, and Earth, and the histories of other places of God's creation and divine acts.

Highly advanced religions, generally within this context can be described as those which have evolved beyond certain geographic and parochial perspectives, and which accept and integrate knowledge of the sciences. They are global or even trans-global in character. A less advanced religion may be compared to those most familiar to humans akin in certain human societies,[127] typically exhibit diverse mythologies, are oral without written texts, are limited linguistically, culturally, and geographically, maintain a focus on the present life rather than otherworldly; consider the motives and works of gods to be often mysterious and fearsome, and contain cyclical time rather than historical or progressive theological impetus and finality. A simpler description can be made between natural religion, that discovered by unaided reason, versus a supernaturally revealed religion. As evolution stresses the importance of higher forms of life, more advanced religions should provide fuller content to divine works and understanding.

The distinction between true and false religions may be described in this context as that between supernatural and natural religions. This may be determined from the perspective of what is unmistakably supernatural of a supernatural religion versus what supernatural content is absent in a natural religion (although natural religions can contain divine truths). The constants of supernatural religion can first be attributed to moral absolutes for creatures which must be maintained in any bona fide religion. While an extraterrestrial religion may differ in its revealed realization of divinity, modality of communication, and functions of divine persons or divinity, it cannot propose a fundamental theology antithetical to Christian morality and ethics and be considered by Christian doctrine divinely inspired. All creatures in their nature as creatures must be understood to have their absolute goal holiness as a supernatural end. A supernatural religion likely would maintain belief in an omniscient, omnipotent, benevolent divine being who creates and sustains all in material and spiritual existence; that God orders and sets physical laws governing the universe; that he creates rational, intelligent beings with free will to know him and share in his life; that he provides creatures the free choice to serve him and the opportunity for a special elevation to greater participation in divine life; that those choosing not to serve may be removed from the divine presence; that God can initiate to engage his creatures in relationship with himself, and that God exists as a transcendent and immanent being. God throughout divine revelation is chiefly concerned with the holiness and devotion of his people, as God himself is holy and seeks relationship with his creatures, and all creatures must have a natural

---

127. Animist, totemic, shamanist, henotheist, pantheist, polytheist, monist, ancestral, or naturalistic.

capacity for God. These are a short outline of the fundamental relation of the divinity to creatures, the destiny of creatures, and dignity of creatures and their corollaries that in an extraterrestrial religion would be a central focus in this task. This would seem more important than particular modes of revelation or the perceived identity of divine entities which establish extraterrestrial religions. One measure is to know a religion by its fruits.[128] Denial of the fundamental ideals of love, freedom, the reality of evil of and liberation from sin, the free and transformational nature of grace, the spiritual journey and unity of creature and creator in heaven would indicate a system lacking authentic religion. Lastly, an extraterrestrial supernatural religion would not be falsifiable through reason or science.

Extraterrestrial societies may have a heterogeneous population or a homogenous one, they may have a unified or universal religious system or two or more competing or non-competing religious systems, as a result of differing races, locales, national or collective identities, technological sophistication, languages, or other attributes. Contact would create a massive paradigm shift in religious consciousness as humanity is removed from its central role in recorded history, and the deposit of the entire Christian tradition would be reoriented within the greater setting of other intelligences created outside the human family. It is likely that conscientious extraterrestrials might not desire to impart too much information, particularly of a religious nature, aware this might cause great destabilization of human societies. In fact, benign societies may not attempt to communicate a religious message, aware of its disruptive effect on a terrestrial-bound populace. It would be reasonable, therefore, that benevolent extraterrestrials planning a contact/disclosure event would prepare for the time when our cultural evolution produces scientific, theological, and philosophical perspectives more aligned to theirs so not produce social disruption. Therefore, a religion imparted from a more ancient and/or culturally/technologically superior race need not necessarily cause despair for Christianity. Extraterrestrials, most likely civilizations more advanced among those capable of contacting us, would be in the position of dictating the terms and conditions of contact. Others would likely not be interested in human science or technology, having surpassed our knowledge; instead having interest in our religious beliefs, culture, and arts. Universalist species would likely be aware and concerned regarding our self-destructive tendencies. SETI proponents and others envision human advancement could be achieved by means of information imparted by highly advanced extraterrestrials disposed to assist lesser developed societies; this

---

128. Matt 7:16–17.

has been considered by some as a thinly veiled religious quest couched in scientific language and methodology.

Section D: The Divine Pedagogy

According to the *varied* argument, the historical theological hesitancy to consider extraterrestrials within the corpus of established Christian doctrine is unfounded; the discovery of a second genesis of intelligence does not compromise faith in Christ but rather reveals God's special and unique means chosen for the salvation of a human civilization among other redemptive modes of putative civilizations. This argument represents a development of doctrine in accord with new information from contemporary science, as a denial of the possible existence of extraterrestrial intelligent life given the scale of the known cosmos is incompatible with the traditional teaching of the 'omni-properties' of God. The local and historic particularization of the Christian doctrine of the Incarnation has historically been heralded as the crux of an incompatibility between Christianity and intelligent extraterrestrials. An incarnation on Earth among other types of divine interventions in species does not render God's revelation and redemptive actions to humans and Earth as necessarily void, inferior, or indeed superior to others; rather, the supreme action and manifestation of God in the human person of Jesus Christ is a uniquely special, but isolated and singular activity among a myriad of other divine actions. Jesus is the image of the invisible God,[129] the fullest and most intimate way the Creator has made himself known to humans and as the most exhaustive revelation of the supernatural to the natural; the Incarnation provides humanity some possible indication of divine action in other places. The Earthly Incarnation was a divinely willed, determined theomorphistic event; God became human *for* humans, incarnated as a human for the *deliberate intention* of an ultimate and long-awaited geocentralizing of religion to supplant erroneous precursory fragmented religious constructs. God's untouchable, ineffable spiritual transcendence, characterized by many Earth religions, was utterly transformed into a fully realized, knowable, extraordinarily and fundamentally immanent supernatural and physical presence. Divinity was fully particularized in the human species, born within a specific race, language, religious heritage, cultural and political milieu, and geographic region on a particular planet. The message, personification, and presence of God in Christ and transmitted through the Holy Spirit was divinely willed to become a *world system* for human civilization; that all human individual and collective

---

129. Col 1:15.

thought and behavior would be reoriented in the shared reality of the free gift of divine grace. By means of the Incarnation, the joining the divine and human natures, human civilization would be transformed and reoriented to the divine order and a predetermined supernaturally elevated and graced human earthly life, followed by an ultimate heavenly perfected destiny. The theomorphization of the divinity constituted a final and complete unity between Creator and earthly created intelligence.

In considering the reality of the greater creation of the universe and possible inhabitants in other worlds, God's purpose in Christ is revealed further: to reconcile a people to himself within a multiplicity of a larger host of other intelligent creatures equally created and loved by God, within an incomprehensively vast universal divine plan. The Creator reaches out to his creation with the intention of bringing it back into relationship in the event of sin. A supremely powerful, infinitely benevolent being loves his creation, particularly his intelligent creation, and calls that creation to relationship. A divine person is not limited to one incarnation or mode of manifestation; there is no limit to divine realizations within a vast cosmos.[130] However, what is revealed will remain incomplete as an infinite divine person, being inexhaustible, self-limits in interactions with created reality. However incredible a divine revealing is to us in the Incarnation, we remain limited in our comprehension or knowledge of God's greater plans for the cosmos, but can safely assume they are consistent with that of Earth. The "happy fault" of sin in humanity led to a further revelation of God, with greater mercy, power, presence, and penetration into the human being through the personhood of Jesus Christ. Creation entails a deeper communion and call to ever deeper relationship with the Creator, founded upon love and ultimate unity.

God's purpose in creation is for all beings to participate in divine life. Drawn from our known examples of human beings and the angels, God provides an invitation to participate freely in divine life in a condition of love and obedience. All creatures have an ontological dependence upon the Creator which is universal, absolute, and inescapable as all beings are created for God's glory; their existence manifests the beneficence, power, and fundamental glory of the Creator. All beings have a duty, therefore, to consciously glorify God, and extraterrestrial intelligences would highlight our common heritage as creatures of the same divine Creator. A consistent supernatural goal would exist to share supernatural life with creatures in varying capacities, roles, and places; each according to its own. God is God

---

130. "The power of a divine person is infinite and cannot be limited to anything created." Aquinas, *Summa Theologica*, III, Q. 7, Art. 3.

of creation, of love, gift, invitation, revelation, self-communication, grace, mercy, and salvation. The Logos, the Second Person of the Trinity on Earth is an illustration of one specific divine manifestation and series of acts of a divine person for the specific needs according to the personal and collective human condition. There is no theological incompatibility in the presence of multiple divine 'beings' throughout the universe; the order of creation and divine initiative does not necessarily reach its high point in humanity. There may be other, greater forms of God's revelation in other civilizations.

The *divine pedagogy* is defined as divinity's heterogeneous divine action throughout the universe in his creation of diverse intelligent creatures designed and destined to share in supernatural life. It is God's methodology of leading his intelligent creation to himself in varying modes, personages, revelations, and relations within a vast temporal and spatial material and spiritual universe. It can be expressed as "God, wishing to speak to men as friends, manifests . . . by adapting what He has to say by solicitous providence for our earthly condition."[131] These categories of the divine pedagogy in relation to creatures describe distinct modalities of divine interaction with humans in salvation history:[132] God is *Invitational*: the divinity, respectful of free will, calls to intimate relationship, adapts his invitation to individual's culture and social setting, and takes into account the entirety of the human being. *Incarnational*: in humans God creates a unity of divine person and creature; through this medium divinity reveals itself through creation, word, and action. For Homo sapiens, God becomes Christocentric and supremely immanent. *Relational, familial, and communal*: the divinity's primary means of communion with humans is communicated and realized through the matrix of relationship; scriptural metaphors provide examples of God acting in a manner consistent with that of intimate human relations, primarily familial. *Structured, systematic and comprehensive*: divinity reveals itself through a historical process in accord with cultural, social, and spiritual development. Divine action is consistent and rational; revelation is coherent and cumulative over historical epochs; God's activity and message engages the entire person or community, for purposes of complete unity of creature and Creator. *Perpetual*: with an internally consistent scripture and tradition, divine truths are further realized in a historical progression. These pedagogical categories can be applied whereas humans serve as one example of God's action among an array of divine pedagogies in other civilizations. Christians believe that divinity has manifested, revealed, and redeemed

---

131. *General Directory of Catechesis*, n. 146.

132. Basic categories developed at an international catechetical conference on teaching methods in Rome 2009 (unpublished). I have adapted these to express potential divine activity among putative extraterrestrial beings and civilizations.

Divine Covenants and Economies of Salvation

Among the possibility of a range of divine incarnations among species, we can further consider the revelation in Christ as the Incarnation of the God-man an archetype of *general* divine activity, but not necessarily normative for extraterrestrial beings as multiple thinkers suggest.[134] Alternate economies of salvation in diverse biologies, societies, and epochs can result in heterogeneous economies and histories of salvation. Accordingly, incarnation of the Second Person would not be the only realization of divine relation to creatures. This has been demonstrated within Earth's own history in the canon of scripture which provides an historical accounting of modalities of divine interaction among progenitors, patriarchs, and Hebrew people. The Old Testament record reveals a pattern of divine election through singular mediators or representatives within a series of covenants, *each providing a separate economy of salvation*, wherein God adapts modes and elements for a divine relationship with human beings throughout historical stages of their spiritual and social development. The creation account concludes with the Sabbath, which is understood as the sign of God's covenant;[135] the creation of humans in the image and likeness of God implies the original divine relation to humans was one of close kinship.[136] This original, divinely instituted relationship to the early humans can be understood as a 'cosmic covenant' embedded in ancient Israel in the Torah and second Temple sources.[137] In Eden, God provided original gifts of reason, free will, companionship, and authority over Earth; the narrative portrays humans created in a graced, love relationship with God, with a single condition of obedience.[138] The Jewish tradition affirms the creation narratives as the canonical source and primordial form of God's two-fold

---

133. "The history of religion shows that this historical interpretation of transcendental, supernatural revelation comes down to us in such a way that various histories of religion arise in different places in the world and at different times in the history of humanity... Nothing really happens in the realm of the categorical which does not also happen in the history of every other people." Rahner, "Revelation," 444–49.

134. Mascall, Zubek, Pittenger, Congar, and Delio.

135. Exod 31:16–17.

136. Gen 1:26, 28.

137. Balentine, *The Torah's Vision of Worship*. Jub 36:7; "Song of the Cosmic Oath" in 1 Enoch 69:13–22. See Murray, *The Cosmic Covenant*.

138. Gen 2:17.

everlasting covenant: wherein the cosmos is consecrated as covenant on the seventh day, and the first people in the marriage covenant.

The biblical witness offers us the following sequence of divine interaction with humans wherein varying economies of salvation and modes of action are demonstrated. The Adamic covenant represents the original, divinely instituted creation and supernatural, unaltered relation with humans. Theologians term the state of early humans as Original Justice to describe a humanity untainted and undamaged by sin and its effects; the economy of salvation of the primordial period is exceedingly simple and natural, and portrays humans having direct, unfettered access to the divinity with a single prohibition.[139] The Noahic covenant was established as a postlapsarian alliance with one family and all living creatures on Earth, after the deluge in response to widespread sin. The covenant was awarded due to Noah's faithful obedience,[140] and had universal scope: it renewed the created Earth and its inhabitants and reestablished the divine relationship to the cosmic order at creation. Humanity is reestablished with the original callings, obligations, and privileges of Adam. God promises never to punish the world by flood again. The covenant took the form of a unilateral, unconditional, and permanent royal grant, affirmed symbolically by the Rainbow.[141] The Adamic covenant is canceled; a new economy of salvation is granted to a single righteous family due to the love and obedience of Noah. A common grace is provisioned to the survivors and an opportunity and time for God's divine plan for humanity to be carried out.[142] The divine mandate to multiply and fill the earth, subdue the animals, and care for all living things is reaffirmed. Later, to place limits on human propensity for domination and pride, and so mitigate the effects of sin on Earth, God divided the human race into a plurality of nations after the destruction of the tower of Babel.[143] However polytheism, idolatry, and other abominations flourished among those nations, serving to thwart the originally intended relationship renewed through Noah.

In the Abrahamic covenant God elects Abram, originating from a pagan family, to be father of a chosen people singularly devoted to true divine

---

139. Although it needs to be recognized that this has to be read in combination with a scientific understanding of human evolution.

140. Gen 6:8–9; 7:5; Heb 11:7.

141. Gen 9:13–17.

142. Sanctity was possible for those who lived according to the covenant with Noah, waiting for Christ to "gather those scattered abroad" (John 11:52).

143. Rom 1:18–25. The division of nations and disunity at Babel due to sin and polytheism and idolatry threatened this "provisional economy."

relationship; God swears an oath,[144] recognized as the fullest expression of the Abrahamic covenant.[145] As a reconfigured covenant of the previous two, the first portion of the covenant with God includes three specific promises to Abraham: a great nationhood, a great name, and universal blessing.[146] The ritual of the passing of pieces of animals in Genesis 15 denotes that Abram will suffer the fate of the cut animals in any failure to fulfill covenantal obligations with the Lord. During this ritual, Abram is promised a great nationhood, numerous descendants, and land;[147] the covenant is sealed by circumcision and his name lengthened to denote its greatness. The covenant takes the form of a unilateral, unconditional royal grant in return for devotion and recognition of a monotheistic God in a land surrounded by pagan polytheists. After the test with Isaac,[148] Abraham is given again the promise of universal blessing and a promise of many descendants, foremost that his descendants will become a nation from which all nations will be blessed.[149] The purpose of the Abrahamic covenant is to sow the seeds of a divine plan of redemption for all nations[150] through a future messiah; responsibility for keeping the covenant lies with one man Abraham. After patriarchs Isaac, Jacob, and Joseph, God provided as a sign of faithfulness: the deliverance of the chosen people from slavery and as further demonstration of blessings of the descendants of Abraham.

The original Mosaic covenant instituted on Mount Sinai was a different covenant: Israel would be raised to a unique, special status as a royal priesthood and holy people singularly devoted to God, contingent on honoring their obligations to the covenant.[151] This included fidelity to the tables of the Law,[152] civil regulations, participation in a sacrificial blood ritual acceptable by God (signified by sprinkling on the altar and people[153] and a

---

144. Gen 22:15–18.

145. Exod 32:13; Deut 4:31; 7:12; 8:18; Luke 1:72–73; Acts 3:25; Heb 6:13–17.

146. Gen 12:2–3.

147. Gen 15:5, 18–21.

148. God's test of Abraham to sacrifice his son Isaac, among other things, was to indicate child sacrifice known among neighboring polytheistic religions was not acceptable to the true God, and hence they worshipped false gods.

149. Gen 22:15–18; Abraham was the beginning of the calling together of nations "the father of all nations." "In you all nations on earth shall be blessed" (Gen 17:5; cf. Gal 3:8).

150. Gen 12:1–3; 17:1–8; 18:19; 22:1–8; 26:4; 28:13; Rom 4; Gal 3:7–9, 26–29.

151. Exod 19:5–6.

152. Exod 20.

153. Exod 24:6, 8.

shared familial meal.[154] The golden calf incident shortly thereafter broke this covenant. The Sinai covenant was renewed on appeal by Moses. Interestingly, Moses's discourse with the Lord reveals God's desire to destroy the Hebrews due to their sin of the golden calf; and willingness to make yet *another* great nation with Moses as leader; demonstrating the possibility of a new divinely instituted plan and separate economy of salvation for the perpetuation and salvation of humanity.[155] Upon appeal by Moses, the Sinaitic covenant was reconfigured with added preconditions as a direct result of the early violation: the Levites will replace the firstborn in priestly service,[156] and cultic regulations are instituted, concluding with blessing and curses as consequences to fidelity or infidelity. After later repeated failure to live up the requirements of the reconstituted Sinai covenant, especially the defection at Baal-Peor,[157] further punishments were provided, including delimitations of the priesthood.[158] After the golden calf incident, God mandated rules to contend with the Israelite predilection for idolatry; it was necessary to train the people in habits of conscience, obedience, and deference to Yahweh. There was no holiness without obedience; as God had in the original Sinai covenant intended a renewed intimacy between himself and each Israelite, now Moses became mediator between God and a disobedient people, indicating another variation of divine relationship. God in Leviticus instituted a new economy of salvation: a remedial-pedagogical-sacrificial religion to teach obedience; placed on a guilty nation which required a symbolic atonement by holocausts and burnt offerings.[159] These offerings did not justify or make holy, but served as a medium, containing external signs of an interior sacrifice; in the rituals the virtues of religion were writ large, physical, and graphic to create spiritual discipline for a people demonstrably not ready for a fuller relationship with God. The Deuteronomic reconfigured covenant, a result of continual infidelity to the Mosaic covenant, imposed even more regulation. As Moses had become the lawgiver, permission for a monarchy, warfare, usury, divorce, and remarriage was provided as a result of the Hebrew predilection for being "stiff necked."[160] These allowances were far from the ideal of those belonging to

154. Exod. 24:10–11.
155. Exod 32:10.
156. Num 8:16–18.
157. Deut 4:3; Josh 22:17; Ps 106:28.
158. Num 25:1–15.
159. There was no divine precept for sacrifice given to the Israelites until the golden calf. The command for animal sacrifice was to specifically sacrifice those animals worshipped by the Egyptians in order to wean the Hebrews of their 'addiction' to idolatry.
160. Deut 17:14–20; 20:16–18; 23:20; 24: 1–4.

a royal priesthood, but were concessions as a result of persistent obstinacy which required a protracted, legalistically burdensome ordering of Israelite society.[161] The Mosaic covenant was a suzerain, bilateral, and conditionally temporary covenant structured as a vassal treaty compared to the Sinai covenant. The Mosaic Law, designed as a guardian[162] as described by Paul, was to teach righteous standards and identify sin for a people seemingly incapable of avoiding it, and to direct the Hebrew people from the sinful behavior of their polytheistic neighbors. They were provided special guidance by means of the tables of the Law, the Ark of the Covenant, the tabernacle, and divine presence in the pillars of cloud and fire. God spoke directly to Moses above the cherubim of the Ark and manifested in columns of smoke and fire. Therefore, variations in the Mosaic covenant as a result of disobedience created a changed divine-human relationship, and separate economies of salvation in nature and form resulted in dramatic adaptations from those implemented through former covenants.

While Noah and Abraham were deemed righteous by God,[163] those subjected to the Mosaic and Deuteronomic Laws were required to offer animal sacrifice, to demonstrate communion by means of gifts to acknowledge the supreme domination of the Lord. These removed sins which prevented Israelites from fulfilling obligations of the covenant. The sacrifice did not justify, but placed upon Israel an abiding need for symbolic atonement by holocausts and burnt offerings; each symbolizing the repentance and reconciliation of the offerer. In contrast, the sacrifices of Abel, Noah, and Abraham were a type of natural family religion distinct from the sacrificial cult instituted by Moses, where sacrifice was the principal means of ratifying, renewing, and repairing the relational bond between God and his people.[164] Therefore, sacrifice before the law came from the heart of man; and sacrifice under the law came under from God's commandment; each revealing a separate pedagogy of worship and divine relation. The Davidic covenant raised the nation of Israel to a kingdom with the promise of an everlasting throne.[165] This was provided by God in terms of a unilateral, unconditional royal grant to King David and his royal descendants. The Davidic covenant provided the gift of divine sonship for those anointed heirs[166] and centrality of wor-

---

161. Jesus later taught these concessions were necessary due to the Hebrew's "hardness of heart," far from the ideal planned by God on Sinai. Matt 19:8–9.

162. Gal 3:24.

163. Jas 2:23–24.

164. Durken, *The New Collegeville Bible Commentary*, 120.

165. 2 Sam 7:13–16.

166. 2 Sam 7:14; Pss 2:6–9; 89: 26–27.

ship in Jerusalem temple.[167] It solidified the Kingdom and the liturgy of the Jerusalem Temple as the one true Kingdom of God on Earth and the manifestation of God's reign on Earth. After division and decline of the Davidic unified kingdom, prophets referenced the violations of the Mosaic covenant which triggered the covenantal curses[168] and the Israelite people were provided hope of a new, everlasting covenant for all people.[169] The Davidic covenant provided yet another type of economy of salvation centered on Temple worship and a relationship of divine sonship.

The everlasting covenant of Christ reveals a Jesus who provides a transition from the Old to the New Testaments by using explicit kinship language rather than explicit covenant language. It is a unilateral, unconditional, everlasting royal grant, announced during Israel's captivity.[170] The institution of the Eucharist redefined the Passover meal;[171] the New Covenant is connected to the church and the new action of a Holy Spirit sanctifies, guides, and elevates Christians in supernatural grace to serve and worship God according to the greatest commandment, negating the need for codified Mosaic Law. Serving as the fulfillment of Abrahamic and Davidic covenants, the New Covenant positions the Israelite nation as the father of all nations and center of true worship and salvation for all people by means of the foretold messiah who brings the Good News.[172] This new Kingdom of God serves as a transformed Davidic kingdom, made in the supernatural temple of Christ.[173] The New Law of love is the new commandment which encapsulates and fulfills all the old prescriptions of the Pentateuch. Paul contrasts the old and new covenants; he notes that the new covenant in Christ fulfills the promises and terms of the grant-type Abrahamic covenant, considered more rudimentary than the vassal-type Mosaic covenant containing laws and curses. The Everlasting Covenant therefore, is gifted to humanity as a new economy of salvation centered on the gift of a divine-human Incarnation and his redemptive sacrifice, allowing for the free gift of individual divine grace and individual human response. Proper relationship with God, mediated through Christ allows for the personal salvation of individual people to achieve unity with the

---

167. 1 Kgs 8:41–43; Isa 2:1–4; 56:6–7.
168. Lev 26:14–46; Deut 28:15–68.
169. Isa 2:2–4.
170. Isa 59:20–21; Jer 31:31–34; Ezek 36:24–31; 37:26–38; Heb 8:8–12; 10:15–18.
171. Matt 26:26–29.
172. Gen 22:18.
173. John 2: 19–20.

divinity in a supernatural state surpassing even that provided to Adam. No human sin can nullify the Everlasting Covenant.

The earthly covenantal history of divine initiatives was designed to establish a kinship with creatures;[174] despite human resistance and persistent failure to observe the conditions for perpetuity of divine covenants, God's continued historical interventions led to the establishment of larger groups of a chosen people, from the Adamic (first people), Noahic (family), Abrahamic (tribe), Mosaic (nation), Davidic (kingdom), and church (international). All revelation within these economies is intrinsically connected. The final, everlasting covenant serves as a restoration of the original Adamic relationship, transformed, however and elevated by new grace in Christ.[175] In this way the covenantal language and structure is transformed into that of the familial relationships between father and son,[176] bride and groom,[177] and even as close friends.[178] The Hebrew and Christian religions manifested on a social and historical continuum, leading to a restoration and cosmic reordering of humans to God and the culmination and final revelation in Jesus Christ.[179] Humans are elevated and transformed through the divine-human Christ, who is both high priest and sacrificial victim of the new covenant.[180] Much of the Lord's activity in the Old Testament is in initiating, maintaining, and meting out rewards and punishments according to divine covenants. The Noahic and Davidic covenants were divine grants with conditional elements; the Abrahamic and Mosaic were dynamic: the Abrahamic covenant was established as a kinship covenant[181] but later as a vassal type[182] and finally as a grant type.[183] The Mosaic covenant was first gifted as a kinship[184] but reconfigured after the calf incident as a vassal type[185] and again more severely in Deuteronomy.

---

174. There is no extrabiblical evidence of a supreme deity entering into a covenant with humans. Ancient Near East secular covenants between human parties were widespread; only Israel is known to have had divine covenants.

175. Gen 1:26, 28; 5:3; Luke 3:38.

176. Rom 8:15; Gal 4:6; 1 John 3:1.

177. Eph 5:21–32; Rev 21:2, 9, 22:17.

178. John 15:12–17.

179. Gen 1:26, 28; 5:3; Luke 3:38.

180. Luke 22:20; John 6:53–58.

181. Gen 15.

182. Gen 17.

183. Gen 22.

184. Exod 24:1–8.

185. Exod 34—Lev 26.

In each instance, God accommodated humans by means of individual covenants to engage humanity along its behavioral, familial, sociological, situational, and relational trajectory with divinity. Christians acknowledge the Adamic, Noahic, and Abrahamic covenants and the Mosaic tabernacle-centered cultic sacrificial worship as divinely inspired, with varying revelations and modalities of relationship with the divinity. Among them, each community recognized and worshipped the same God within its own situational particularities and in accordance with divine requirements, resulting in varying economies of salvation. In each case, God accommodated creatures according to the overarching divine desire for their survival and salvation. In reference to the ancient Hebrews, Irenaeus stated that humanity was *nuper factus*, a "newly made" being, childlike and requiring education.[186] Therefore, God's pedagogy was to enter into covenants with humans to gradually persuade, rather than coerce toward spiritual maturity.[187] So according to Paul, the Mosaic Law was imperfect and did not make people perfect, and lacked grace necessary for a person to fulfill it. But it provided a sufficient economy of salvation to serve as a custodian, or *paidogogos* until the time of Christ.[188] Therefore, there can be a variation in *fides quae*; there are similarities and differences in theophany, revelation, modality of presence, message, praxis, and economies of salvation between the Hebrews and Christian community, however both serve the same Creator.[189] The Judaic

---

186. Orbe, "Homo Nuper Factus": *En torno a s. Ireneo, Adv. haer, IV,38,1', Gregorianum* 46 (1965), Steenberg, 'Children in Paradise: Adam and Eve as "infants" in Irenaeus of Lyons', 481–84 in *Journal of Early Christian Studies* 12.1 (2004), 20.

187. "The education of the human race, represented by the people of God, has advanced, like that of an individual, through certain epochs, or, as it were, ages, so that it might gradually rise from earthly to heavenly things, and from the visible to the invisible. This object was kept so clearly in view, that, even in the period when temporal rewards were promised, the one God was presented as the object of worship, that men might not acknowledge any other than the true Creator and Lord of the spirit, even in connection with the earthly blessings of this transitory life . . . It was best, therefore, that the soul of man, which was still weakly desiring earthly things, should be accustomed to seek from God alone even these petty temporal boons, and the earthly necessities of this transitory life, which are contemptible in comparison to eternal blessings, in order that the desire even of these things might not draw it aside form the worship of Him, to whom we come by despising and forsaking such things." St. Augustine, *City of God*, Book X, 14.

188. Gal 3:23–25. The Epistle to the Hebrews states with the death of Jesus the Hebrew system of sacrifices was fulfilled and destroyed. James, *Sacrifice and Sacrament*, 118.

189. "That which is called the Christian religion existed among the ancients . . . from the beginning of the human race until Christ came in the flesh, at which time the true religion which already existed began to be called Christianity." Augustine, *Retractationes*, 1.13.3.

and Christian covenants, therefore, provide examples of modulating and varying relational, credal, cultic, and soteriological architectures within possible extraterrestrial divine-creature relationships. The nature of each relationship is dependent upon the divine prerogative, creaturely response, historical trajectories, and situational particularities, taking example from our own religious histories. Extraterrestrials, therefore, might be considered to share in certain short or long-term relational dynamics with divinity in accordance with divine interactions demonstrated according to the scriptures. The modus of divine election, typically centered on one individual on behalf of a community appears common as a hierarchical model for transmitting divine messages and acts. Jewish and Christian history has clearly indicated that God tends to choose, at least in human civilizations, holy people as his emissaries, with a preference for the uneducated, the poor, and the simple. Throughout the scriptures, it seems divinity waits for a certain individual to be born to inaugurate its plans. In summary, divine action among extraterrestrial civilizations can be expected to share in these aspects in relation to creatures as demonstrated above: God's interaction with creatures demonstrates divine freedom in designing covenant, worship, and economies of salvation; contingency and flexibility of plans in accordance with creaturely action; patience with regard for certain persons to respond or act for a special purpose or mission in furtherance of the divine will; a preference for certain individuals, families, group, or nation as the recipient of divine gifts and relationship; a revealing of divine presence according to varying modalities in relationship with creatures; varying modes of divine communication; and definitive divine acts designed to protect and redeem creatures; each action serving the original and sole design of the Creator in his desire of loving relationship with creatures.

Acknowledging there is considerable debate regarding the accuracy and historicity of certain events in the period describing covenantal history, theologically these are considered representative of important stages of the revelation and relationship between the Hebrew people and their deity. For Christians these histories are important benchmarks for understanding the divine-human relationship which transpired over a long historical period, within a variety of circumstances and personages and culminating in the Christ event. In regards interpreting scripture, Aquinas described several interpretive methods of scripture.[190] Vatican II's *Dei Verbum*, on the *Dogmatic Constitution on Divine Revelation* outlines an approach to scripture

---

190. Aquinas, *Summa Theologica*, I, Q. 1, Art. 10. Aquinas delineated several meanings hidden within scripture. The allegorical (hidden theological meaning), literal, scientific, historical critical, anagogical (heavenly sense), tropological (significance of an individual's behavior).

encouraging biblical criticism, affirms divine authorship, with attention devoted to the content and unity of the whole of scripture. The covenants serve as revelation, memory, and guidepost to understand divinity's will to engage humans in loving relationship in a variety of modes, circumstances, and historical epochs, where examples of divine action are varied; principally, that God is a Lord of history, unlike neighboring polytheists and works towards a goal of human-divine unity. God acts consistently and seeks out humanity in relationship, as understood by the sacred writers, seen from a historical perspective. Each account of divine covenant in the Old seeks to teach an important theological truth, while together they maintain a unity moving toward fulfillment in the New. The economy of salvation as revealed in the Old Testament, and each punctuated by divine covenants, although imperfect and provisional, are necessary and giving witness to the whole divine pedagogy of God's love of humans in varying circumstances. Revelation is, therefore, historical, conditional, and particularized accordingly. The Pentateuch is based on ancient traditions, providing a history of divine acts and theological truths, and represent some of the earliest traditions of the Israelite people. The history of Israel is interpreted as salvation history, a history of man's knowledge of the true God. The Pentateuch is a book of promises; whereby election and promise are ratified by covenant—which brings the Law, which constitutes the foci of the books. However, the theological significance of covenants takes precedence over exact historical narratives. Even if historical accuracy is in some ways removed, they provided a collective theological interpretation of a divine pedagogy within a large expanse of time to an ancient and primitive people. God communicated through the writers to express, through the medium of human words, a supernatural teaching on the person of God, humankind, and the relationship between them. The Papal encyclical *Humani Generis* was promulgated in 1950 to counter among other things, certain scriptural interpretive errors which threatened to undermine authentic teaching. It also encouraged further research on hermeneutical techniques to render a more accurate understanding of the Old Testament:

> Just as in the biological and anthropological sciences, so also in the historical sciences there are those who boldly transgress the limits and safeguards established by the Church. In a particular way must be deplored a certain too free interpretation of the historical books of the Old Testament. Those who favor this system, in order to defend their cause, wrongly refer to the Letter which was sent not long ago to the Archbishop of Paris by the Pontifical Commission on Biblical Studies. This letter, in fact, clearly points out that the first eleven chapters of Genesis,

although properly speaking not conforming to the historical method used by the best Greek and Latin writers or by competent authors of our time, do nevertheless pertain to history in a true sense, which however must be further studied and determined by exegetes; the same chapters, (the Letter points out), in simple and metaphorical language adapted to the mentality of the people but little cultured, both state the principal truths which are fundamental for our salvation, and also give a popular description of the origin of the human race and the chosen people. If, however the ancient sacred writers have taken anything from popular narrations (and this may be conceded), it must never be forgotten that they did so with the help of divine inspiration, through which they were rendered immune from any error in selecting and evaluating those documents.[191]

As the accounts of the early Hebrew books can contain popular narrations utilizing metaphorical language and other literary methods to reveal and instruct divine truths; the remainder of patriarchal history must be considered in the same manner. Divine activity and central events among the Hebrews must be acknowledged to make Christianity comprehensible,[192] although the precision and manner of narrative may vary accordingly as conditioned by historical methodology and context. Therefore, respect for the Hebrew books as divine inspiration is maintained despite manner of expression, and that certain theological and religious truths were intended to be communicated within the medium of human methods.

An economy of salvation by definition is a comprehensive divine action that provides a healing response to creaturely sin. It is a free movement of a supernatural love-gift which (for humans) involves divine revelation, communication, theophany, and grace; historically interwoven with a corresponding creaturely free response of love, action, and faith for the purpose of restoration of the divine-creature love relationship.[193] According to the Christian understanding of divine justice, sin creates imbalance in the spiritual and material worlds by injury to the divine honor in relation to creatures intended for divinely willed ends. An economy of salvation is the divine effort to restore that balance. This is a subject which cannot be explicated in full detail here, but only a guide on general parameters determined consistent with the supernatural revelation of the attributes of the Godhead and intelligent creatures. One form of economy of salvation could

---

191. Pope Pius XII, *Humani Generis*, 38.

192. The denial of the person of Abraham for example, would render much of Jewish history moot, as Jesus himself attested. "Before Abraham was, I AM" (John 8:58).

193. According to the Christian tradition for humans.

be a historically contingent movement of creatures towards the perfection of natural attributes and eventual divine unity. Other types of salvation histories are dependent upon the physical, morphological, and theological anthropology of sentient beings, and divinity's action upon those individually and collectively. Certain of these may be considered equal in dignity/desirability to incarnation and to each other. A starting point in consideration of variant economies of salvation among extraterrestrial species is our human revelation of a Trinitarian reality of the Godhead. Within the *varied* hypothesis of extraterrestrial religion is the argument for a diversity of religious expressions among creatures, flowing from a definitive revelation according to a distinct economy of revelation. Although religions will be diverse, all supernatural economies will be inspired by a Trinitarian God, while not necessarily revealed or conceptualized as Trinitarian (e.g. ancient Judaism). A basic framework for possible economies of salvation would entail fundamentals of the natures and acts of divine persons and creatures. Within the matrix of the divine-creature relationship there are absolutes and finitude; divine free acts and power and creaturely choice. In divinity there is personhood, omni-properties, the love-acts of creation, grace, redemption, and unification with creatures. With creatures there is personhood, freedom, will, mind, body, soul, suffering, death, and the reality of sin. All these provide structure to consider non-human economies of salvation. As mentioned above, economies of salvation can in important ways also depend and differ on the varying theological anthropologies of creatures, historical trajectories, and acts of the divinity.

Supernatural economies could be categorized according the *mode* of interaction or encounter which divinity uses to relate to creatures. These forms would be those which are mediated, where a medium is introduced to serve as a representative/gifter; and non-mediated, where divinity encounters creatures directly without intermediaries. Among those mediated are certain mediator types, of *persons* (e.g., creature-representative,[194] divine incarnation, supernatural apparition/representative);[195] *objects* (e.g.,

---

194. A member of the host race, or a separate, more advanced and spiritually advanced race from another planet or location, or a group within a race who serves as administrators of grace and redemption for another species or group.

195. The dynamic dialectic of the historically hidden, silent divinity interacting with humanity may contrast with a more direct, unambiguous, existential knowledge of divine presence, communication, and relationship in other societies, where "knowing" supplants "believing." The divine hiddenness known to Earth religion, where indirect experience of divinity remains the standard, other religions may have total concealment, relative transparency, or a full revealing to the individual or community, or some combination of the three.

nature, church, *acheiropoieta*);[196] and *events* (e.g., miracle, interior locution, revelation, ecstatic state).[197] Among those non-mediated are direct divine person-to-creature, known by the examples of Adam, Noah, and Abraham talking directly with God, and that between the angels and God, and could include certain mystical religions discussed below. Mediated economies appear more complex. As mentioned according to *person*-mediators, they can be creature-representatives, such as Moses or David; an incarnated divine person as is Christ; or in a non-human example, a supernatural representative such as an angel who carries out certain divine acts or messages to creatures, or the Holy Spirit whose action is normally invisible and imperceptible. In *objects*, divinity can be mediated through nature (although more a system than an object) where divinity is revealed and communicated in general revelation; through church, where the treasure house of graces and divine teaching is administered, and where the divine economy for humans is broken down into its pastoral and ecclesial components; or by means of an object such as *acheiropoieta*, which in a putative extraterrestrial economy could serve a dual purpose as supernatural presence and a communication medium of divine messages; or a physical, constructed (or natural, such as Mt. Sinai) object such as the Ark of the covenant, which becomes 'divinized' and a source of divine presence and power. By *events*, in external or communal forms, mediation could be accomplished by miracles, as a one-time historical event, or in series of supernatural acts which demonstrate the presence and message of a divinity; or interiorly, by means of interior locutions within individuals or collectively, through revelations such as a collective illumination of conscience, or by means of ecstatic states whereby divine messages are provided or miracles are performed. Among these forms, mediations of person do not necessarily require objects or supernatural events; mediation of objects does not necessarily require the presence of persons or events, and mediations by means of events would not necessarily require the perceivable presence of divine persons or objects. Alternatively, it is possible that any singular medium or combinations of these could manifest in extraterrestrial economies of salvation as evidenced in our own history.

As illustrated earlier, scripture details several varying human economies of salvation, each version a divine modification and response to human behavior, and principally as necessary segments of a larger divine scheme of human salvation. Within and among these types, they can be

---

196. *Acheiropoieta*, in Medieval Greek, is rendered, "made without hand," a physical artifact produced solely by supernatural means.

197. These types fall more under the category of non-mediated economies; but are obviously present in mediated economies as well.

further distinguished by types of *process*, for example, as a cosmic-single condition (angelic and Adamic); among those covenantal, there is unilateral gift (Noahic, Abrahamic); provisional-sacrificial-pedagogical (Mosaic, Deuteronomic); and substitutionary atonement (Messianic). According to mediation of relation, they can be grouped as follows: angelic (non-mediated, cosmic, unilateral, provision of supernatural gifts, single-conditional gift of heavenly beatitude, custodial role in spiritual care of humans); Adamic (non-mediated, earthly, unilateral, provision of preternatural gifts, single-conditional gift of supernatural life with divinity, custodial role in care of Earth); Noahic (non-mediated, covenantal-unilateral, unconditional gift of divine election, divine mandate as custodian of Earth); Abrahamic (non-mediated, covenantal-unilateral, unconditional gift of divine election, promise and blessing); Mosaic (mediated [by Moses], covenantal-provisional-remedial-sacrificial-pedagogical, conditional promise of divine election with blessings/punishments); Deuteronomic (symbolic sacrificial atonements, reconfigured Mosaic with additional imposed regulations as a result of continued disobedience); and Messianic (mediated by Christ-Holy Spirit/Church, covenantal-substitutional atonement, conditional gifts of grace and eternal salvation).

Although the human economy has been mediated by persons it is not necessary that extraterrestrials require the same type. Other economies may contain many, all, or few of the elements known to the Judeo-Christian religion, including an overall architecture which can entail pedagogy, punishments, gifts, forms of remembering, teaching, and revelation. All these elements may exist in economies fitting for creatures, whether having lost an original graced creative condition, a lost graced condition following forgiveness, or unfallen creatures gifted with an economy to further their spiritual development. Each economy outlined constitutes its own paradigm of divine relation to creatures which is unique and accommodates the particularities of their ontological and situational reality. Following these, extraterrestrial economies can be considered according to the frameworks and elements provided in human history while taking into account important differences among creatures.

J. Patout Burns outlines two economies of salvation from the patristic period, the Greek form developed by Origen and Gregory of Nyssa, which taught a universal availability of salvation as a gradual development of the soul from birth to beatitude, and emphasized asceticism as a primary means of perfection. In this sense the Greek form emphasized the importance of the continuity of developmental natural processes. This contrasted with the Latin theology best expressed by Augustine, of salvation made available principally through the Church, by fulfilling divinely-imposed conditions for participation in Christ's redemption, where grace available through divine interventional acts in the Church provided the primary means of salvation. In this sense the Latin tradition is centered on a divine redemptive interventionist model.[198] According to Burns, the process of each economy can be understood according to categories of divine and human acts. These may be continuous (a series of stages, each following the last) or discontinuous (prior stages are not required for certain acts); or developmental (each stage builds upon the last) or interventionist (external caused by divinity, no stages are required, not cooperative with subject).[199] A developmental process allows for only co-operative graces. An interventionist process requires only operative graces. These can also be hybridized.[200] Accordingly, Gregory's process of spiritual growth is continuous, developmental, and is achieved by co-operative graces. It has categories of continuity (purification of the body) and discontinuity (the operative interventions of Christ), development and co-operative (cleansing of the soul by grace) and intervention (assistance of the Holy Spirit).[201] Augustine's model emphasized the discontinuity of creaturely efforts and instead stressed the power of divine intervention, with salvation only available through the Church - original sin can only be removed by baptism, available only in the Church which has a constitutive role. The Holy Spirit creates a bond of peace and unity of the Christian community, which mediates Christian virtues by teaching and sacraments. Grace transforms the flesh and provides the soul with the fullness of charity in the vision of God.[202] These categories can be fruitfully utilized in consideration of the modes, elements, processes, and ecclesial forms of extraterrestrial economies in future exotheological studies.

The created capacities, histories, and environments of extraterrestrials would serve to modify elements of the functions of creaturely acts coupled with divine movements of operative and co-operative grace; the duality of divine gift and creaturely response determines future actions of divinity which motivate to satisfy divine justice, of which loved creatures are an integrated part. In the Christian system, all mediations transmit operative grace or elicit a co-operative creaturely response to grace in one degree or another. On their own, creaturely efforts do not produce supernatural merit without the action of operative and co-operative grace, whose transformational

198. Burns, "The Economy of Salvation," 598–601.
199. Burns, "The Economy of Salvation," 600.
200. Burns, "The Economy of Salvation," 601.
201. Gregory of Nyssa, *De Vita Moysis*, 3.12–4.18, 116.21–23. See Burns, "The Economy of Salvation," 606–7.
202. Augustine, *De Spiritu et littera*, 36, 64; Augustine, *De Natura et Gratia*, 38, 45; Augustine, *De Perfectione Iustitiae Hominis*, 6, 14.

nature and action is central in the task of an economy of salvation, and which could vary extensively in conjunction with creatures by virtue of a likely great diversity in created capacities. There can be economies of general or universal salvation, available to creatures regardless of capacities and situation, versus economies which provide grace and salvation for a special privileged group, as in divine election. In other cases, there might be a hybrid of these combining a universal economy (for example, an economy devoid of special revelation, with mediation known through nature and general grace, living according to one's conscience) and a specialized economy (person mediated, supernatural grace with divine teaching). Other extraterrestrial economies may involve inexhaustible combinations of mediations and operations of supernatural grace to creatures which each serve to satisfy divine love and justice. These cannot be limited to Christian forms, and may produce ecclesiologies with recognizable elements and others completely alien. Each economy will have an inner logical consistency, taking into full account divine will, action, and creaturely means and ends. Economies of a universalist type may take non-mediated forms as they are not conditioned by mediators which can be limited historically and situationally; economies of a mediated, specialized type can transform, by historical progression into universalist types, as in Judeo-Christianity.[203] Economies can be provisional, contingent upon creaturely actions, or until the fulfillment of certain divine plans or promises; they can be probationary, entailing typically binary choices for creatures (as with the angels shortly after their creation), or incrementalized as with humans during their life-cycle.[204] Ecclesiastically, they can be mediated hierarchically as with the patriarchs, through the governance of a mediating church or religious group; or non-hierarchical such as the prophets in disclosing divine plans. Each economy can interact with creatures according to internal (individualized) and external (communal) modalities; they are inherently congruent and contingent by virtue of supernatural corollaries which match creaturely nature and action in a variety of circumstances. For example, divine election appears as a form of mediation as a result of widespread sin; therefore, the more sin, the more mediated;

---

203. Universalized economies can be religious-based or manifest as non-religious means of movement towards spiritual evolution to unity with divinity, as in natural religions.

204. Angels, being wholly incorporeal, their knowledge gained by pure intuition rather than by use of senses. Their interactions with humans are incidental, rather than essential to their activity. Humans, by contrast have knowledge which is discursive, beginning with sense and involving a movement from non-knowing to knowing through sense experience and the active intellect. Angels are not necessary but contingent beings, as are humans, existing from God's willful act to supernaturally elevate them to their proper end.

as a result, the earthly economy was historically progressive in a trajectory from unmediated action to the mediated.

If the divine-human earthly analog is reliable as a (albeit limited) model for extraterrestrial economies, other composite rational beings of body/soul could exhibit patterns of relation with divinity known to us, while others might be wholly unknown. These categories of divine actions and creaturely response can be as stated, contingent on factors such as biology, history, theological anthropology, divine will/action and creaturely action. Extraterrestrial economies would likely be highly dependent on the theological anthropology of creatures, as demonstrated in the angelic economy versus the human. Insofar as creaturely action, extraterrestrials need not reject a messiah or other divine representative to fulfill a particular economy of salvation. God is capable of producing the necessary supernatural redemptive effect in other types of historical epochs, persons, and circumstances. It could be argued that divinity may simply forgive certain transgressions out of divine mercy without need for individual or collective sacrifice.[205] The smallest suffering or act on the part of a divine, infinite being could satisfy justice, as Christ's sacrifice can be seen a supererogatory act of love to satisfy the divine honor and restore humanity in proper relationship to God. It can be questioned whether in extraterrestrial societies supernatural acts to satisfy divine justice only deal with past sin or also future sin, as in Christian formulation, where justifying grace is provided by sacraments which forgive past sin, and contend with future sin by the provision of grace; or whether extraterrestrials might be provided with one opportunity for repentance, as known in early church baptismal practices.

Extraterrestrial supernatural mediated manifestations and communications may take a form where a divine being makes itself present on a continual basis, without recourse to inspired writings, sacramental life, priesthood, or hierarchical church structure containing a set of teachings designed to instruct the faithful. Demonstrable supernatural miracles may only be necessary for beings prone to ignorance and a limited vision resulting from loss of grace. Such presence could radically alter the form and function of an economy and trajectory of a particular salvation history. Among non-mediated economies, a mystical religion may manifest as invisible graced and interior states without a historical progression of revelation or theological development; the economy could be completely personalized, non communal, non species-specific, or planetary based. It may produce a type of salvation history characterized in great contrast to our own. Mystical

---

[205] Among Roman Catholics, equivalent to a plenary indulgence; the most well-known example being the *Divine Mercy* devotion.

religions could possibly exist as forms perhaps considered superior to organized religions given its individualized and authoritative nature of revelation. As such, divinity could manifest itself in a purely introvertive union and communicate with creatures without the use of conventional language nor sacraments, scriptural documents, liturgy, or hierarchical authoritative religious structures to serve as a medium through which one encounters the divine. In this case, historical theological development might be unnecessary or irrelevant as a result of received ineffable mystical knowledge, and public supernatural communications unnecessary given individual high capacity of an advanced state of holiness and communion with divinity. Therefore, economies manifesting from the divine will are inherently contingent, characterized by variables demonstrated by divine prerogative, creaturely response, historical trajectories, and situational particularities, exhibiting varying relational and soteriological architectures.

Economies as described in the extraterrestrial context are comprehensive divine actions which represent the summary of creaturely states and behaviors in conjunction with the operations of divine will. Aquinas argued for the fittingness of the Incarnation for humans, as the fullest possible revelation of divinity to creatures,[206] but asserted there were other ways to redeem humanity other than incarnation (which he did not explore). In accordance with this, he stated, "Since God is the universal cause of all things, it is necessary that he aim chiefly at what is useful for the entire universe of things. But the assuming of human nature pertains only to what is useful for man. Therefore, it was not fitting that if God should have assumed a foreign nature, that he would have assumed only a human nature."[207] By incarnation God chose the best way for humans according to our particular nature, history, acts, and divinely intended supernatural end as creatures. In this way, incarnation was not required for humans as there could have been other *equivalent* ways to satisfy the requirements of divine justice, although the Incarnation/Cross and was the most fitting solution in accordance with creaturely need and divine love. Therefore, it may be stated that divine action in the redemption of creatures does not require incarnation; what can be considered by creatures as most desirable and dignified in divine acts with creatures is the particular action which God chooses, as he chooses them.

There are metaphysical and epistemological implications to the *varied* view as well, in how creatures' theological anthropology may exist as a result of a universe containing diversified life forms. One of these can be

---

206. Aquinas, *Summa Theologica*, III, Q. 1, Art. 1.
207. Aquinas, *Summa Contra Gentiles*, IV, ch. 54.

to consider an individual descended from two species, and heir to varying economies of salvation. There are many biological impediments to inter-species reproduction, let alone those of inter-genus among putative biological extraterrestrials. Generally, variant species reproducing sexually would require compatible sexual organs, complementary pheromones, and the means to overcome many other evolutionary mechanisms producing reproductive isolation critical for speciation,[208] as well as pre- and post-zygotic reproductive isolating mechanisms.[209] As a hybrid individual contains half if its chromosomes from each parent, the absence of a necessary gene or presence of a varying one, or difference in the number of chromosomes can arrest normal development and cause non-viability or sterility.[210] However, in the example of hybridized individuals resulting from artificial genetic manipulation, it can be theorized a highly advanced technological society could produce such individuals among divergent species. Theologically, a hypothetical individual derived from two species could be theorized to inherit two distinct economies of salvation.[211] Given our model of humanity with a single soteriological mode according to Christianity, it is a system integrated to speciation as a result of divinity's joining itself to human nature in order to elevate and redeem it through grace. The extent of this economy is biological (meaning humans and their nature) rather than spatial (the physical extent of the Earth) or temporal (within the time continuum of our species). It might be supposed that an incarnation-based soteriology, at least that known in the human example, is species-specific whereas a non-incarnational economy would not necessarily have creature-type boundary or restriction, and could be applicable or available to more than one species; thus our incarnational human model should not be necessarily projected onto other species.[212] Economies of salvation can in important ways depend and differ on the varying theological anthropologies of creatures, historical trajectories, and divine will. Within the Christian model, (which may be considered limited in this context) can be theorized several different states of beings, each corresponding to diverse soteriological arrangements. For sake of this particular question, a homologous supernatural economy of

---

208. Barton and Bengtsson, "The Barrier to Genetic Exchange," 357–76.

209. Mayr, *Animal Species and Evolution*.

210. Strickberger, *Genética*, 874–79.

211. The example used here is of a "rational animal," like humans, composed of an immaterial soul and physical body forming a distinct single nature, but who are differentiated from humans in their accidentals. See *Catechism of the Catholic Church*, 365.

212. A fundamental argument of the *varied* hypothesis against the *multiple* soteriological view.

salvation can be considered,[213] with a similar creature embodying a wounded or imperfected nature with the natural powers subject to it (to include a natural capacity for God) among each species. A creature descended from two species could be composed of a binary of combined natures which may result in one (new) nature, composed of body and soul. In the Thomist theology being explored here, such a creature could potentially have access to either mode of redemption as each has as its product a supernatural religion derived from the same divine source. The operations of actual and sanctifying grace and a creature's reception to it within a duality of supernatural religions can differ; however each mode could enable the elevation to supernatural life and redemption as its subject is one soul. As grace and nature interrelate due to the necessary contingencies of creatures,[214] the nature of the subject would receive supernatural grace according to the particularities of its unique nature. Grace by its operations heals, elevates and completes creatures according to natures and within each, their personal particularities. This model is applicable to a being in reception of two incarnation-based economies as well as that combining an incarnation-based with a non-incarnational-based economy, or dual non-incarnational-based economies. For a hybrid being inheriting that of a natural religion in conjunction with a supernatural religion, the supernatural would take precedence over the natural. Such beings in this case could practise a syncretistic religion or choose one religion; the inheritance of grace from a supernatural religion would supersede a purely natural system.

In the greater cosmic scheme, the creation of intelligences and ultimate calling into intimate relationship with God are fundamental and the self-determinative choice of destiny a necessary consequence of rational beings. Extraterrestrials may have various created capacities, gifts, divine probations and testing, and general outcomes in use of their free will in comparison to humans. In consideration of putative intelligent extraterrestrials, we must consider a range of possible natural/supernatural/preternatural combinations and their natural and supernatural destinies given the unlimited creative freedom of the Creator.

---

213. Although Judaism and Christianity are each considered supernatural religions, they are not equivalent in their modes of grace and scope of redemptive activity.

214. Aquinas, *Summa Theologica*, I, Q. 1, Art. 8. resp. 2. *Gratia non tollit naturam, sed perficit.*

Economies of Revelation

Origen described God's redemptive work as a transcendent action which gradually through time takes effect in every realm of creation but which, nevertheless, needs to find corporeal expression in a particular place on a particular occasion.[215] As demonstrated in the exposition and implementation of divine covenants, the specific divine pedagogy in the divinity's revelation to human beings is in a series of progressive steps; communicated to human beings in stages of supernatural revelations that reached its summation in the person of Jesus Christ. There is a harmony, uniformity, and consistent pattern to God's actions in relationship with creatures revealed on Earth; divine presence and action is mediated in its many forms and modalities. Divine revelations and revealings are always mediated as God is an infinite, inexhaustible being; in a vast universe divine mediations are polymorphous. The divinity makes obvious to creatures what was non-evident, and knowledge from what was pure mystery for the ultimate purpose of creaturely salvation and divine unification. Aquinas observed that, "The right way to manifest the unseen things of God is through things that are seen, and this is the purpose of the whole world."[216] The great volume of varying modalities of supernatural theophanies, manifestations, and divine action in scripture alone provides ample evidences for the immense variety of ways God has available to accomplish his ends for humans within salvation history. These provide indicators of what could be expected of supernatural presence and action among extraterrestrial civilizations for purposes of creaturely edification and redemption.

Theophany and divine presence takes many forms according to Hebrew and Christian scriptural accounts and functioned within diverse contexts in distinct ways: God is manifested in divine covenant;[217] a pillar of fire and cloud;[218] in the tables of the law;[219] and in the divine nearness in the tent of meeting and the Jerusalem temple, unparalleled in history. God was manifested according to the Hebrew scriptures at creation, at the Sinaitic theophany;[220] through actions and presence over the cherubim in

---

215. Lyons, *The Cosmic Christ*, 214.

216. Aquinas, *Summa Theologica*, III, Q. 1, *sed contra*.

217. Gen 9:11, 13; Deut 4:13; 7:9; 31:8; Ps 103:17–18; Heb 8:6; 9:15; 13:20–21; Exod 19:5; 34:28; Job 31:1.

218. Exod 13:21; 14:19–20; 16:9–10, 42; 19:16; 40:34, 36–37.

219. Exod 31:18; 32:15; 34:1, 4, 29; Deut 4:13; 5:22; 9:10–11, 15; 10:1, 3; 2 Chr 5:10.

220. An invisible creative spirit (cf. Gen 1); after the creation of life on Earth and Adam and Eve, described as a physical being "moving about the garden at the breezy time of the day" (Gen 3:8); interacting with first humans (Gen 3:8–19; 4:9–15; 6:13; 7:1;

the ark of the covenant;[221] as a personal being, expressed anthropomorphically and fraternally by Christ as Father;[222] to Jews and his adopted Gentile

---

8:15; 9:1–8, 18). through the intermediaries of angels at the destruction of Sodom and Gomorrah (Gen 18—23). An angel of the Lord speaks the words of God to Abraham in the first person, and the voice of God is heard through the angel, as a manifestation of God himself (Gen 22:12). As an angel of the Lord, a fire flaming out of a bush (Exod 3:2–6). God's presence and power can be known by a storm; as glory of light and brightness (Exod 19:16; 20:18; Judg 5:2–31). In the Sinaitic revelation, God descends and appears upon Earth, accompanied by thunder and lighting, a fiery flame reaching to the sky; and loud notes of a trumpet; the mountain smokes and quakes; out of the smoke and flames a voice reveals the Decalogue (Exod 19:16–25; cf 16:10). The Lord reveals to the seventy elders who accompany Moses, Aaron, Nadab, and Abihu on the mountain, described where "under his feet appeared to be a sapphire tilework, as clear as the sky itself" (Exod 24:10). To the entire Israelite community, the glory of the Lord is seen as a consuming fire on the mountaintop. God's appearance and proximity to humans in the Old Testament invoke awe and fear, and frequently include lighting, thunder, earthquakes, storm winds, brightness, and darkness to demonstrate divine power and presence (Exod 24:9–18; Deut 4:11–12, 33–36; 5:4–19); God states to Moses, "You cannot see my face; for no one shall see me and live" (Exod 33:20). However, God speaks with Moses "mouth to mouth" and "as a man would speak to his neighbor," in clear sight; perhaps a privilege to provided to the six-winged angels of the Seraphim, whose two wings cover their faces so to not look upon the divinity (cf. Num 12:6-8; Exod 33:11; Deut 34: 10; also Gen 17:1; 18:1; Exod 6:2–3; 24:9–11; Num 12:6–8). Pentateuchal narratives contain manifestations in which God appears in "glory," as "messenger," and as a "face." Divinity is represented in a host of expressions appears as a physical being to Hagar (Gen 16:9–13); to Abraham and Sarah at Mamre as three men (Gen 18:1–33); to Jacob at Peniel (Gen 32:24–43); again at Mount Moriah as an angel (Gen 22:11–14); Moses in the burning bush (Exod 3:2–4:17); God principally appears to individuals, and in rare moments groups of people.

221. The Ark's design was instructed by God himself. The ark was carried approximately eight hundred meters in advance of the Hebrew army, and it displayed its powers by parting the waters of the Jordan river, similar to that of the Red Sea, and was present at the destruction of Jericho. God was said to speak to Moses in a cloud appearing "from between the two Cherubim" on the ark's lid (Exod 35:22). Moses was instructed by God not to enter the holy place within the veil enclosing the ark without the cloud of the Lord on the mercy seat, lest he shall die (Lev 16:1–2). The ark served as a leader of the Israelites in their wanderings in the desert and force of their power in warfare, visibly apparent during the day as a column of cloud and in the night as a pillar of fire (Exod 13:21–22). The ark's capture by the Philistines and its placement within the altar of Dagon resulted in a serious of plagues and maladies upon the people of Gath and Ekron (1 Sam 4:8–12), and subsequently returned to the Israelites on a cart yoked by two cows, which made its way to the Israelite town of Beth-Shemesh, coming into the possession of the Bethsames. Due to their lack of proper respect for the ark by opening the lid and peering inside, a large number of Bethsames fell dead. After Solomon's temple was dedicated and the ark placed therein, the temple was filled with a cloud, "for the glory of the Lord had filled the house of the Lord." (1 Kgs 8:10–11; 2 Chr 5:13–14).

222. God is not a remote, indifferent, or malevolent being but rather an intimate, benevolent, and personage deeply involved with human lives. Provider of the Sabbath and Torah to the chosen people; both life-giver and law-giver. The divine personage of God

children;[223] manifested or is represented by an angel of God;[224] as a disembodied voice;[225] in visions and dreams,[226] and through prophecy.[227] The prophets Isaiah and Ezekiel received their commissions in the midst of glorious manifestations of the divine presence; Isaiah saw God only as a glorious robe, the hem and train of which filled the entire temple of heaven.[228] Ezekiel envisioned the divine throne appearing as a chariot, accompanied

---

as Father not only of the Hebrews, but all human beings as promised through Abraham, takes an active interest in human activity as a human father would his own children, human fatherhood is modeled after the divine fatherhood. As father, the divinity is emphasized as having ultimate authority, protectiveness, and powers that far exceed human conception. "Father" implies masculine characteristics. In referring to the divinity, Jesus used male pronouns, and affectionately called him "Abba," a word exclusively used within the family context, to indicate the intimate relationship God desires with human beings. In the giving of the law, God the father maintained an exclusive, covenantal father-child relationship with the Hebrews, who received his laws, acted as stewards of his prophecies and miracles, maintained a special claim of divine election, and received his guidance and special blessings. Jesus revealed the Lord of Israel, of Abraham, Isaac, and Jacob as "father" of the human race, as *principium or origo* of divinity is in the Father. The New Testament reveals God as father in eternal, exclusive, and intimate familiar relationship to his son Jesus, and that fatherhood is inherent to God and the example by which human fatherhood is understood. God was father to the Israelite nation by sovereign election over and above other peoples of the period, and father of Christians. Fatherhood is considered in a more literal sense and substantive sense, requiring the Son as means for accessing the Father, making for more a metaphysical than metaphorical interpretation. See Goshen-Gottstein, "God the Father in Rabbinic Judaism and Christianity." See also Exod 4:22; Deut 8:5; 14:1; 32:6; Isa 63:16; 64:8; 2 Sam 7:13–14; 1 Chr 17: 12–13; Hos 11:1; Mal 2:10; Pss 68:5; 89:26–27; Prov 3:11–12; 103:13; Jer 3:19; 31: 9; Matt 3:17; 5:48; 6:9; 7:11; 11:27; 12:50; 23:9 26; Mark 1:11; 3:35; Luke 2:9; 3:22; 8:21; 12:29–32; John 1:12–13; 5:17–18; 17:11; 20:17; 1 John 2:13; 3:1; Col 1:12; 1 Cor 8:6; 2 Cor 1:3; 3:26; Rom 8:17; 15:6; Gal 3:26, 29; 4:5–7; Eph 1:3; 3:14–15; 4:3–6; Acts 17: 24–28; Heb 2:11; 12:5–6, 9; 1 Pet 1:3; Tit 3:7; 1 Pet 1:17.

223. John 1:3, 12; Gal. 4:4–7. See Scott, *Paul's Way of Knowing*, 159–60; Rom 8:23; 9:4; Eph 1:5.

224. Gen 16:7–9; 18:1–2; Exod 3:2–6; Josh 5:14; Judg 2:1–5; 6:11, 20–25.

225. Isa 6:8; 64:4; Exod 19:9; Num 7:89; Deut 4:12, 33, 36; 5:23; Dan 8:16; 10:9; John 12:29; Heb 3:7, 16; 4:7; 2 Pet 1:18.

226. Gen 15:1–21; 16:7–13; 18:1–33; 20:3–7; 28:12–17; 40:12, 13, 18–19; 41:25–32; Dan 2:16–23, 28–30; 7:1–2; 1 Sam, 3:15, 21; Isa 5:1–5, 15; 6:1–13; Amos 7: 1–6; 8:1; 9:1; Jer 1:11, 13; Ezek. 8–11; Zech. 2:1–5. And early Christian Church: Acts 16: 9–10; Dan 7—8; Rev 4:2–3, 12, 17).

227. Isa 38:1, 4–5; 40:3; 55:10–11; Deut 18:22; 1 Kgs 8:15–21, 23–24; 11:29–39; 12:15; 13:1–3, 21–22; 14:7–13; 15:29; 16:1–7, 34; 20:13–21; 17:14, 35–36; 21:20–29; 22:17; 2 Kgs 1:6; 7:1–2; 15:12; 19:20–37; 20:1, 4–6, 17–18; 21:10–15; 22:15–20; 23:16–18, 30; 24:2; 25:1–7; 2 Chr 6:4–11, 14–15; 36:17, 21; Hab 1:6–11; Jer 21:3–7; 29:10; 31:15; 32:3–5; 39:1–7; Ezek 12:12–14; Dan 9:2; Joel 2:28–32; Josh 6:26; Matt 2:17–18; 3:1–3; 21:11; Mark 1:2–4; 6:4; 8:27–28; Luke 3:1–6; 7:16; 18:31; 24:44; 25:17–19; John 4:19; 1 Cor 12:10; 13:1–2; 14:3–5, 26, 28–29; Rom 12:3–6; Eph 12:10; 1 Pet 4:10; Acts 2:14–21; 3:18; 11:27–28; 15:32; 21:9.

228. Isa 6:1–7; Ezek 2:1—3:3; Rev 4:2.

by a great cloud and ceaseless fire, surrounded by an amazing brightness.[229] In his vision of the cherubim, he saw the divine being, having the likeness of a man, whose upper body is shining, and lower surrounded by flames.[230] Later, Christ's physical presence in the Incarnation went beyond the divine self-communication in the creation of the material world and divine presences among the Hebrews. In the messianic age, God is first presented to humanity as a helpless infant;[231] as the Word-Logos;[232] as the Incarnation of the God-man Jesus;[233] and as an active force after the ascension of Christ in the person of the Holy Spirit;[234] and through the speaking in tongues.[235]

---

229. Ezek 1:1–3.

230. Ezek 10:20.

231. Ps 72:11; Isa 7:14; 9:6; Matt 1:23; 2:1–12; Luke 1:39–55; 2:7, 15–20; John 1:9–14; 3:16, Gal 4:4.

232. John 1:1–14. The Logos before the creation of the Earth, the preexistent and transcendent and incarnate Word and self expression of God in the Second Person of the Trinity (John 1:1–14). Christ is referred to as "the power of God, the wisdom of God" (1 Cor 1:24); "the wisdom of God" (2 Cor 4:4).

233. Lord and brother, the highest and fullest manifestation of the divine, taking the nature and form of a specific species. God self-communicates his own divine life, joining in a single creature the human and divine natures. Through this divine self-manifestation within human nature we have witnessed the fullness of human nature, of a being given fully over to God (John 8:20; 14:6, 9; 1 John 2:1; 5:1).

234. In the canonically Old Testament texts the Spirit is known in the creation of the universe (Gen 1:2) and producing prophecy (Gen 41:38; Num 24:2; 1 Sam 10:10; Hos 9:7; Mic 3:8; Ezek 2:2; 3:24). In later periods, the Spirit is active in messianic and eschatological activity, bringing salvation to Israel, and pouring itself upon Israel's leaders, prophets, and people (Isa 11:1–10; 42:1; 61:1; 32:15; 44:3; Ezek 39:29; Joel 2:28). It was commonly held by those of later pre-New Testament Judaism that the Spirit had departed the community due to disobedience and sin but would be return upon the restoration of Israel. Jesus's birth and ministry begins through the action of the Holy Spirit, and it figures prominently in the final commissioning of the apostles to baptize in the Holy Spirit (Matt 10:17–22; 12:17–28; 28:18–20), its descending in physical form at Pentecost, and in the spectacular outpouring of the Spirit in Acts (2:1–4; 4:28–31; 8:15–17; 10:44; 19:6). For Paul, the Holy Spirit resides in the individual Christian as well as the leadership in its guidance of the Church. Paul described nine specific gifts of the Holy Spirit (1 Cor 12—14: wisdom, knowledge, healing, miracles, prophecy, discernment of spirits, speaking in tongues, the interpretation of tongues, and love), and its seven fruits (Gal 5: 22–23: love, joy, peace, patience, kindness, generosity, faithfulness, gentleness, and self-control). According to Paul, the Spirit helps those who cannot pray properly and is integral to the salvation of all creation, and takes the place of carrying out the miracles and works of Christ among the early Christian community after the Ascension. Among the Old and New Testaments is seen a progression of the activity of the Holy Spirit, where God's Spirit manifests among the Hebrews indirectly and singularly among its prophets and seers, and directly and communally within the infant church. The Old Testament refers to a wind, breath of life, or divine inspiration of the prophets (Pss 33:6; 147:18).

235. 1 Cor 14:1–5, 18–19, 22–25, 27–28, 39–40; Acts 2:1–4; 19:16; John 16:13–15; Eph 5:18.

God is present under the forms of bread and wine in the Eucharist;[236] in the form of the person encountered on the road to Emmaus;[237] as a man on the shore at the Sea of Tiberius,[238] to Mary Magdeline,[239] as a glorified being envisioned during the transfiguration.[240] Divinity acted in the revelation to St. Peter;[241] the apparition of Christ to St. Paul;[242] in the risen Christ to the apostles;[243] in the form of the Trinitarian indwelling of God's presence after Jesus's ascension, and in the descent of the Holy Spirit at Pentecost.[244] Numerous other divine action and modalities exist within tradition and salvation history. Divine action is manifested through various miracles, whether spiritual or physical healings, Eucharistic miracles, incorruptibles, and exorcism;[245] by divine message: as public revelation[246] in scripture and expressed in Church teaching in special revelation; as private revelation by means of angelic intermediaries, through nature, Marian and other apparitions, interior locutions, visions, dreams, ecstasies, mystical revelations, 'presence,' and mediation of saints; and throughout ordinary[247] and extraordinary divine grace known in sacraments, sacramentals, teachings, and by the good works of graced people.

An inventory of these divine modalities and actions provides a topology of divine revelations in other putative intelligent species; divinity which purposefully creates with the sole intention of relationship reveals to creatures within, and as evidenced on Earth, beyond their natural capacity for knowing the divine. As revealed in the Jewish and Christian testaments, the

236. Matt 26:17–30; Mark 14:22–24; Luke 13:26; 22:19–20; John 6:35, 51, 53–57; 13:1–4; Acts 2:42, 46–47; 20:7; 1 Cor 10:16–17, 21–22; 11:20–34. As Christ is present in the Eucharist, we see him in the "forms" of bread and wine; however, that physical reality is transcended by a reality of God present, unprecedented in the history of salvation.

237. Luke 24:30–31.

238. John 21:1.

239. At first recognized by Mary as another person. John 20:11–18.

240. Matt 17:1–8; Mark 9:2–8; Luke 9:28–36; 2 Pet 1:16–18.

241. Matt 16:13–15.

242. Acts 9:1–19.

243. John 11:25–26; 20:8–9; Mark 16:6; Luke 24:6–7; 1 Thess 4:14; 1 Pet 1:3; 3:21; Matt 20:18–19; 28:5–6; 1 Cor 15:1–11; 2 Cor 5:14–15; Rom 6:5–6; Heb 13:20–21; Acts 26:22–23; Rev 20:6, 12–13.

244. Acts 2:1–13, 41–42.

245. Luke 4:41; 13:32; Mark 1:2, 23–26, 39; Matt 12:28; 17:18.

246. Aquinas taught that all public revelation ceased with the death of St. John the Apostle. Private revelation cannot surpass, correct, improve, fulfill, complete, or perfect public revelation. *Catechism of the Catholic Church*, 66–67.

247. Ps 19:1; Rom 1:18–20. God can first be known through the light of natural reason.

divinity is a personal being and not merely an impersonal creative or sustaining force in the universe. It may be expected that extraterrestrial theologies might acknowledge the Creator as a personal being, as intelligent and non-intelligent beings are inherently and naturally relational on a variety of levels; humans not only desire relationship but require them to function individually and collectively.[248] Therefore, revelation of God as a personal being, composed of three divine persons who exist in dynamic, divine relationship is consonant with his earthly creation. As a personal being, God initiates action out of love to create, maintain, or redeem creatures. In each historical manifestation God reveals some aspect of himself.[249] Divine purpose determined the manner and type of divine manifestation/revelation: divine plan, leading, encouraging, punishing, communicating, displaying power, prophesizing, preparing, teaching, assuring, confirming, proving, forgiving, and redeeming creatures, among others. God uses multisensory methodologies characterized through teaching and action. Pre-eminent in the divine pedagogy in relation to humans, and presumably other intelligent creatures, is mediation; all supernatural actions made to creatures are transmitted through finite manifestations and action, while there remains the apocalyptic hope of the unmediated presence of God in the end times.[250]

In comparison to the Hebrew Scriptures, the testimony of the New Testament demonstrated a new divine initiative with an accompanying new economy of revelation with the advent of the Messianic age and the divine Incarnation.[251] Mediated through the Church, continued divine action in tradition so described made apparent an entire new set of grace-filled divine actions known in revelatory events.[252]

248. There is a true necessity of the human nature for union with the supernatural order, not merely a capacity.

249. Martin Buber and Franz Rosenzweig, in their translation of the Hebrew Bible into German, in the account of Exodus 3:14 of Moses' encounter with the Lord in the burning bush, indicate God's proper name in Hebrew, YHWH, transliterated as "the Lord." They interpreted the biblical iteration in which the name is described in Hebrew *eheye asher ehyeh* as "I will be there howsoever I will be there," meaning God is present in the mode of presence which he chooses, among an innumerable modalities possible in the universe. See Buber, "Gottesfinsternis," 503–603.

250. 1 Thess 1:10; 2:19; 3:13; 4:13–5:11, as in the afterlife.

251. This included a new knowledge and artistic portrayal of God in the Christian world. Whereas in ancient Judaism the use of divine images were condemned, the Incarnation of God in a human being introduced a new, accurate, and acceptable economy of divine images.

252. These are revelations according to recognized mystics: works, teachings, and miracles ascribed to holy saints, approved Marian apparitions and messages, locutions, and healings, teachings of the doctors of the Church, and incorruptible saints, among others.

Accordingly, extraterrestrial civilizations could demonstrate a new, unknown economy of revelation consonant with our own, containing none, some, or all of the supernatural elements familiar to humans. Divine acts in Earth history are oriented, without exception, towards the restoration, maintenance, and final fulfillment of the human-divine relationship. Other civilizations on planets, event timelines, and historical trajectories might display divine action or a series of supernatural actions according to the same divine will for creatures. The historical "playing out" of the interchange of events between creature and Creator in other societies would be dependent, using Earth as an example, on God's free action and creaturely response. As illustrated in the history of covenants, each was initiated in response to human sin, and each remained valid until its reoccurrence required the modification or reformulation of a new covenant. Therefore, a trajectory of divine revelations and forms of covenant or divine-creature relation can be dependent on creaturely response; what God does in civilizations is tethered to free individual and communal behavior as a result of the originally and inescapably instituted divine-creature relationship. What one party does in relationship (in the case of the Hebrews, contractual) invariably affects the other(s). However, extraterrestrials may not reject a "saviour" or messiah figure in their own histories; the patterned history of sin, forgiveness, covenant, and breaking of covenant of the ancient Hebrews need not find parallels in other peoples. Divine action, revelation, and salvific acts might be as legitimate as those known to humans, and need not have distinct parallels in human civilization. Within the expanded context of extraterrestrial intelligences created by God, we must consider anew divine action, relations, and manifestations according to our terrestrial accountings and consider them as *analogical* representations of other, similar and dissimilar actions, relations, expressions, and appearances of God or other divine beings in other civilizations, rather than a singular manifestation of divinity within a solitary creation of intelligent species in the cosmos.

Syncatabasis

In considering the pedagogy of divine initiatives in relationship with extraterrestrials, and taking into account the biblical record and Christian tradition, the hermeneutic of God's "accommodation" in his self-revelation with creatures according to each particular and varying condition within an historical epoch is evident. Divinity modifies self-revelation to the unique conditions of creatures by "*syncatabasis*." God "particularizes" his manifestation and communication according to individual species' biological, historical, psychological, social, cultural, and even individual circumstances, to create a synthesis of internally and externally consistent interactions within the divine relationship; as expressed in the two testaments, incorporating covenants, theophanies, communications, and culminating with a divine-corporeal redeemer. John Chrysostom describes this aspect of the divine pedagogy as *"attemperatio"*:

> In sacred scripture, therefore, while the truth and holiness of God always remains intact, the marvelous "condescension" of eternal wisdom is clearly shown, "that we may learn the gentle kindness of God, which words cannot express, and how far He has gone in adapting His language with thoughtful concern for our weak human nature."[253]

Each divine revelatory and definitive action known in scripture and tradition is an example of a continuous *syncatabasis* for the exposition and medium of supernatural work within and for creatures. This is a central principle of the *varied* argument of exotheology. The Church expresses this pedagogy as "God, wishing to speak to men as friends, manifests ... by adapting what He has to say by solicitous providence for our earthly condition."[254] God meets each in their own place, "Being a work at once common and personal, the whole divine economy makes known [to humans] both what is proper to the divine persons and their one divine nature."[255] *Syncatabasis* in human and, in presumably extraterrestrial civilizations present a union of the divinity, divine action, physical place, time, and person(s), which create the form, method, and conditions for a divine-creature encounter for purposes of the most fitting revelation, edification, and supernatural unity with creatures. Humans are engaged by divinity in a manner consonant with their innate, divinely created capacities, subjected to history, culture, and personal and collective circumstances. Any revealing does not complete or exhaust all which can be known of God, and is accomplished in a series of progressive steps in accordance with human ability to understand and respond. Maimonides stated that the ideal religion originally gifted on Mt. Sinai was to include neither temple nor sacrifice.[256] However after the "original sin" of the Hebrews over the golden calf, a pedagogical-remedial-sacrificial religion was instituted in place of pagan idols to wean the Hebrews away from the Egyptian idolatry

---

253. Chrysostom, *"In Genesis"* 3, 8 (Homily l7, 1) PG 53, 134; *"Attemperatio"* (in English "Suitable adjustment") in Greek *"synkatabasis."*
254. *General Directory of Catechesis*, n. 146.
255. *Catechism of the Catholic Church*, n. 259.
256. Maimonides, *Guide for the Perplexed*, 1190, III:32.

and towards fitting worship.[257] Therefore, divinity accommodated the divine plan in response to the Hebrew rejection of divine authority and relationship; however God did not desire to institute the pedagogical Mosaic Law in perpetuity as it was provisional in nature; an "imperfect" set of laws was enacted due to the limitations of humans, and a more perfect teaching and messenger would come at the proper time. As John Chrysostom wrote, "God Therefore, countenanced sacrifices which he did not want, in order to assure the success of what He really wanted."[258] According to the *Glossia Ordinaria*, "The Law, indeed, like their Teacher, stipulated that one offer sacrifices to God, in order to avoid sacrifices to idols, as they would be occupied with these lawful sacrifices. The Mosaic Law was a divine concession made to the Hebrews in exchange for their destruction after the sin at Sinai. God "accommodated" the original divine plan, substituting a lesser model of divine relation, in order to have it fulfilled in the new law during the messianic age.[259] Nevertheless, He made the sacrifices holy by letting them foreshadow the mysteries of the New Covenant."[260] Therefore, the divine accommodation in the historical trajectory of salvation history on Earth required that God not reveal to the Hebrews the mysteries of the Christian faith until the proper time. Further, divine accommodation is evident in the creation of the canon of scripture; divinity influences its creation by means of fallible humans to accomplish a divinely willed objective: divine truth is communicated through humans so to be understandable to humans. The Incarnation is the supreme example of *syncatabasis*, the summation of all prior divine accommodations.[261] In the Incarnation God is present in the human condition, revealed in the most perfect and appropriate manner for humans, as human. This is the condescension, or more accurately, considerateness of the divinity in its engagement in the messianic age. By Incarnation human weakness and history were taken into full account, but in no way comprised the integrity of divine truth. Justin

---

257. Martyr, *Dialogue with Trypho*, ch. 19. Also Tertullian explained, "God did not want them for Himself, but He was moved by His solicitude for a people given to idolatry and disobedience. He wished to attach them to Himself by arrangements similar to those in force in contemporary paganism, but with a view to turning them from their idolatry. Furthermore, He prescribed that the sacrifices be offered to Himself, as if He desired them, in order that the people not sin by offering sacrifices to idols." Tertullian, *Contra Marcion*, Book II, 18.

258. Chrysostom, *On Is.* 1.4 (PG 56, 19).

259. Martyr, Justin. Adv. Haer. IV.

260. *Glossa Ordinaria*, "Lex ergo, quasi paedagogus eorum praecepit Deo sacrificare ut in hoc occupati abstinerent se a sacrificio idolatriae. Tamen sanctificavit sacrificia quibus mysteria significantur futura." PL 113, 344–46 (early 12th century).

261. In extraterrestrial civilizations, this accommodation can take a vast range of modalities and forms, to include incarnation.

Martyr expressed it thus, "As in the historical Incarnation the eternal Word became flesh, so in the Bible God's glory *veils itself* in the fleshly garments of human thought and human language.²⁶² The entire soteriological system of the crucifixion and death, and resurrection of Jesus are all modalities in which God accommodated the needs of human civilization to be reconciled to the Father. A more recent instance *par excellence* of divine pedagogical "*Attemperatio*" is the Guadalupian *acheiropoieta* artifact of 1531, through an image intended to transform the persons and relations between Indian and European in post-conquest Mesoamerica by means of a codex, incorporating symbolic language to transmit Christian teaching in consonance with indigenous Aztec theology.²⁶³ The image provides evidence of a new, previously unknown divine pedagogy which served as transmitter and translator of Christian to Indian; in effect recasting and relegating indigenous gods under the headship of the Christian God.

Therefore, every divine action is necessarily made and modulated as *syncatabasis*; due to God being an infinite, vastly incomprehensible reality to creatures who are subjected to a spatial/temporal and situational matrix, whether terrestrial or extraterrestrial. Extraterrestrial societies would experience divine action, taking the form of theophanies, manifestations, and other acts, as has been illustrated for humans throughout Judeo-Christian history. A divinely instituted and inspired religion, with humanity as an example can evolve in its central elements of worship, community, doctrine, and praxis in accordance with divine initiative and creaturely response, and consonant with the principle of *syncatabasis* and spiritual and cultural evolution. Therefore, religious types and expressions among extraterrestrial civilizations can be manifested along an array of potentialities which are determined in large part on contingent divine actions and correspondent creaturely responses.

The Exotheological Metanarrative

The *exotheological metanarrative* is defined as the macro-scale divine life, religious formulations, theological architecture, and divine relations encompassing creation within a universal context among a diversity of

---

262. Chase, *Chrysostom*, 17.

263. Exhaustive accounts on Guadalupe can be found in these texts: Royer, *The Franciscans Came First*; O'Leary, *Our Lady of Guadalupe*; Cawley, *Guadalupe from the Aztec*; Rengers, *Mary of the Americas*; Lynch, *Our Lady of Guadalupe and Her Missionary Image*; Franciscan Friars of the Immaculate, *A Handbook on Guadalupe*; Guadalupe, *The Seven Veils of Our Lady of Guadalupe*.

intelligent beings; and of the transterrestrial and universal heterogeneous interpretation of mundus-centric parochial perspectives on unresolved questions regarding creatureliness and its role within an incomprehensibly vast and unexplored creation. The exotheological metanarrative postulates a cosmic, plenary unity of divine action and salvation history is possible, even likely, and that the history of sin and salvation contained in the Old and New Testaments, and further demonstrated through tradition constitutes a record of divine action on one particular planet, within a single race of beings, in a particular epoch, among a vast array of other civilizations. The lack of scriptural prohibitions, the fundamental Christian teaching of an omnipotent and omnipresent Trinitarian Creator, and the unknowable freedom of God within an expanding human perspective of his visible creation further weakens the hypothesis for a single divine revealing on a solitary inhabited planet in the cosmos. Fundamentally, whether or not realized, all creation, life, and principally intelligent life is created for the purpose of participating in divine life. Terrestrial religions, upon discovery of an advanced extraterrestrial race or races should prepare for the possibility that there exists an external, cosmic metanarrative beyond that of human awareness but which encompasses a host of inhabited planets, our galaxy, a cluster or superclusters of galaxies, or potentially the entire cosmos and all intelligent life forms contained therein. This metanarrative could embrace primeval information which include the sciences, philosophy, historical chronicles, and importantly, theological data; which knowledge may be the result of accumulated data from one particular race or a compendium of a variety of disparate civilizations. A reasonable hypothesis can be drawn that the universe has produced and will continue to produce innumerable histories analogous or non-analogous to ours, and that Homo sapiens occupy one link within a vast collection of heterogeneous civilizations which extend throughout the universe. Given this possible context, it is necessary to consider that terrestrial theologies be understood as a singular example and potential subset of what may exist in a broader context of other, intelligent non-human religions, and of the transcendence of divine action and plans beyond our geocentric and anthropocentric concerns.

God is not an inert being but an active force and his being is realized in unlimited and unknown ways by means of creation, sustenance, and the raising up of intelligent creatures to his love and life. He is an infinite source of potential and actual beings, motivated out of love and generosity to create other intelligences to share his life. His will is manifested through external realizations in the inorganic, organic, material and spiritual intelligences. Given the aforementioned possible extraterrestrial typologies, it is possible that there exists a vast continuum of beings in a variety of states, relationship to the Creator, and of varying capacities and functions. We may expect that there exist creatures to which God has revealed himself in varying measures, however less, equal, or greater than our own. The nature of divine relationship to creation is the love-gift, the first gift the act of creation, of being, then of consciousness, and of intelligent creatures, self-consciousness and awareness of the existence and relationship to the Creator. All rational creatures are necessarily free, capable of discerning and choosing between good and evil, between serving the Creator or refusing to serve, and are tested prior to ultimate union in divine life. The *varied* form of exotheology is not neo-dominionistic, nor imperialist but acknowledges the total validity of the corpus of the Christian message of the unique Incarnation, Redemption, and salvation through the God-human Jesus Christ; while acknowledging the legitimacy of the varied forms of divine self-communication according to their individual manifestations and provision of the necessary means to accomplish the union or reunion of creatures with the Creator. A cosmic redemptive plurality does not impede humans from expounding and legitimizing our particularized and terrestrial salvation history within the Judaic and Christian traditions. Salvation history can be as long as the universe, perhaps extending into other universes. Humans, therefore, possess insights into a minute, compartmentalized and particularized portion of a potentially infinity of divine activity.

Individual religious truths among worlds and societies are not relative; although particularized, each valid religion responds to a plurality of divine manifestations which must be internally coherent as God is an ultimate unity. The distinctiveness of human experience with the divinity in relation to other races renders our experience unique within a continuum of unique experiences. Contact with extraterrestrial life would highlight our common heritage as creatures of the same divine Creator. We recognize similar grains of truth within various Earth traditions, many of which were independently founded and developed. Through our interactions with other civilizations, we learn the relationship between God and his creation, and recognize commonalities in how God has acted throughout various intelligent civilizations. Given the age of the universe estimated at ≈ 13.8 billion years, and the advent of the genus Homo sapiens occurring only a short time ago in cosmic terms, it follows that humanity may occupy a position at the lower end of sophistication within a large spectrum of extraterrestrial civilizations which populate solar systems with parent stars much older than our sun. The most-cited claimed conflict of Christianity with the new cosmology was its long-standing assertion to the particularity of salvation in Christ, which transcends a single, remote, and isolated human civilization of a mere few thousand years, to include the remotest regions of an utterly incomprehensible vast universe composed of billions of galaxies, hosting possible civilizations that pre-existed humanity for millions, if not billions of years. However, the reality of divine creation and self-manifestation well before and after our historical epoch cannot be excluded from possibility. Given our incomplete physical and intellectual perspectives within a transient terrestrial human history, human religious claims are to an extent limited to our present knowledge of the universe. After approximately six thousand years of documented human history, we are only just now beginning to piece together our true place within creation. According to the revelations of paleontology, biology, and geochemistry, humanity has been demonstrated as a recent and perhaps transient manifestation in cosmic time. The early Christian church and Christian theology has spent two millennia in delving into and integrating the historical events, teachings, and personages of the New and Old Testaments as one narrative of the divine intervention within the human species. Similarly, if a second genesis is discovered within an unknown and foreign divine manifestation and salvation history, theology may require decades or centuries in understanding and synthesizing its message and modus in conjunction with ours. Christianity remains an 'open system,' that is, open to new types of information whether, from the human disciplines or intelligent extraterrestrials. Christianity is enriched through the historical process of accommodating, as well as assessing new information according to its own particular world view while retaining its veracity and internal integrity among other truth claims and new knowledge among a variety of disciplines. The discovery or disclosure of extraterrestrial existence would provide a new opportunity for accommodations of new modes of thought, providing further enrichment of Christian theology, in continuity with previous historical developments.

EXOTHEOLOGY AND TRADITIONAL CHRISTOLOGICAL FORMULATIONS 265

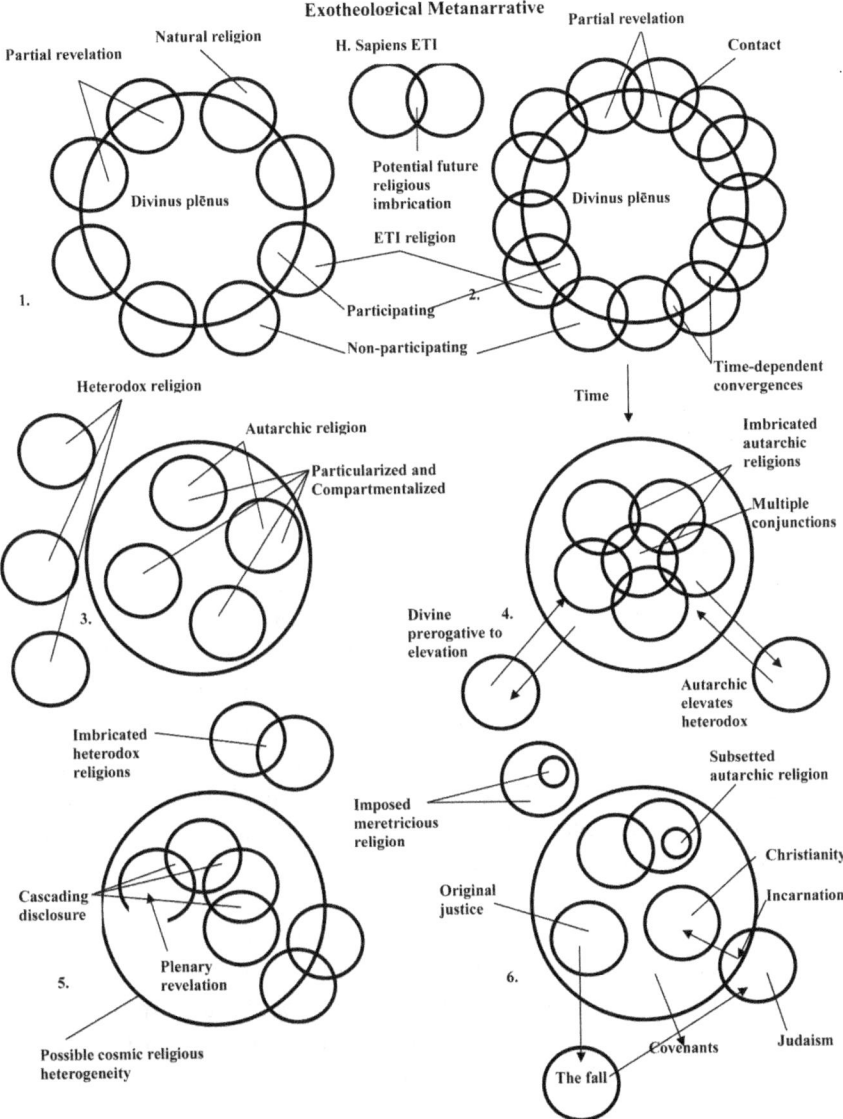

Exotheological Metanarrative Diagram

The Venn diagrams above illustrate a perspective of the *varied* view of the relations between possible religions among intelligent civilizations in the universe. Each large circle represents the domain of plenary supernatural life. Divinity initiates and reveals a religious system according to the principles of divine prerogative and *syncatabasis* within intelligent civilizations. The *Divinus plēnus* represents the plenum of divine action in a material universe, in which individual divinely instituted religions share in supernatural life. Each small circle represents, for simplicity, an individual extraterrestrial civilization containing a single, independent supernatural or natural religious tradition. At top center is shown a future potential contact event of humans with extraterrestrials, each possessing a divinely instituted and revealed religion; in this case, each religion could share some theological resonance, while not sacrificing its uniqueness and specialness in conjunction with another supernatural religion. Figure 1 indicates several divinely initiated extraterrestrial religions, however each containing divine and mundane elements; ancient Judaism would serve as an example of a supernatural religion containing remedial and mundane features, such as Mosaic Law, as a result of sin. Each civilization in Figure 1 can be categorized as fallen, or conversely, in need of redemption or divine completion and perfection. Whether sin in societies originates as a singularity or manifests as a collective phenomenon, each religious tradition subjected to an original loss of original divine relationship awaits further revelation or divine action for a restoration of an original supernaturally created unity with divinity. In this sense, each religion contains incompleteness, as a species in need of restoration to divine relationship require further revelation, grace, and supernatural completion; as such, each religion possesses participating and non-participating aspects in supernatural religion. Figure 2 represents contact events among civilizations and individual religions of the same types; some resulting in syncretistic beliefs in compatible and contrasting elements of the supernatural and natural, or non-syncretized, imbricated traditions maintaining shared recognition of the divine nature of each. Figure 3 represents autarchic, particularized, and compartmentalized supernatural religions according to planetary or planetary systems, or other domain. Christianity is an example of a fully realized, divine religion, existing separately but equally among other foreign, but valid supernatural extraterrestrial religions. Heterodox religions, typically arising out of nature or represented by distortions such as those known in certain Earth societies, exist outside the divine plenum, and are devoid of supernatural presence or action. Figure 4 indicates an instance of elevation of a heterodox or natural religion by means of divine prerogative; in other cases a supernatural religion can serve as a means of elevation for a heterodox population, as illustrated in the Christianization of indigenous peoples of the New World. Civilizations possessing autarchic religions and capable of contact, whether through long-range communication or direct contact in time may become imbricated; with some becoming syncretistic and others maintaining individuality and separateness. There could be multiple conjunctions of this kind, alliances created where each acknowledges and recognizes the divine and supernatural character and legitimacy of others. Figure 5 represents the possibility of a confluence of two (or more) heterodox religious traditions; it also indicates the possibility of an autarchic religion receiving a fuller disclosure of divine revelation beyond present knowledge, resulting in a cascading disclosure among linked or imbricated divine religions. All divinely instituted religions can be heterogeneous and participate in supernatural life, but cannot exhaust divine revelation. Figure 6 represents the example of Judaic-Christianity within this larger context; presupposing an original divine relationship between God and humans, a sin event or events, resulting in a fracturing of the divine-human relationship. The divine initiative of election and covenant, which ultimately resulted in the Mosaic tradition, reinstituted that relationship while falling short of that originally designed. A reconstituted divine relationship achieved by means of the Mosaic Law provided renewed access to the divine life and relationship; and was brought to completion with the Incarnation: the divine initiative to reintegrate the Nation of Israel, and all humanity into a fully realized, autarchic religion of Christianity. Figure 6 also indicates an instance of a supernatural religion in guiding a lesser developed civilization in true divine relationship; and conversely the instance of a heterodox civilization imposing a meretricious religion on a lesser advance or developed civilization. Divine initiative is always active in creating, sustaining, leading, teaching, redeeming, and calling creatures to greater relationship in fulfillment of the original divine intent of creation.

Paradigm of the Varied Type

Christianity, affirmed

The *varied* type paradigm asserts the human, terrestrial account of salvation in Jesus Christ as an authentic, divinely instituted religion among an unknowable number and types of economies of salvation among extraterrestrial civilizations, and represents the final evolution in the development of doctrine with regard to Christianity's role among potential outside intelligences. The Christian religion is not compromised by the existence of extraterrestrial intelligences, as Jesus's special role in the human economy of salvation is not affected; there need be no direct relation to human history of the historical God-man Jesus with other, external divine acts other than they be divinely inspired as truth cannot contradict truth; God cannot deny himself.[264] As Jesus provides complete knowledge of God sufficient for human beings of our particular epoch; other, continuous and conforming, but unknown divine truths can be disclosed to other sentient species. God is knowable whilst eternally unknowable; in encountering a supernatural extraterrestrial religion humans would not find themselves completely alienated as divinity exists as an ultimate unity. Therefore, there is no need of the particularized Christian religion with its exclusive ties to geography, history, and Person be expanded to a vast universe of undoubtedly varying environments, creatures, and plurality of histories. The elements of covenant, grace, sacrifice, and redemption of our divinely instituted religion are circumscribed in an anthropological-historical-geocentric matrix; and cannot be theologically or practically adapted to extraterrestrial creatures. The silence of scripture on the subject of intelligent extraterrestrials, therefore, should not be considered as evidence of their absence, but rather evidence of the Judeo-Christian narrative and teachings as a *human religion* whose revelation and economy of salvation are neither designed nor intended for exportation to other species. Christianity is the sole possession of humanity as God's special and individual revelation and divine gift to and for humans—so special that divinity was incarnate in the human species to demonstrate and complete the divine will of human salvation. Earthly salvation history therefore must be considered as one action within a much larger, vast divine plan for creatures. The revelation to humanity according to the Old and New Testaments is affirmed within a greater context of other, valid divine manifestations, revelations, and actions.[265] God's greatest accommodation of humanity is in scripture, its "prime analog . . . [in] that (other) Incarnation of the Word in the person of Jesus."[266] Jesus Christ is by no means affected by past, present, or future divine relationships with intelligent extraterrestrials; the Christ as redeemer and mediator is affirmed as the unique manifestation and personification of God to Homo sapiens. The supernaturally instituted, divine revelation and relation to humans cannot be falsified within a context of other intelligences as it remains divine.

264. 2 Tim 2:13. "Truth cannot contradict truth" (Pope Leo XIII, "*Providentissimus Deus*"); "Two truths cannot contradict each other" (Galilei, *Letter to Madame Christina of Lorraine*, 120. Also expressed in *Lettter to Castelli* (1613), 107, as, "It is impossible for two truths to contradict each other." The concept, however, predates Galileo, and was expressed by Averroes and Aristotle; see Guessoum. *Islam's Quantum Question*.

265. "Christian faith cannot accept 'revelations' that claim to surpass or correct the revelation of which Christ is the fulfillment" (*Catechism of the Catholic Church*, 67). Christ's revelation is super-abundant in revealing the meaning and purpose of human life. New revelations from extraterrestrials should not be allowed to impact the human mode and message of salvation.

266. John Chrysostom in Hill, *Reading the Old Testament in Antioch*, 39.

The Christian religion and its terrestrial foundations are reaffirmed within the context of inestimable divine manifestations of God, as the unrepeatable and unique testament to divine love within a specific supernatural realization willed by a divine Creator in the human species. As a child is born, loved, and raised in the natural setting of family and home, in God's *attemperatio*, Judaism and Christianity were instituted, loved, and evolved within the context of the human family and home of Earth—each enters the greater 'world' amidst potentially equivalent, although varying models and persons of love and relationships. Humanity in the encounter with an extraterrestrial 'other' may achieve adulthood and affirm its rightful place as a divinely created, graced, loved, and redeemed species within a continuum of God's other realizations of divine love and beauty. The exotheological metanarrative of divine *attemperatio* reveals a Christianized humanity *recontextualized* as a particularized, species-specific religion and spiritual reality among other beings and revelations; there is no annihilation of doctrine or substantial modifications to Christian teaching or praxis; rather a 'relocating' of human-Christianism within a greater matrix of divine presence and actions among intelligent beings in a vast universe. Paradoxically, rather than the feared subjugation of human religion, knowledge, and experience in the event of contact with intelligent beings, highlighted will be humanity's uniqueness and specialness within a larger theater of varied divinely created intelligent beings, each possessing unique salvation histories, and producing cultures and societies amidst galactic timescales and distances.

Therefore, this thesis proposes these merits of the *varied* view of exotheology: that the unlimited freedom and omnipotent will of God to create intelligent beings throughout the cosmos is not antithetical to Christian orthodoxy, rather, it strengthens and expands human notions of God's beneficence and magnificence, and that the fullness of revelation, the free offer of grace, and the salvation won and gifted to humanity by Christ can be similarly realized in pluriform hypostatic unions, diverse theophanies, and unknown supernatural actions of the divine will. This view acknowledges and recognizes the diversity of divine manifestations and means of communicating, sanctifying, and redeeming humanity according to Hebrew and Christian traditions, and argues their fittingness as terrestrial analogs for a possible vast continuum of divine expressions and relationships within extraterrestrial civilizations. It posits that divine disclosure in general and special revelation occur on a historical and environmental trajectory as evidenced in Earthly religious systems, and regards the biblical record and Christian tradition as a paradigmatic exemplar of other possible, yet undiscovered supernatural interventions in non-human societies.

Conclusion

The history of philosophical, theological, and scientific inquiries of the plurality of worlds, and the directly related subject of intelligent extraterrestrials represents attempts spanning nearly two millennia to accommodate or 'locate' non-human rational beings within an internally and externally coherent framework inclusive of each discipline. Theological considerations about outside intelligences were challenged by advances in science, predominantly telescopic technology beginning with Galileo, which inaugurated a progressive de-centering of humanity within an increasingly expanding cosmos. The modern era exponentially intensified this process with new insights into the nature and physical extent of the universe and the discoveries of exoplanets with the *Hubble* and *Kepler* telescopes; each revealing in more profound detail the actual context of humanity within an incomprehensibly vast creation. Hesitatingly and incompletely, theologians offered solutions to reconcile the capital doctrines of the Incarnation and Redemption of Christ within the potentiality of other, divinely created extraterrestrial intelligences in this expanded setting. These theological approaches, spanning centuries in efforts to resolve the quandary of divine relation to humans and extraterrestrials, resulted in the reflexive effort to maintain the closed, classical isolationism of the *exclusivist* model; the strained, discursive tethering of a 'cosmic' Christ of the *inclusivist* model; the narrow, presumptuous imposition of the *multiple* model; and among the least developed—but most generative, the beauty, freedom, and versatility of the *varied* model.

Fundamentally, the *varied* form acknowledges and respects the omniproperties of the Creator in fashioning worlds, beings, and relationships in accordance with the inscrutable divine will. It argues that Earth, the human being, its civilization, and central religious identity and legacy of Christianity provide a measure of attestations of the modalities and realizations of divine work elsewhere in creation. It recognizes a divine pedagogy evidenced in the canon of scripture and tradition as a means to understand supernatural acts with creatures; each representative of divine action, relation, revelation, and creaturely response in extraterrestrial civilizations, and each intended to be understood as *analogical* rather than a singular series of events between Creator and creature on a solitary planet. It asserts all divine acts, and the supreme, unique act of the Incarnation of the God-man Christ in human civilization as one modality and representative of other, potentially equivalent divine acts of redemption, unification, and finalization of creatures in civilizational epochs in accordance with the principles of divine prerogative and *attemperatio*. Therefore, discovery or contact with extraterrestrial intelligences, rather than necessitating an abandonment or major reformulation of Christian doctrine, represents the final phase in the *natural* process of an evolutionary reorientation of humanity as a species and Christianity as a divinely instituted *human religion, recontextualized* within a panoply of organic divine actions, processes, and expressions in the universe; expanding our knowledge of the extent of God's action, and providing further insight into the unique phenomenon of the human being as a particular manifestation of the divine will.

The Earth within the Milky Way galaxy moves in a cluster of other galaxies, within a larger group of galaxies, within a larger collection of superclusters. NASA in 2013 estimated that forty billion habitable planets exist in our galaxy, in an observational universe containing hundreds of billions of galaxies; astronomers have calculated the potentiality of a hundred quintillion Earth analogs,[1] locales of potential independent sacred traditions in the universe where there may exist intelligent life. New worlds are evolving and old ones are becoming uninhabitable. There is structure, order, and purpose in the physical universe; as in the human mind, the message of Christianity, and testimony of divine work on Earth. Different epochs of human and alien civilizations are entirely possible; our planet will continue to be habitable for six hundred million to one billion years,[2] well beyond what many believe the time of fulfillment of the Christian eschaton. Our home as humans is the Earth, the universe is our ultimate horizon; Christianity will remain the domain of human beings regardless of our cosmic context. Our vocation as humans does not set us apart from our fundamental relatedness to an entire universe and community of beings. Christianity exists as an integral part of a vast, incomprehensible whole in the larger scope of God's entire creation and divine work.

1. Choi, "New Estimate for Alien Earths."
2. O'Malley-James et al., "Swansong Biospheres," 99–112.

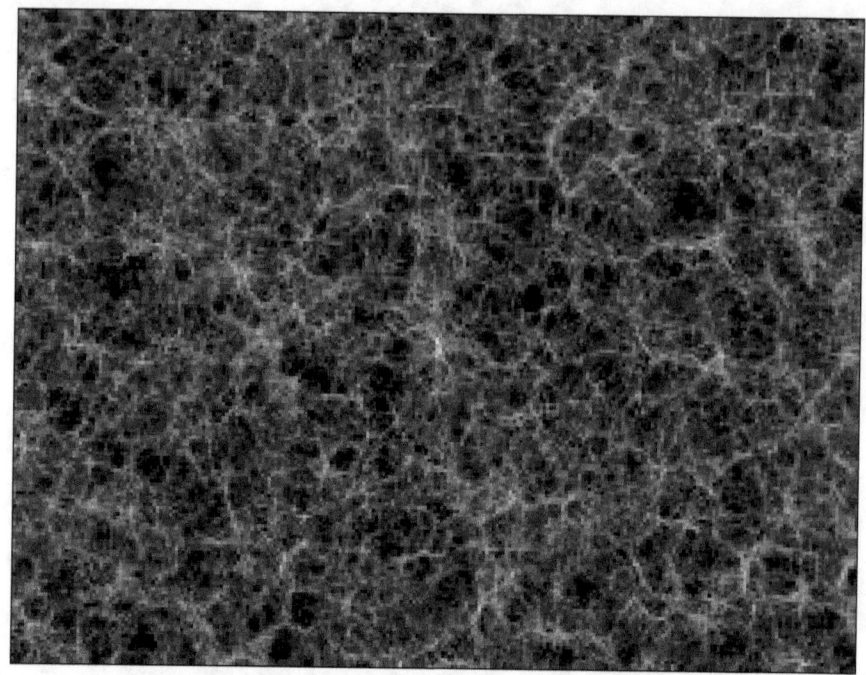

Simulated large scale structure of the universe, showing the distribution of galaxies and interstellar space. (Credit: The Millennium Simulation: V. Springel et al. 2005, Nature 435, 629)

This book provides a propaedeutic for a continuing exotheology, a natural theology to understand divine presence and action within intelligent civilizations in the absolutest context. It is within such a framework that exotheology can inform other theologies, by accommodating a larger, broader perspective in considering the milieu of humanity in this greater setting. The advances in space exploration provide a new frontier for theology in the study of creation and role of humanity within it. Exotheology is destined to be a companion to scientific explorations of the universe, the search for exoplanets and their environments, and astrobiology in its quest for life in other worlds, simple or complex. As science is tasked with reaching out to explore space and the integration of its findings into corpus of human knowledge; Christian theology has an equivalent responsibility in the exploration of God's creation where consummations of the divine will are manifested. Exotheology follows discovery.

Historical Cosmology and Theological Timeline

Period	Proponents	Scientific Cosmology	Scientific Development
30,000 B.C.	Primitive man	Magic universe[1]	First calendars[2]
4,000-800 B.C	Sumerian, Assyro–Babylonian, Minoan, Chinese, Norse, Celtic, Mayan, Egyptian, Hebrew, etc.	Mythic universe[3]	Celestial poles, zenith, meridian, ecliptic, planets, zodiac, heliacal setting & rising, celestial equator, solstices, equinoxes, Saros cycle[4]
6th c. B.C	Thales, Anaximander, Pythagoras, Plato	Pythagorean system[5]	Eclipse predictions, angular measurement[6]
4th c. B.C– 3rd c. A.D	Mediterranean world	Aristotelian universe	Spherical moon, relative distances of Sun and moon[7]
Late 4th c. B.C— 1st c. A.D.	Mediterranean world	Epicurean universe	Precession, star catalog, Earth size

Scientific World View	Theological Cosmology	Theological Canon	Theological World View
Geocentric, flat earth, surrounded by watery abyss Earth sole planet in universe	Anthropocentric Animistic	Grounded in the superstition of benevolent and demonic spirits inhabiting all living and non-living things	Geocentric, flat earth, surrounded by watery abyss Singularity of worlds
Geocentric, flat earth, finite sphere of fixed stars Earth sole planet in universe	Anthropocentric Pantheistic Monistic	Mythic gods served and protected humans, astrology, creation stories	Geocentric, flat earth, finite sphere of fixed stars Singularity of world, Creation myths
Geocentric, spherical earth, finite sphere of fixed stars Plurality of worlds, Atomistic[13]	Anthropocentric Pantheistic	Greek pantheon governed movements of heavenly bodies	Geocentric, spherical earth, finite sphere of fixed stars Singularity of worlds
Geocentric, spherical earth, finite sphere of fixed stars Singularity of worlds	Anthropocentric Pantheistic Monistic	Accepted in Judeo-Christian-Islamic world	Geocentric, spherical earth, finite sphere of fixed stars Singularity of worlds
Geocentric, spherical earth plurality of worlds, infinite plurality of worlds, infinite universe, Atomistic	Anthropocentric Atheistic	Condemned in Judeo-Christian-Islamic world	Geocentric, spherical earth, finite sphere of fixed stars, Singularity of worlds

Period	Proponents	Scientific Cosmology	Scientific Development
3rd c. B.C.–2nd c. A.D.	Roman Empire	Stoic universe	—
2nd c.–16th c	Ptolemy	Ptolemaic system[8]	Epicyclic theory
Early Middle Ages	Europe	Medieval (mythic) universe[9]	—
High Middle Ages 11th–14th c.	Anselm	Empyrean (Stoic)[10]	—
1540–1690	Copernicus	Copernican system	Telescope, unseen stars, lunar mountains, Jovian satellites, Venetian phases, sunspots, moon as an independent world
1596–1687	Descartes	Cartesian system	Saturnian rings (Huygens), planetary elliptical orbits, Kepler's laws, speed of light

Scientific World View	Theological Cosmology	Theological Canon	Theological World View
Geocentric, spherical earth, finite sphere of fixed stars, surrounded by infinite void Singularity of worlds	Anthropocentric Monistic Pantheistic	Accepted in Judeo-Christian-Islamic world	Geocentric, spherical earth, finite sphere of fixed stars, surrounded by infinite void Singularity of worlds
Geocentric, spherical earth with eccentrics, epicycles, equants within finite sphere of fixed stars, singularity of worlds	Anthropocentric Monistic	Accepted in Judeo-Christian-Islamic world	Geocentric, spherical earth, finite sphere of fixed stars, surrounded by infinite void Singularity of worlds
Geocentric, flat earth surrounded by watery abyss Singularity of worlds	Anthropocentric Monistic	Accepted in Judeo-Christian world	Geocentric, flat earth surrounded by watery abyss Singularity of worlds
Geocentric, spherical earth surrounded by sphere of fire occupied by God, Singularity of worlds	Anthropocentric Monistic	Condemnation of 1277 of Aristotelian cosmology[14]	Geocentric, spherical earth, surrounded by sphere of fire occupied by God, Singularity of worlds of worlds
Heliocentric, spherical earth, surrounded by finite sphere of fixed stars, plurality of worlds	Anthropocentric Monistic	RCC imprisonment of Galileo, forced to recant. Died under house arrest. RCC acknowledges error in 1993. Execution of Giordano Bruno in 1600 for asserting plurality of worlds heresy.	Geocentric, spherical earth, finite sphere of fixed stars Singularity of worlds
Vortices,[15] plurality of worlds, Indefinitely extended universe Plurality of worlds	Anthropocentric Monistic	Strongly opposed by RCC, accepted in Reformation countries	Geocentric, spherical earth, finite sphere of fixed stars Singularity of worlds

Period	Proponents	Scientific Cosmology	Scientific Development
1687–1750	Newton	Newton-Euclidian universe	Explanation of tides, precession of Earth's axis, paths of planets and comets, calculus, universal gravity, relative/absolute motion, properties of light, reflecting telescope
1750–1840	Wright, Kant	Hierarchical universe[11]	Milky Way system, other nebulae; photography, spectroscopy, star, nebulae composition and their radial velocities, Bodes rule, astrophysics, Doppler effect
1837–1901	Herschel	Victorian universe[12]	Age of universe (disagreed with scripture), finite speed of light, nature of nebulae, identification of globular clusters, distance indicators, interstellar light absorption
1917–1929	Einstein, Sitter	Einsteinian universe	Special and general relativity theory, space-time
1927–present	Lemaitre	Big Bang	Red shift, curved space, black holes, first discovery of exoplanets in 1995

Scientific World View	Theological Cosmology	Theological Canon	Theological World View
Geocentric, spherical earth, plurality of worlds, infinite universe (Epicurean)	Anthropocentric Monistic	Idea of infinite universe rejected by Church as incompatible with scripture of creation in time	Geocentric, spherical earth Singularity of worlds
Galactocentric, galaxy composed of billions of stars surrounded by other, similar galaxies, each with its own "supernatural galactic center" Plurality of worlds	Anthropocentric Monistic	Idea of infinite universe rejected by Church as incompatible with scripture of creation in time	Geocentric and galactocentric, finite universe Singularity of worlds
Galactocentric, surrounded by star clusters and nebulae, beyond an infinite void similar to Stoic view. Plurality of worlds	Anthropocentric Monistic	RCC rejected notion of plurality of worlds, with Earth as sole habitat of life. Until 1930s affirms scriptural accounting of Earth's age of approximately 10,000 years.	Geocentric and galactocentric, finite universe Singularity of worlds
Homogeneous, isotropic, centerless and static universe with curved space. Plurality of worlds	Anthropocentric Monistic or Atheistic	Theological accommodation to evidence of extreme antiquity of Earth	RCC accepts homogeneous, isotropic model, retains singularity of worlds view
Expanding centerless universe from singularity. Plurality of worlds	Anthropocentric Monistic or Atheistic	RCC affirmed big bang's notion of God creating a point in time, in accord with scripture.	Geocentric, expanding universe from singularity. Plurality of worlds

Table Notes

1. A product of early animism in primitive social groups, the projection of anthropocentricity upon the living and non-living world created a pantheistic perspective motivated by indwelling spirits who acted as controlling forces. The magic universe served as a precursor to astrological and provided explanations for unknown processes on earth and in the heavens. Control over these forces was gained by way of prayers, supplication, sacrifice, and gifts to the spirits, benign and demonic.

2. Thousands of artifacts dating as far back as 30,000 BC indicate early record-keeping by the early peoples across Europe, Africa, and the Soviet Union. Many accounted for the days between moon phases and other celestial patterns, and later, after the domestication of animals and agrarian social development, became important for seasonal predictions. Accumulated astronomical experience revealed the correlation between the cycle of seasons and the apparent movement of the Sun, which aided in the understanding of the exact number of days in the year and defining exact dates for celestial events. Mesopotamian calendars were in development between 4000 and 3000 BC.

3. A natural evolution from the magic universe, animistic and pantheistic indwelling and free spirits were redefined within a more transcendent framework, becoming powerful mythic gods who personified abstractions of thought and language. The beginning of the age of theism, it developed approximately ten thousand years ago in the early city-states of Sumeria, Assryo-Babylonian, Greek, Chinese, Norse, Celtic, Mayan, and others. From the mythic universe originated creation myths, most prominent those of the Sumerian epic *Enuma Elish*, Egyptian, Indian Vedic texts, Greek, and within the ethical religions of Confucianism, Buddhism, and Zoroastrianism.

4. Astronomical discoveries developed by Babylonic study of the heavens, motivated by astrolatry and astromancy. Explanation of these processes were guided by mythic principles rather than scientific rationale.

5. Pythagoras was first to propose a spherical Earth, and put it in the center of a spherical universe composed of perfect spheres. Later Pythagoreans proposed that the Earth moved, as the Moon and other planets, around a distant center. His system predates Copernicus' correct model of the solar system by approximately two thousand years.

6. Angular measurement system, developed by the Greeks, from which developed Euclid geometry and the Pythagorean theorem. Angular measure provided a means to accurately locate celestial objects, revealing a detailed systematics of the movement of heavenly bodies.

7. Aristarchus (ca. 310–230 BC) devised a system of measurement based upon geometry of the Moon's orbit, phases, and eclipses, by which he closely calculated the relative sizes and distances of the Sun and Moon. His findings led to the conclusion that the Sun, not Earth must be the central body in the solar system, which remained unconfirmed until the Copernican Revolution.

8. Ptolemy (2nd c. BC) an astronomer and mathematician at the Museum of Alexandria, sought to bring the Aristotelian system in accord with astronomical observations to account for planetary retrograde motion and changed apparent luminosity with the added idea of the equant, or "equalizing point" a non-central point about which a planet orbits at a constant angular rate. Postulated earlier by other astronomers were the idea of rotating spheres, called epicycles, to account for planetary retrogression

and progression and varied luminosity. Ptolemy's *Almagest* sought to harmonize, with the equants, a precise picture of the observed motions of the plants in a geocentric universe, which endured for 1400 years until overthrown by the works of Copernicus, Kepler, and Galileo.

9. The Early Middle Ages brought the fall of the Roman Empire in the 5th century, and all scientific and intellectual pursuits decayed under the rule of barbaric tribes. Cosmological perspectives reverted to the mythic polarization of heaven and hell, a flat Earth in the form of a rectangular tabernacle surrounded by an abyss of water. Remnants of ancient knowledge survived in Persia, Syria, and Byzantium.

10. St. Anselm of Canterbury (11th c.) described the universe as the Earth surrounded by a sphere of stars, surrounded by a realm of purest fire where God dwelt. As astronomical knowledge expanded, and with it, the theorized extent of the universe, as well as the abode of the creator.

11. Proposed by Thomas Wright, Kant, and Johann Lambert, in attempt to explain the nature of nebulae observed beyond the disk of the Milky Way galaxy. Although galactocentric, Wright argued for a multitude of "centers of creation" within each galaxy populated by other, unknown intelligent life forms. Conjectured were that galaxies were contained within a still larger cluster of galaxies, and these clusters within even larger clusters.

12. The standard cosmological theory during the reign of Queen Victoria of Great Britain (1837–1901), geocentric and galactocentric, maintained a one-galaxy universe surrounded by small "islands" or nebulae and ultimately an unknown, endless Stoic void.

13. Anaxagoras, and later Leucippus (6th c BC) proposed the Atomic theory of causality, whose universe was composed of an infinity of atoms, composed of similar substance although different in shape and size, within a void infinite in extent. Atoms moved freely about the void, eternally colliding and aggregating to form celestial bodies, and decaying over time, to later coalesce to form other bodies.

14. Aquinas' accommodation of Aristotelian philosophy resulted in conflicts with Christian theology that required modification, given Aristotle's finite universe and argument that God could not move the Earth if he willed or create other worlds if he willed. These constraints on the power of God were rejected in the Condemnations by Etienne Tempier, bishop of Paris, in order to affirm God's omnipotence, eternity, and infinity.

15. Descartes proposed, in contrast to Aristotle's finite universe, and in acknowledgement of the necessity in preserving God's infinity, an "indefinite" universe governed by natural laws whereby all forces act via direct physical contact, as he found forces acting without direct material contact were contrary to reason. In this system, planetary orbits resulted from vortical motions of interplanetary matter and pressure was exerted by the force of swirling fluids, so that each solar system acted as a vortex connected to an endless expanse of solar systems.

Bibliography

"17 Billion Earth-Size Alien Planets Inhabit Milky Way." Space.com. January 7, 2013. http://www.space.com/19157-billions-Earth-size-alien-planets-aas221.html.

"1,901 New Kepler Candidates." NASA. February 28, 2012. http://kepler.nasa.gov/news/nasakeplernews/index.cfm?FuseAction=ShowNews&NewsID=190.

Adams, George. *Lectures*, Barthold Heinrich Brockes, *Irdisches Vergnügen in Gott* in 9 volumes (1680–1740) Memphis, Tennessee: General Books LLC, 2012, Vol. I, p. 435; vol. IV, p. 244.

Adams, Marilyn. "Christ as God-Man, Metaphysically Construed." In *Oxford Readings in Philosophical Theology*, edited by Michael Rae, 239–63. Oxford: Oxford University Press, 2009.

———. "The Metaphysics of the Incarnation in Some Fourteenth-Century Franciscans." In *Essays Honoring Allan B. Wolter*, edited by William A. Frank and Girard J. Etzkorn, 21–57. St. Bonaventure: Franciscan Institute, 1985.

———. "What's Metaphysically Special about Supposits? Some Medieval Variations on Aristotelian Substance." *Aristotelian Society Supplementary Volume 79* (2005) 15–52.

Adriani, Oscar, et al. "The Discovery of Geomagnetically Trapped Cosmic Ray Antiprotons." *The Astrophysical Journal* 736.29 (2011) 79–87.

"A Long Childhood Is of Advantage: Synchrotron Reveals Human Children Outpaced Neanderthals by Slowing Down." Max-Planck-Gesellschaft. November 15, 2010. https://www.mpg.de/617475/pressRelease20101111.

Alexander, Victoria. *The Alexander UFO Religious Crisis Survey: The Impact of UFOs and Their Occupants on Religion*. Las Vegas: Bigelow Foundation, 1994.

Allen, Gavin. "Did Jesus Die for Klingons Too?" Dailymail.co.uk. October 4, 2011. http://www.dailymail.companyuk/sciencetech/article-2044730/Did-Jesus-die-Klingons-Christian-Weidemanns-speech-100-Year-Starship-Symposium.html.

Allen, John, Jr. "This Time, the Catholic Church is Ready." *National Catholic Reporter*. February 27, 2004. http://natcath.org/NCR_Online/archives2/2004a/022704/022704e.htm.

Allen, James. "Supernova Effects." NASA/GSFC. February 2, 1998. https://www.mutah.edu.jo/eijaz/supernova.htm.

Almar, Ivan. "The Consequences of a Discovery: Different Scenarios." In *Progress in the Search for Extraterrestrial Life*, edited by Seth Shostak, 499–505. San Francisco: Astronomical Society of the Pacific, 1995.

Angelo, Joseph. *The Extraterrestrial Encyclopedia*. New York: Facts of Life, 1991.

An Expanded View of the Universe- Science with the European Extremely Large Telescope E-ELT Science Office. http://www.eso.org/sci/facilities/eelt/science/doc/eelt_sciencecase.pdf.

Anonymous. "Messages from Space." *America: A Catholic Review of the Week* 111 (1964) 770–71.

———. "No Room for Christian Faith." *Sign: A National Catholic Monthly Magazine* 36 (1956) 14.

Aristotle. *On the Heavens*. Translated by Keith Chambers Guthrie. Loeb Classical Library. London: Harvard University Press, 1953.

Ashkenazi, Michael. "Not the Sons of Adam: Religious Responses to ETI." *Space Policy* 8 (1992) 341–50.

Asimov, Isaac. *Extraterrestrial Civilizations*. New York: Crown, 1979.

Association of Religion Data Archives. "The World." http://www.thearda.com/internationaldata/regions/profiles/Region_23_1.asp.

Aquinas, St. Thomas. *Commentary on Aristotle's "On the Heavens."* Translated by Pierre Conway and F. R. Larcher. Unpublished but circulated in photocopied form, 1963–64.

———. *Commento al Corpus Paulinum: Expositio et Lectura Super Epistolas S. Pauli, ad Ephesios*. Vol. 2. 9th ed. Turin: Marietti, 1986.

———. *Compendium Theologica*. In vol. 1 of *Opuscula Theologica*, edited by Raymond A. Verardo. Rome: Marietti, 1954.

———. *Expositio In Libros Aristotelis De Caelo et Mundo*. Edited by Raymond M. Spiazzi. Turin: Marietti, 1952.

———. *In Aristotelis Librum De Anima Commentarium*. Italy: Marietti, 1959.

———. *In Duodecim Libros Metaphysicorum Aristotelis Expositio*. Edited by Raymond M. Spiazzi. Rome: Marietti, 1950.

———. *Quaestio Disputata de Anima*. In vol. 2 of *Quaestiones Disputatae*, edited by Pio Bazzi. Turin: Marietti, 1965.

———. *Quaestio Disputata de Spiritualibus Creaturis*. In vol. 2 of *Quaestiones Disputatae*, edited by Pio Bazzi. Turin: Marietti, 1965.

———. *Scripta super libros sententiarum magistri Petri Lombardi*. Paris: Lethielleux, 1929. Albany: Global Academic, 1997.

———. *Scriptum super libros Sententiarum*. Albany: Global Academic, 1997.

———. *Summa Contra Gentiles*. London: University of Notre Dame Press, 1955.

———. *Summa Theologica*, III, QQ. 1–59. London: Catholic Way, 2014.

———. *Summa Theologica*. Fathers of the English Dominican Province. New York: Benziger Brothers, 1947.

———. *Super Epistolas S. Pauli, ad Hebraeos. Lectura*, vol. 2. Edited by Turin R. Cai. Rome: Marietti, 1953.

———. *Super Evangelium Ioannis Lectura* 1:4, Turin: Marietti, 1952.

Arendzen, J. P. *Whom Do You Say-?: A Study in the Doctrine of the Incarnation*. New York: Sheed and Ward, 1941.

Athanasius, St. *On the Incarnation*. Edited and translated by Penelope Lawson. New York: Macmillan, 1981.

Atkins, John, et al. *The RNA World: The Nature of Modern RNA Suggests a Prebiotic RNA World*. Plainview: Cold Spring Harbor Laboratory, 2006.

Augustine, St. *The City of God*. New York: The Modern Library, 1950.

———. *De Genesi adv. Man*. Augustine Institute, Villanova University, 1980.

———. *De Natura et Gratia*. Paris: Societe Philosophique, 1953.
———. *De Perfectione Iustitiae Hominis*. London: Forgotten Books, 2018.
———. *De Spiritu et Littera*. Ann Arbor: University of Michigan Library, 1914.
———. *Letters*. Translated by John George Cunningham. Germany: Jazzybee Verlag, 2015.
———. "Literal Meaning of Genesis." *Ancient Christian Writers*. Vol. 1, Mahwah: Paulist, 1982.
———. *Retractationes*. Fathers of the Church Patristic Series. Translated by Mary Inez Bogan, RSM. Washington, DC: Catholic University of America Press, 1999.
Aune, David. "Cosmology." In *The Westminster Dictionary of the New Testament and Early Christian Literature*. Westminster: John Knox, 2003.
Aunger, Robert. *Darwinizing Culture: The Status of Memetics as a Science*. Oxford: Oxford University Press, 2000.
Bailey, Cyril, ed. and trans. *Epicurus: The Extant Remains*. Oxford: Clarendon, 1926.
———. *The Greek Atomists and Epicurus*. New York: Russell & Russell, 1964.
Baker, Kenneth. *Jesus Christ—True God and True Man: A Handbook on Christology for Non-Theologians*. South Bend: Saint Augustine's, 2013.
Balabanski, Vicky. "Critiquing Anthropocentric Cosmology: Retrieving a Stoic 'Permeation Cosmology' in Colossians 1:15–20." In *Exploring Ecological Hermeneutics*, edited by Norman Habel and Peter Trudinger, 151–59. Atlanta: Society of Biblical Literature, 2008.
Balducci, Corrado. "UFOs and Extraterrestrials - A Problem for the Church?" *UFO Evidence*. http://www.ufoevidence.org/documents/doc814.htm.
Balentine, Samuel. *The Torah's Vision of Worship*. Minneapolis: Fortress, 1999.
Barbour, Ian. *Issues in Science and Religion*. New York: Harper and Row, 1966.
———. *Religion and Science, Historical and Contemporary Issues*. San Francisco: Harper and Row, 1997.
Barrow, John D. *The Anthropic Cosmological Principle*. Oxford: Clarendon, 1986.
Barton, Nick, and Bengtsson, Bengt Olle. "The Barrier to Genetic Exchange Between Hybridizing Populations." *Heredity* 57.3 (1986) 357–76.
Basalla, George. *Civilized Life in the Universe: Scientists on Intelligent Extraterrestrials*. Oxford: Oxford University Press, 2006.
Battaglia, Debbora. *E.T. Culture: Anthropology in Outerspaces*. Durham: Duke University Press, 2005.
Batalha, Natalie. "Exploring Exoplanet Populations with NASA's Kepler Mission." *Proceedings of the National Academy of Sciences of the USA* 111 (2014), 12647–54.
Baxter, Anthony. "Chalcedon, and the Subject of Christ." *Downside Review* 107 (1989) 1–21.
Baxter, Steven. *The Science of Avatar*. London: Orbit, 2012.
Baylor Institute for Studies of Religion. *American Piety in the 21st Century: New Insights to the Depth and Complexity of Religion in the US*. September 2006. http://www.baylor.edu/content/services/document.php/33304.pdf.
Beal, Heather. "Thermophiles" In *Encyclopedia of Environmental Microbiology*, vol. 3, edited by Gabriel Bitton. Hoboken: Wiley, 2002.
Becker, Jurgen. *Paul: Apostle to the Gentiles*. Louisville: Westiminster, 1993.
Bennett, Charles, et al. "Nine-Year Wilkinson Microwave Anisotropy Probe (WMAP) Observations: Final Maps and Results." *The Astrophysical Journal Supplement Series* 208 (2013) 20. http://arxiv.org/abs/1212.5225.

Bentley, Richard. *A Confutation of Atheism from the Origin and Frame of the World.* London: St. Paul's Church-yard, 1693.

Benton, Michael J. *When Life Nearly Died: The Greatest Mass Extinction of All Time.* London: Thames & Hudson, 2003.

Berger, Peter. "Reflections on the Sociology of Religion Today." *Sociology of Religion* 62 (2001) 443–54.

Bertka, Constance. "Astrobiology in a Societal Context." In *Exploring the Origin, Extent, and Future of Life,* edited by Constance Bertka, 1–18. Cambridge: Cambridge University Press, 2009.

———. "Christianity's Response to the Discovery of Extraterrestrial Intelligent Life: Insights from Science and Religion and the Sociology of Religion." In *Astrobiology, History, and Society: Life Beyond Earth and the Impact of Discovery,* edited by Douglas A. Vakoch, 329–40. Heidelberg: Springer, 2013.

———, et al. "Workshop Report: Philosophical, Ethical, and Theological Implications of Astrobiology." *American Association for the Advancement of Science.* Washington, DC, 2007.

Bettenson, Henry, ed. and trans. *The Later Christian Fathers: A Selection from the Writings of the Fathers from St. Cyril of Jerusalem to St. Leo the Great.* Oxford: Oxford University Press, 1972.

Bianciardi, Giorgio, et al. "Complexity Analysis of the Viking Labeled Release Experiments." *IJASS* 13 (2012) 14–26.

Beadle. George W. "The Place of Genetics in Modern Biology." Eleventh Annual Arthur Dehon Little Memorial Lecture. Cambridge: Massachusetts Institute of Technology, 1959.

Bieri, Robert. "Humanoids on Other Planets?" *American Scientist* 52.4 (1964) 425–58.

Billingham, John, et al. *Societal Implications of the Detection of an Extraterrestrial Civilization.* Mountain View: SETI, 1999.

———. "Summary of Results of the Seminar on the Cultural Impact of Extraterrestrial Contact." In *Bioastronomy '99: A New Era in Bioastronomy,* edited by Guillermo A. Lemarchand and Karen J. Meech, 667–75. San Francisco: Astronomical Society of the Pacific, 2000.

———. "Who Said What: A Summary and Eleven Conclusions." In *If SETI Succeeds: The Impact of High Information Contact,* edited by Allen Tough, 33–39. Bellevue: Foundation For the Future, 2000.

Birks, Thomas Rawson. *Modern Astronomy.* London, 1850.

Bless, Robert. *Discovering the Cosmos.* Herndon: University Science, 2012.

Blish, James. *A Case of Conscience.* New York: Ballantine, 1958.

Blumenberg, Hans. *Die Vollzähligkeit der Sterne.* Frankfurt am Main: Suhrkamp, 2000.

Bonaventure. *Commentaria in quatuor libros sententiarum.* Quaracchi: Collegium S. Bonaventurae, 1882.

Bonting, Sjoerd. "Theological Implications of Possible Extraterrestrial Life." *Zygon* 38 (2003) 587–602.

Borowski, Steve. "Comparison of Fusion/Anti-Matter Propulsion Systems for Interplanetary Travel." *Technical Memorandum 107030.* San Diego: NASA, 1987.

Bostrom, Nick. *Superintelligence: Paths, Dangers, Strategies.* Oxford: Oxford University Press, 2014.

Bouvier, Audrey, and Meenakshi Wadhwa. "The Age of the Solar System Redefined by the Oldest Pb-Pb Age of a Meteoritic Inclusion." *Nature Geoscience* 3 (2010) 637-41. https://doi.org/10.1038/ngeo941.
Bowler, Peter. *Evolution: The History of the Idea*. Berkeley: University of California Press, 1989.
Bracewell, Ronald. *The Galactic Club: Intelligent Life in Outer Space*. San Francisco: HW Freeman, 1975.
Brady, Ignatius. "The Declaratio seu Retractatio of William of Varuouillon." *Archivum franciscanum historicum* 58 (1965) 394-416.
———. *William of Vaurouillon, O. Min Miscellanea Melchior de Pobladura 1*. Rome: Institutum Historicum O.F.M. Cap., 1964.
Brazier, Paul. "C. S. Lewis: The Question of Multiple Incarnations." *The Heythrop Journal* 55 (2014) 391-408.
———. *C. S. Lewis—On the Christ of a Religious Economy, 3.1: I. Creation and sub-creation*. Eugene: Pickwick, 2013.
Breed, David. "Ralph Wendell Burhoe: His Life and His Thought." *Zygon* 26 (1991) 397-428.
Breig, Joseph. "Man Stands Alone." *America: A Catholic Review of the Week* 104 (1960) 294-97.
Brewster, David. *Memoirs of Life, Writings, and Discoveries of Sir Isaac Newton*. Boston: Adamant Media Corporation, 2001.
———. *More Worlds than One: The Creed of the Philosopher and the Hope of the Christian*. London: J. Murray, 1894.
Brockes, Barrthold Heinrich. *Irdisches Vergnügen in Gott*. 9 vols. Berlin: Contumax GmbH & Co., 2019.
Broms, Allen. *Our Emerging Universe*, Garden City: Doubleday, 1961.
Brooke, Lindsay. "A Universe of Two Trillion Galaxies." Royal Astronomical Society. January 16, 2017. https://m.phys.org.
Bruce, Frederick. *The Epistles to the Colossians, to Philemon, and to the Ephesians*. Grand Rapids: Eerdmans, 1984.
Bruno, Giordano. *De l'infinito, Universo et Mondi, Dialogs IV-V*. Montevarchi: Harmakis Edizioni, 2018.
———. *De immenso*. Books VI-VIII. Paris: Belles Lettres, 1995.
———. *De la Causa, Principio et Uno, Cause, Principle, and Unity*. Translated by Jack Lindsay. Castle Hedingham: Daimon, 1962.
———. *Le Cena de le Ceneri*. Edited by Giovanni Aquilecchia. Torino: Einaudi, 1955.
Bruns, J. Edgar. "Cosmolatry." *Catholic World* 191 (1960) 286.
Brook, John. "Natural Theology and the Plurality of Worlds: Observations on the Brewster-Whewell Debate." *Annals of Science* 34 (1977) 221-86.
Buber, Martin. "Gottesfinsternis." In *Werke: Erster Band. Schriften zur Philosophie*, edited by Martin Buber, 503-603. Heidelberg: Schneider, 1962.
Burgess, Andrew. "Earth Chauvinism." *Christian Century* 93 (1976) 1098.
Bultmann, Rudolf. *Theology of the New Testament*. New York: Scribners, 1951-1955.
Burke, Patrick T. "Theology and Anthropology." In *The Word in History*, edited by Patrick Burke, 1-23. New York: Sheed & Ward, 1966.
Burns, J. Patout, SJ, "The Economy of Salvation: Two Patristic Traditions." *Theological Studies*, 37.4 (1976) 598-601.

———, trans. and ed. *Theological Anthropology.* Sources of Early Christian Thought. Philadelphia: Fortress, 1981.

Burque, Abbe Francois Xavier. *Pluralité des Mondes Habités Considérée au Point de Vue Négatif.* Montreal: Cadieux & Derome, 1898.

Baugh, Carlton, and Carlos Frenk. "How are Galaxies Made?" *PhysicsWeb.* May 1999. http://physicsweb.org/article/world/12/5/9.

Campanella, Tommaso. *Apologia pro Galileo.* Frankfort: Impensis Godefridi Tampachii, 1602.

Carr, Derek. "Take Me to Your Leader." *Homiletic and Pastoral Review* 65 (1964) 255–56.

Carroll, Bradley. *An Introduction to Modern Astrophysics.* London: Pearson, 2013.

Carroll, Sean. *Spacetime and Geometry: An Introduction to General Relativity.* London: Pearson, 2013.

Castello, Don Nello. *Cosi Parlo Padre Pio.* Vicenza, La Casa Sollievo della Sofferenza, 1974.

Catholic Bishop's Conference of England and Wales and Catholic Bishops' Conference of Scotland. *The Gift of Scripture: A Teaching Document of the Bishops Conferences of England and Wales, and of Scotland.* London: Catholic Truth Society, 2005.

Catholic Encyclopedia, vol. 1, New York: Robert Appleton, 1909.

Cawley, Martinus. *Guadalupe from the Aztec.* In *Anthology of Early Guadalupan Literature.* Lafayette: Trappist Abbey of Our Lady, 1968.

Cayrel, Roger, et al. "Measurement of Stellar Age from Uranium Decay." *Nature* 409 (2001) 691–92.

Chrysostom, John. *Homilies on Colossians.* http://www.documenta-catholica.eu/d_0345-0407-%20Iohannes%20Chrysostomus%20-%20Homilies%20on%20Colossians%20-%20EN.pdf.

———. *"In Genesis"* 3, 8 (Homily l7, 1). Patrologia Graeca 53. Paris: Migne, 1860.

———. *On Isaiah 1.4.* Patrologia Graeca 56. Montrouge: SEU Petit, 1859.

Cowan, J. J., et al. "The Chemical Composition and Age of the Metal-poor Halo Star BD+1703248." *The Astrophysical Journal* 572 (2002) 861–79.

Catechism of the Catholic Church. 2nd edition. Città del Vaticano: Libreria Editrice Vaticana, 2002.

Chalmers, Thomas. *Astronomical Discourses.* New York: Carter and Brothers, 1871.

———. *Discourses on the Christian Revelation Viewed in Connexion with Modern Astronomy.* New York: Robert Carter and Brothers, 1855.

Chandrasekhar, Subrahmanyan. *Mathematical Theory of Black Holes.* Oxford: Oxford University Press, 1999.

Charleton, Walter. *Physiologia Epicuro-Gasssendo-Charltoniana: Or a Fabrik of Science Natural, upon the Hypothesis of Atoms, "Founded by Epicurus, Repaired by Petrus Gassendus, Augmented by Walter Charlton."* 1654. https://philpapers.org/rec/CHAPEO-2.

Chase, Frederick. *Chrysostom: A Study of Biblical Interpretation.* London: Aeterna, 2015.

Choi, Charles. "New Estimate for Alien Earths: 2 Billion in Our Galaxy Alone." *Space.com.* March 21, 2011. https://www.space.com/11188-alien-earths-planets-sun-stars.html.

Chow, Denise. "5 Rocky Planets Revealed by NASA's Kepler Spacecraft." Space.com. January 7, 2014. http://www.space.com/24180-rocky-planets-mini-neptunes-aas223.html.
———. "Discovery: Cosmic Dust Contains Organic Matter from Stars." Space.com. October 26, 2011. http://www.space.com/13401-cosmic-star-dust-complex-organic-compounds.html.
Clarke, Arthur C. *Childhood's End*. New York: Del Rey, 2015.
Clifford, Steven J. *Astrotheology: For the Cosmic Adventure*. Techny: Divine Word, 1969.
Clooney, Francis. *Beyond Compare: St. Francis de Sales and Sri Vedanta Desika on Loving Surrender to God*. Washington: Georgetown University Press, 2008.
Cockell, Charles S., and Marco Lee. "Interstellar Predation." *Journal of the British Interplanetary Society* 55 (2002) 8–20.
Cohen, Bernard, ed. *Isaac Newton's Papers and Letters on Natural Philosophy*. Cambridge: Harvard University Press, 1958.
Colombo, Giulio, and Elio Giorello. *L'intelligenza dell' Universo*. Casale Monferrato: Piemme, 1999.
Congar, Yves. "Has God Peopled the Stars?" In *Wide World My Parish: Salvation and Its Problems*, 184–85, 188. Baltimore: Helicon, 1961.
———. "Non-Christian Religions and Christianity." In *Evangelization, Dialogue, and Development*, 144–51. Rome: Gregoriana, 1972.
———. "Preface." In André Feuillet, *Le Christ Sagesse de Dieu après Les Épitres Pauliniennes*. Paris: Gabalda, 1966.
———. "Theologian of Grace in a Vast World." In *Yves Congar: Theologian of the Church*, edited by Gabriel Flynn, 371–400. Louvain: Peters, 2005.
Connell, Francis, J., CSSR. "Flying Saucers and Theology." In *The Truth About Flying Saucers*, edited by Aime Michel, 255–58. New York: Pyramid, 1967.
Conner, Sam, et al. "A Spectrum of Views on ETI." Reasons to Believe. http://www.reasons.org/articles/a-spectrum-of-views-on-eti.
Consolmagno, Guy. *Brother Astronomer: Adventures of a Vatican Scientist*. New York: McGraw-Hill, 2000.
Conway, Bertrand. "The Question Box." *Catholic Messenger* 82 (1964) 10.
Cook, Jia-Rui. "Clay-Like Minerals Found on Icy Crust of Europa." December 11, 2013. http://www.nasa.gov/jpl/news/europa-clay-like-minerals-20131211.html.
Copernicus, Nicolaus. *De Revolutionibus*. Nuremberg: Johannes Petreius, 1583.
Corbally, Christopher. "What if there were Other Inhabited Worlds?" *Studies in Science and Theology* 5 (1997) 77–88.
Cornford, Francis. *Plato's Cosmology: The Timaeus of Plato*. New York, Hackett, 1937.
Council of Trent. Session VI, "On Justification." http://www.thecounciloftrent.com/ch6.htm.
Cosmovici, Cristiano Batalli, et al. *Astronomical and Biochemical Origins and the Search for Life in the Universe*. International Astronomical Union Colloquium 161. Bologna: Editrice Compositori, 1997.
Coyne, George V., SJ. "The Evolution of Intelligent Life on Earth and Possibly Elsewhere: Reflections from a Religious Tradition." In *Many Worlds: The New Universe, and the Theological Implications*, edited by Steven Dick, 177–88. Philadelphia: Templeton, 2000.
Craig, William. "Flint's Radical Molinist Christology Not Radical Enough." *Faith and Philosophy: Journal of the Society of Christian Philosophers* 23.1 (2006) 55–64.

Crampton, Josiah. *Testimony of the Heavens to Their Creator: A Lecture to the Enniskillen Young Men's Christian Association*. Dublin: Parish Register Society of Ireland, 1857.

Cranfield, Charles. "Some Observations on Romans 8:19–21." In *Reconciliation and Hope: Essays on Atonement and Eschatology*, edited by Robert Banks, 224–30. Grand Rapids: Eerdmans, 1974.

Crisp, Oliver. *God Incarnate: Explorations in Christology*. London: T&T Clark, 2009.

———. "Multiple Incarnations." In *Reason, Faith, and History: Philosophical Essays for Paul Helm*, edited by Martin Stone, 219–38. Cambridge: Cambridge University Press, 2008.

Cross, Richard. *The Metaphysics of the Incarnation: Thomas Aquinas to Duns Scotus*. New York: Oxford University Press, 2005.

Crowe, Michael. *The Extraterrestrial Life Debate, 1750–1900: The Idea of a Plurality of Worlds from Kant to Lowell*. Cambridge: Cambridge University Press, 1986.

———. *The Extraterrestrial Life Debate: Antiquity to 1915. A Source Book*. Notre Dame: University of Notre Dame Press, 2008.

———. "History of the Extraterrestrial Life Debate." *Zygon* 32 (1997) 147–62.

Crowe, Sean, et al. "Atmospheric Oxygenation Three Billion Years Ago." *Nature* 501 (2013) 535–38.

Crysdale, Cynthia. "God, Evolution, and Astrobiology." In *Exploring the Origin, Extent, and Future of Life*, edited by Constance Bertka, 220–41. Cambridge: Cambridge University Press, 2009.

Csaba, Kecskes. "Evolution and Detectability of Advanced Civilizations." *Journal of the British Interplanetary Society* 62.9 (2009) 316–19.

Cusa, Nicholas. *Of Learned Ignorance*. Translated by Fr. Germain Heron. New Haven: Yale University Press, 1954.

Cyril of Alexandria. *Quod unus sit Christus*. LFC 47. http://www.tertullian.org/fathers/cyril_christ_is_one_01_text.htm.

Dales, Richard C. *The Intellectual Life of Western Europe in the Middle Ages*. Washington, DC: Catholic University of America Press, 1980.

Darling, David. *The Extraterrestrial Encyclopedia*. New York: Three Rivers, 2000.

———. "Variety of Extraterrestrial Life." The Encyclopedia of Science. http://www.daviddarling.info/encyclopedia/E/etlifevar.html.

Davids, A. "Justin Martyr on Monotheism and Heresy." *Dutch Review of Church History Nieuwe Serie* 56.1 (1975) 210–34.

Davidson, Paul. "The Structure of Heaven and Earth: How Ancient Cosmology Shaped Everyone's Theology." *Is That in the Bible?* August 17, 2019. https://isthatinthebible.wordpress.com/2019/08/17/the-structure-of-heaven-and-earth-how-ancient-cosmology-shaped-everyones-theology/.

Davies, Paul. *Are We Alone?: Philosophical Implications of the Discovery of Extraterrestrial Life*. New York: Basic, 1995.

———. "Biological Determinism, Information Theory, and the Origin of Life." In *Many Worlds: The New Universe, and the Theological Implications*, edited by Steven J. Dick, 15–28. Philadelphia: Templeton, 2000.

———. "ET and God." *The Atlantic Monthly* 292.2 (2003) 112–18.

———. *God and the New Physics*. New York: Simon and Schuster, 1983.

Davis, Charles. "The Place of Christ." *The Clergy Review* 45 (1960) 706–18.

Davis, John. "Search for Extraterrestrial Intelligence and the Christian Doctrine of Redemption." *Science and Christian Belief* 9 (1997) 21–34.
Davis, Leo D. *The First Seven Ecumenical Councils (325–787): Their History and Theology*. Theology and Life 1. Wilmington: Glazer, 1988.
Davison, Andrew. "Christian Systematic Theology and Life Elsewhere in the Universe: A Study in Suitability." *Theology and Science* 16.4 (2018) 447–61.
Dawkins, Richard. "Universal Darwinism." In *Evolution from Microbes to Men*, edited by Derek Bendall, 403–25. Cambridge: Cambridge University Press, 1983.
Deacon, Terrence W. *The Symbolic Species: The Co-evolution of Language and the Brain*. New York: Norton & Co., 1997.
Deane-Drummond, Celia. "The Alpha and the Omega: Reflections on the Origin and Future of Life from the Perspective of Christian Theology and Ethics." In *Exploring the Origin, Extent, and Future of Life*, edited by Constance Berkta, 96–112. Cambridge: Cambridge University Press, 2009.
De Concilio, Januarius. *Harmony between Science and Revelation*. Whitefish: Kessinger, 1889.
De Duve, Christian. *Vital Dust: Life as a Cosmic Imperative*. New York: Basic, 1995.
De Fontenelle, Bernard. *Conversations on the Plurality of Worlds*. 2nd edition. London: Thomas Caslon, 1767.
———. *Entretiens sur la Pluralite des Mondes*. Sydney: Wentworth, 2018.
De Lérins, Vincent. *Commonitorium*. London: Aeterna, 2016.
De Lubac, Henri. *Augustinianism and Modern Theology*. New York: Herder and Herder, 1965.
———. *Surnaturel*. Paris: Aubier, 1946.
De Maistre, Comte Joseph. *Soirées de Saint-Pétersbourg*. Anvers: Société Catholique pour le Royaume des Pays-Bas, 1821.
De Montiqnuez. *Theeorie chretinne sur la Pluralite des Mondes, Archives Théologiques*. Paris: Ménard et Desenne, 1866.
De Vaurouillon, Guillaume. *Quattuor Liborum Setentiarum Compendium Venerabilis Pais Fratris Guiermi*. Basel: Langerdorf, 1510.
Deacon, Terrence. *The Symbolic Species*. New York: W. W. Norton, 1997.
Deamer, David. "A Giant Step Towards Artificial Life?" *Trends in Biotechnology* 23.7 (2005) 336–38.
Del Colle, Ralph. *Christ and the Spirit: Spirit-Christology in Trinitarian Perspective*. New York: Oxford University Press, 1994.
Delano, Kenneth J. *Many Worlds, One God*. Hicksville: Exposition, 1977.
Delio, Ilia, OSF. "Christ and Extraterrestrial Life." *Theology and Science* 5.3 (2007) 253–60.
———. "Cosmic Christology in the Thought of Zachary Hayes." *Franciscan Studies* 65 (2007) 107–20.
———. "Saint Bonaventure, Francis Mayron, William Vorilong, and the Doctrine of the Plurality of Worlds." *Speculum* 12 (1937) 386–89.
Dennett, Daniel. *Darwin's Dangerous Idea*. New York: Simon and Shuster, 1996.
Denzinger, Heinrich. *Enchiridion symbolorum, definitionum et declarationum de rebus fidei et morum*. 43rd ed. Edited by Peter Hünermann. San Francisco: Ignatius, 2012.
Denzler, Brenda. *The Lure of the Edge: Scientific Passions, Religious Beliefs, and the Pursuit of UFOs*. Berkeley: University of California Press, 2001.

Descartes, Rene. *Principles of Philosophy*. Amsterdam: Louis Elzevir, 1644.
Dethier, Vincent G. "Life on Other Planets." *Catholic World* 198 (1964) 250–51.
DeVito, Carl. *Science, SETI, and Mathematics*, New York: Berghahn, 2014.
Dexter, Miriam Robbins. "Proto-Indo-European Sun Maidens and Gods of the Moon." *Mankind Quarterly* 25.1,2 (1984) 137–44.
Dick, Steven. *The Biological Universe: The Twentieth Century Extraterrestrial Life Debate and the Limits of Science*. Cambridge: Cambridge University Press, 1996.
———. "Bringing Culture to Cosmos: the Postbiological Universe." In *Cosmos and Culture: Cultural Evolution in a Cosmic Context*, edited by Steven Dick and Mark Lupisella, 468. Washington, DC: NASA, 2013.
———. "Consequences of Success in SETI: Lessons from the History of Science." In *Progress in the Search for Extraterrestrial Life*, edited by Seth Shostak, 521–32. San Francisco: Astronomical Society of the Pacific, 1995.
———. "Cosmotheology Revisited: Theological Implications of Extraterrestrial Life." 2005. https://bdigital.ufp.pt/bitstream/10284/778/2/287-301Cons-Ciencias%2002-5.pdf.
———. *Life on Other Worlds. The Twentieth Century Extraterrestrial Life Debate*. Cambridge: Cambridge University Press, 1998.
———. *Many Worlds: The New Universe, Extraterrestrial Life, and the Theological Implications*. Edited by Steven Dick, Philadelphia: Templeton, 2000.
———. "Origins of the Extraterrestrial Life Debate and Its Relation to the Scientific Revolution." *Journal of the History of Ideas* 41.1 (1980) 4–6.
———. "Plurality of Worlds." In *Encyclopedia of Cosmology*, edited by Norriss Hetherington, 89, 501–12. New York: Garland, 1993.
———. *Plurality of Worlds: The Origins of the Extraterrestrial Life Debate from Democritus to Kant*. Cambridge: Cambridge University Press, 1982.
———, and James Strick. *The Living Universe, NASA, and the Development of Astrobiology*. New Brunswick: Rutgers University Press, 2004.
Dobzhansky, Theodosius. "Darwinian Evolution and the Problem of Extraterrestrial Life." *Perspectives in Biology and Medicine* 15.2 (1972) 157–75.
Drake, Frank. "Extraterrestrial Intelligence." *Science* 260.5107 (1993) 474–75.
Dressler, Alan, and Frogel, Jay. "GSMT AND JWST: Looking Back to the Future of the Universe." https://www.nsf.gov/mps/ast/aaac/reports/gsmt-jwst_synergy_combined.pdf
Duffy, Stephen. *The Dynamics of Grace: Perspectives in Theological Anthropology*. New Theology Studies 3. Collegeville: The Liturgical Press, 1993.
Duhem, Pierre. *Etudes sur Leonard de Vinci*. 3 vols. Paris: Hermann, 1906–1913.
———. *Le Systéme du Monde*. Paris: A. Hermann, 1958.
Durken, Daniel. *The New Collegeville Bible Commentary: In One Volume*. Collegeville: Liturgical, 2017.
Durkheim, Emile. *The Elementary Forms of Religious Life*. Edited and translated by Karen Fields. New York: Free, 1995.
Duverger, Christian. "La Flor Letal: Economía del Sacrificio Azteca." *Fondo de Cultura Económica* (2005) 83–93.
Dvorak, John. *Mask of the Sun: The Science, History, and Forgotten Lore of Eclipses*. New York: Pegasus, 2017.
Dwight, Timothy. *Theology Explained and Defended*. Middletown: Clark and Lyman, 1818.

Dyson, Freeman. "Search for Artificial Stellar Sources of Infrared Radiation." *Science* 131.3414 (1960) 1667–668.

———. "The Search for Extraterrestrial Technology." In *Perspectives in Modern Physics*, edited by Robert Marshak, 641–55. New York: John Wiley & Sons, 1966.

Dyson, G. Darwin. *Among the Machines: The Evolution of Global Intelligence*. Cambridge: Perseus, 2012.

Ebrard, Johann. *Der Glaube an die Heilige Schrift und die Ergebnisse der Naturforschung*. Hamburg: Gräfe und Unzer, 1861.

Edwards, Denis. "Resurrection of the Body and Transformation of the Universe in the Theology of Karl Rahner." *Philosophy and Theology* 18 (2006) 357–83.

"The European Extremely Large Telescope." ESO. http://www.eso.org/public/images/ann12096a/.

Eggen, Olin, et al. "Evidence from the Motion of Old Stars that the Galaxy Collapsed." *The Astrophysical Journal* 136 (1962) 748–66.

Eiseley Loren. "Is Man Alone in Space?" *Scientific American* 189.7 (1953) 80–86.

Eltester, Walther. *Eikon im Deuen Testament*. BZNW 23. Berlin: Topelmann, 1958.

Emery, N., and N. Clayton. "Comparative Social Cognition." *Annual Review of Psychology* 60 (2009) 87–113.

"ExoMars: ESA and Roscosmos Set for Mars Missions." European Space Agency. March 14, 2013. http://www.esa.int/Our_Activities/Space_Science/ExoMars_ESA_and_Roscosmos_set_for_Mars_missions.

"Extrasolar Planet Detected by Gravitational Microlensing." NASA. Last updated March 3, 2021. https://exoplanets.nasa.gov/resources/53/extrasolar-planet-detected-by-gravitational-microlensing/

"The Extraterrestrial is My Brother." Interview with Fr. Funes. *L'Obsservatore Romano*. May 14, 2008. https://www.nytimes.com/2008/05/14/world/europe/14iht-vat.4.12885393.html.

Feinberg, Gerald, and Robert Shapiro. *Life Beyond Earth: The Intelligent Earthlings Guide to Life in the Universe*. New York: Morrow and Company, 1980.

Fiorenza, Francis, and John Galvin. *Foundational Theology: Jesus and the Church*. New York: Crossroad, 1984.

———. *Systematic Theology: Roman Catholic Perspectives, Vol. 1*. Minneapolis: Fortress, 1991.

Fimmel, Richard, et al. *Pioneer Odyssey: Encounter with a Giant*. Washington, DC: NASA, 1974.

Finney, Ben. "The Impact of Contact." *Acta Astronautica* 21.2 (1990) 11–21.

Finocchiaro, Maurice. *Galileo on the World Systems: A New Abridged Translation and Guide*. Berkeley: University of California Press, 1997.

———. *Retrying Galileo*. Berkeley: University of California Press, 2007.

Fisher, Christopher, and David Fergusson. "Karl Rahner and The Extra-Terrestrial Intelligence Question." *The Heythrop Journal* 47.2 (2006) 275–90.

Fitzgerald, Randall. *The Complete Book of Extraterrestrial Encounters*. New York: Collier, 1979.

Flammarion, Camille. *L'Humanité dans L'univers*. In *La pluralite des Mondes Habites*, 33rd ed. Paris, 1885.

———. *Memoires Biographiques et Philosophiques D`un Astronome*. 1st ed. Paris: Ernest, 1911.

———. *La Multiplicite des Mondes Habites*. Vol. X. Paris: Seuil, 1969.

Flint, Thomas. "Molinism and Incarnation." In *Molinism: The Contemporary Debate*, edited by Ken Perszyk, 187–207. Oxford: Oxford University Press, 2012.

———. "The Possibilities of Incarnation: Some Radical Molinist Suggestions." *Religious Studies* 37.3 (2001) 307–20.

Ford, Lewis. *The Lure of God*. Philadelphia: Fortress, 1979.

———. "Theological Reflections on Extra-Terrestrial life." *Raymond Review* 3.1 (1968) 2.

Franciscan Friars of the Immaculate. *A Handbook on Guadalupe*. New Bedford: Academy of the Immaculate, 2001.

Fraser, Craig. *The Cosmos: A Historical Perspective*. Santa Barbara: Greenwood, 2016.

Freddoso, Alfred. "Human Nature, Potency, and the Incarnation." *Faith and Philosophy* 3.1 (1986) 27–53.

———. "Logic, Ontology, and Ockham's Christology." *The New Scholasticism* 57.3 (1983) 293–30.

Freitas Robert. "Extraterrestrial Zoology." *Analog Science Fiction/Science Fact* 101 (1981) 53–67.

Fritsch, Harald. "Vollendung des Cosmos." In *Vollendende Selbstmitteilung Gottes und seine Schöpfund: Die Eschatologie Karl Rahners*, 508–11. Zustand: Leichte Gebrauchsspuren, 2006.

Fuller, Andrew. *The Gospel Its Own Witness*. Mulberry: Sovereign Grace, 1961.

Gabathuler, Huber. *Jesus Christus Haupt der Kirche-Haupt der Welt*. ATANT. Stuttgart: Zwingli, 1965.

Galilei, Galileo. "Letter to Castelli (1613)." Famous Trials. https://www.famous-trials.com/galileotrial/1025-castelliletter.

———. "Letter to the Grand Duchess Christina of Tuscany: Concerning the Use of Biblical Quotations in Matters of Science." Translated by Maurice A. Finocchiaro, 120. In *The Essential Galileo*. Cambridge: Hackett, 2008.

Galloway, Allan. *The Cosmic Christ*. New York: Harper & Brothers, 1951.

Geister, Phillip. *Aufhebung zur Eigentlichkeit: Zur Problematik Kosmologischer Eschatologie in der Theologie Karl Rahners*. Uppsala: Uppsala University Press, 1996.

General Directory of Catechesis. 1st ed. United States Conference of Catholic Bishops. Washington, DC: USCCB, 1998.

George, Marie. "Aquinas on Intelligent Extra-Terrestrial Life." *The Thomist* 65.2 (2001) 239–58.

———. "Catholic Faith, Scripture, and the Question of the Existence of Intelligent Extra-terrestrial Life." In *Faith, Scholarship, and Culture in the 21st Century*, edited by A. Ramos and Marie George, 135–45. Washington DC: Catholic University of America Press, 2002.

———. *Christianity and Extraterrestrials? A Catholic Perspective*. New York: iUniverse, 2005.

Giordano, Bruno. *De le Causa, Principio e Uno*. 1584. https://link.1er.com/chapter/10.1007/978-90-481-8796-6_8.

Gledhill, Ruth. "Catholic Church No Longer Swears by Truth of the Bible." *The Times* (London). October 5, 2005.

Glossa Ordinaria, PL 113, 344 ff. (early 12th century).

Goldsmith, Donald, and Tobias Owen. *The Search for Life in the Universe*. Sausalito: University Science, 2002.

Gordley, Matthew. The *Colossian Hymn in Context: An Exegesis in Light of the Jewish and Greco-Roman Hymnic and Epistolary Conventions*. Wissenschaftliche Untersuchungen zum Neuen Testament 2.228. Tübingen: Mohr Siebeck, 2007.

Goshen-Gottstein, Alan. "God the Father in Rabbinic Judaism and Christianity: Transformed Background or Common Ground?" *Journal of Ecumenical Studies* 38.4 (2001) 1–34.

Gould, Stephen Jay. "Exaptation: a Crucial Tool for an Evolutionary Psychology." *Journal of Social Issues* 47 (1991) 43–65.

———. "Great Dying." *Natural History* 83 (1974) 22–27.

———. *Rock of Ages: Science and Religion in the Fullness of Life*. New York: Ballantine, 1999.

———. "An Unsung Single-Celled Hero." *Natural History* 83 (1974) 33–42.

Grady, Monica. *Astrobiology*. Washington, DC: Smithsonian Institution, 2001.

Grant, Andrew. "At Last, Voyager 1 Slips into Interstellar Space." *Science News*. September 12, 2013. https://www.sciencenews.org/article/last-voyager-1-slips-interstellar-space.

Grant, Edward. *A Source Book in Medieval Science*. Cambridge: Harvard University Press, 1974.

Grasso, Dominico. "Missionaries to Space." *Newsweek*. February 15, 1960.

———. "La Teologia e la Pluralità dei Mondi Abitati." *Civiltà Cattolica* 103.4 (1952) 255–65.

Graves, Robert Perceval. *Life of Sir William Rowan Hamilton, vol. II*. Dublin: Hodges Figgis, 1885.

Gregersen, Neils Henrik. "Deep Incarnation: Why Evolutionary Continuity Matters." *Toronto Journal of Theology* 26.2 (2010) 173–88.

Gregory of Nazianzus. *Epistle 101*. In *On God and Christ the Five Theological Orations and Two Letters of Cledonius*, edited by Lionel Wickham and Frederick Williams, 155–66. Crestwood: St. Vladimir's Seminary Press, 2002.

Gregory of Nyssa. *De Vita Moysis*. In *Opera exegetica en Exodum et Novum Testamentum*, GNO 7/1, edited by Edidit Musurillo. Leiden: Brill, 1991.

Grinspoon, David. *Lonely Planets: The Natural Philosophy of Alien Life*. New York: ECCO, 2003.

Groppe, Elizabeth. *Yves Congar's Theology of the Holy Spirit*. Oxford: Oxford University Press, 2004.

Grotzinger, John. "Introduction to Special Issue - Habitability, Taphonomy, and the Search for Organic Carbon on Mars." *Science* 343 (2014) 386–87.

Guadalupe, Miguel. *The Seven Veils of Our Lady of Guadalupe*. Goleta: Queenship, 1999.

Gudipati, Murty, and Rui Yang. "In-Situ Probing of Radiation-Induced Processing of Organics in Astrophysical Ice Analogs-Novel Laser Desorption Laser Ionization Time-of-Flight Mass Spectroscopic Studies." *The Astrophysical Journal letters* 756 (2012) 1–23.

Guessoum, Nidhal. *Islam's Quantum Question: Reconciling Muslim Tradition and Modern Science*. London: Tauris, 2011.

Guiseppe, Cocconi, and Philip Morrison. "Searching for Interstellar Communications." *Nature* 184.4690 (1959) 844–46.

Guthke, Karl. "The Idea of Extraterrestrial Intelligence." *Harvard Library Bulletin* 33 (1985) 196–210.

Guthrie, Donald. *New Testament Theology.* Downers Grove: Inter-Varsity, 1981.
Haldane, Elizabeth. *Descartes: His Life and Times.* New York: Cornell University Library, 1905.
Harford, James. "Rational Beings in Other Worlds." *Jubilee: A Magazine of the Church and Her People* 10 (1962) 17–21.
Harrison, Albert, and Steven Dick. "Contact: Long-Term Implications for Humanity." Section II. Bellevue: Foundation for the Future, 2000.
Harrison, Albert, and Kathleen Connell. "Workshop on the Societal Implications of Astrobiology." NASA Ames Research Center, November 16–17, 1999.
Harrison, Albert, et al. "The Role of Social Science in SETI." In *If SETI Succeeds: The Impact of High Information Contact*, edited by Allen Tough, 71–85. Bellevue: Foundation for the Future, 2000.
Harrison, Nonna Verna, and David G. Hunter, eds. *Suffering and Evil in Early Christian Thought.* Grand Rapids: Baker Academic, 2016.
Hart, Herbert. *The Concept of Law.* 3rd ed. Clarendon Law Series. Oxford: Oxford University Press, 2012.
Hartmann, Lars. "Universal Reconciliation (Col. 1:20)." *SNTU* 10 (1985) 109–20.
Hartmann, William. *Astronomy: The Cosmic Journey.* Belmont: Wadsworth, 2011.
Haught, John. "Theology after Contact: Religion and Extraterrestrial Life." *Annals of the New York Academy of Sciences* 950 (2006) 296–308.
Haye, William. *Religion Philosophi: or, The Principles of Morality and Christianity Illustrated from a View of the Universe, and Man's Situation in it.* Farmington Hills: Gale, 2018.
Hayes, Zachary. *Bonaventure: Mystical Writings.* New York: Crossroad, 1999.
———. "Christ, Word of God, and Exemplar of Humanity." *The Cord* 46.1 (1996) 3–17.
———. "Disputed Questions on the Mystery of the Trinity." In *Works of Saint Bonaventure*, Vol. 3, edited by George Marcil. New York: Franciscan Institute, 1979.
———. "Incarnation and Creation in the Theology of Bonaventure." In *Studies Honoring Ignatius Brady*, edited by Romano Almagno and Conrad Harkins, 320–29. New York: Franciscan Institute, 1976.
———. "The Meaning of Convenientia in the Metaphysics of St. Bonaventure." *Franciscan Studies* 34 (1974) 71–100.
Hebblethwaite, Brian. "The Impossibility of Multiple Incarnations." *Theology* 104.821 (2001) 323–34.
———. *Philosophical Theology and Christian Doctrine.* Hoboken: Wiley & Sons, 2008.
Heger, Alexander, and Stan Woosley. "The Nucleosynthetic Signature of Population III." *Astrophysical Journal* 567 (2001) 532–43.
Hegermann, Harald. *Die Vorstellung vom Schopfungsmittler im Hellenistichen Judentum und Urchristentum.* Berlin: Akademie, 1961.
Heidmann, Jean. *Extraterrestrial Intelligence.* Cambridge: Cambridge University Press, 1997.
Helyer, Larry. "Cosmic Christology and Colossians 1:15–20." *Journal of the Evangelical Theology Society* 37.2 (1994) 235–40.
Herrick, James. *Scientific Mythologies: Science and Science Fiction Forge New Religious Belief.* Westmont: InterVarsity, 2008.
Hetherington, Norriss. "Cosmology, Religious, and Philosophical Aspects." In *Encyclopedia of Science and Religion.* New York: Macmillan, 2003.

Hibbs, Thomas. *Knowledge and Faith in Thomas Aquinas*. Cambridge: Cambridge University Press, 1998.
Hill, Robert C. *St. John Chrysostom: Homilies on the Old Testament: Homilies on Isaiah and Jeremiah*. Brookline: Holy Cross Orthodox, 2007.
———. *Reading the Old Testament in Antioch*. Bible in Ancient Christianity 5, Leiden: Brill, 2005.
Holden, Edward. *Sir William Herschel: His Life and Works*. New York: Scribner's Sons, 1881.
Horrell, David, et al. *Greening Paul: Rereading the Apostle in a Time of Ecological Crisis*. Waco: Baylor University Press, 2010.
Horowitz, Norman H. *To Utopia and Back: The Search for Life in the Solar System*. New York: Freeman & Company, 1986.
"How Many Satellites Are Orbiting The Earth in 2017?" Pixalytics. November 15, 2017. https://www.pixalytics.com/sats-orbiting-earth-2017/#:~:text=According%20to%20the%20Index%20of,357%20objects%20launched%20into%20space.
Hoyningen-Huene, Paul. "Kuhn's Development Before and After *Structure*". In *Kuhn's Structure of Scientific Revolutions – 50 Years On*, edited by William J. Devlin and Alisa Bokulich, 185–96. Boston Studies in the Philosophy and History of Science 311. Cham: Springer International, 2015.
"Hubble Reveals Observable Universe Contains 10 Times More Galaxies Than Previously Thought." NASA. October 13, 2016. https://www.nasa.gov/feature/goddard/2016/hubble-reveals-observable-universe-contains-10-times-more-galaxies-than-previously-thought.
"Hubble Goes to the eXtreme to Assemble Farthest Ever View of the Universe." Hubble Site. September 25, 2012. www.hubblesite.org/newscenter/archive/releases/2012/37/fastfacts/.
Hübner, Hans. *An Philemon, An die Kolosser, An die Epheser*. Tübingen: Mohr, 1997.
Huygens, Christaan. *The Celestial Worlds Dicover'd, Or, Conjectures Concerning the Inhabitants, Plants and Productions of the Worlds in the Planets*. Reprint of 1698 edition, London: Cass and Company, 1968.
———. *Oeuvres completes de Christiaan Huygens: Publiées par la Société hollandaise des sciences*. Vol. XXI. La Haye: Nijhoff, 1888-1950.
Impey, Chris, and Johnathan Lunine. *Frontiers of Astrobiology*. Cambridge: Cambridge University Press, 2012.
Irenaeus. *Adversus Haereses*. https://www.newadvent.org/fathers/0103318.htm.
Ito, Miho, et al. "Thermal Stability of Amino Acids in Seafloor Sediment in Aqueous Solution at High Temperature." *Organic Geochemistry* 37 (2006) 177–88.
Jakosky, Bruce. *Science, Society, and the Search for Life in the Universe*. Tucson: University of Arizona Press, 2006.
James, Edwin. *Sacrifice and Sacrament*. London: Thames and Hudson, 1962.
Jenson, Robert. *Systematic Theology: Volume 1: The Triune God*. Oxford: Oxford University Press, 1997.
Jorgensen, Jess, et al. "Detection of the Simplest Sugar, Glycolaldehyde, in a Solar-Type Protostar with ALMA." *The Astrophysical Journal Letters* 757 (2012) 1–13.
Jouan, René. *La question de L'habitabilité des Mondes Étudiée au Point de vue de L'histoire, de la Science, de la Raison et de la Foi*. Saint-Ilan: par Yffiniac, 1900.
Kaku, Michio. "The Physics of Interstellar Travel: To One Day, Reach the Stars." http://mkaku.org/home/?page_id=250.

Kardashev, Nikolai. "Transmission of Information by Extraterrestrial Civilizations." *Soviet Astronomy* 8 (1964) 282–87.
Kasemann, Ernst. *Essays on New Testament Themes*. SBT 41. Naperville: Allenson, 1964.
Kasting, James, et al. "Habitable Zones around Main Sequence Stars." *Icarus* 101.1 (1993) 108–28.
Keill, John. *Introductio ad Veram Astronomiam*. Delhi: Gyan, 2017.
Kereszty, Roch. *Jesus Christ: Fundamentals of Christology*. 3rd ed. Staten Island: Alba, 2002.
Kevern, Peter. "Limping Principles: A Reply to Brian Hebblethwaite on The Impossibility of Multiple Incarnations." *Theology* 105.827 (2002) 342–47.
Kirchhoffer, David, et al. *Being Human: Groundwork for a Theological Anthropology for the 21st Century*. Eugene: Wipf & Stock, 2013.
Kirk, Geoffrey, and John Raven. *The Presocratic Philosophers*. Cambridge: Cambridge University Press, 1983.
Kirkpatrick, Lee A. "Toward an Evolutionary Psychology of Religion and Personality." *Journal of Personality* 67 (2001) 921–52. https://doi.org/10.1111/1467-6494.00078.
Kleinz, John. "The Theology of Outer Space" *Columbia* 40 (1960) 28.
Kolb, Vera. *Astrobiology: An Evolutionary Approach*. Boca Raton: CRS, 2014.
Kracher, Alfred. "Evolutionary Perspectives on Interstellar Communication: Images of Altruism." In *Extraterrestrial Altruism: Evolution and Ethics in the Cosmos*, edited by Douglas Vakoch, 295–308. Heidelberg: Springer, 2013.
Krebs, J., and Hillebrandt, W. "The Interaction of Supernova Shockfronts and Nearby Interstellar Clouds." *Astronomy and Astrophysics* 128 (1983) 411–19.
Kuhn, Thomas. *The Copernican Revolution - Planetary Astronomy in the Development of Western Thought*. Cambridge: Harvard University Press, 1985.
Kunin, Seth. *Religion: The Modern Theories*. Edinburgh: Edinburgh University Press, 2003.
Küppers, Bernd-Olaf. *Information and the Origin of Life*. Cambridge: MIT Press, 1990.
Kutner, Mark. *Astronomy: A Physical Perspective*. Cambridge: Cambridge University Press, 2003.
Kurtz, Jon Hienrich. *The Bible and Astronomy: An Exposition of the Biblical Cosmology, and Its Relations to Natural Science*. Translated by T. D. Simonton. Philadelphia: Linday & Blakiston, 1857.
Kurzweil, Ray. *The Age of Spiritual Machines: When Computers Exceed Human Intelligence*. New York: Penguin, 1999.
———. *Singularity is Near*. London: Penguin, 2005.
Kutner, Mark L. *Astronomy: A Physical Perspective*. Cambridge: Cambridge University Press, 2003.
Kwok, Sun, and Yong Zhang. "Mixed Aromatic-Aliphatic Organic Nanoparticles As Carriers of Unidentified Infrared Emission Features." *Nature* 479 (2011) 80–83.
Kwon, Jung Yul, et al. "How Will We React to the Discovery of Extraterrestrial Life?" *Frontiers in Psychology* 8 (2018) 1–12.
Labandeira, Conrad, and John Sepkoski. "Insect Diversity in the Fossil Record." *Science* 261 (1993) 310–15.
Lai, Tyrone. "Nicolas of Cusa and the Finite Universe." *Journal of the History of Philosophy* 11 (1973) 161–67.
Lalande, Kevin, and Gillian Brown. *Sense and Nonsense: Evolutionary Perspectives on Human Behavior*. Oxford: Oxford University Press, 2011.

Lamm, Norman. "The Religious Implication of Extraterrestrial Life." In *Challenge: Torah Views on Science and Its Problems*, edited by Aryeh Cannell and Cyril Domb, 354–98. Jerusalem: Feldheim, 1978.

Langford, Jerome, OP. *Galileo, Science, and the Church*. South Bend: St. Augustine's, 1998.

Langlois, Ed. "What If We're Not Alone? Scientist-theologians Say Incarnation and Salvation Are Not Global, but Cosmic." *Catholic Sentinel*. January 25, 2017. https://catholicsentinel.org/MobileContent/Faith-Spirituality/Living-Faith/Article/What-if-we-re-not-alone-/4/29/32821.

Law, Sung Ping. "The Regulation of Menstrual Cycle and its Relationship to the Moon." *Acta Obstet Gynecologica Scandinavica* 65.1 (1986) 45–48.

Leitch, Willliam. *God's Glory in the Heavens*. London: Strahan, 1867.

Leiter, Darryl J., and Sharon Leiter. *A to Z of Physicists*. New York: Infobase, 2009.

Lemarchand, Guillermo A. "Detectability of Extraterrestrial Technological Activities." http://www.coseti.org/lemarch1.htm.

Lemarchand, Guillermo A., and Jon Lomberg. "Communication among Interstellar Intelligent Species: A Search for Universal Cognitive Maps." In *Communication with Extraterrestrial Life*, edited by Douglas Vakoch, 371–95. New York: State University of New York Press, 2011.

Leslie, John. *Modern Cosmology and Philosophy*. New York: Prometheus, 1998.

Lestel, Dominique. "Ethology, Ethnology, and Communication with Extraterrestrial Intelligence." In *Archaeology, Anthropology, and Interstellar Communication*, edited by Douglas A. Vakoch, 229–36. Washington, DC: NASA, 2014.

Lewis, C. S. "Faith and Outer Space." *Time* 71 (1958) 37.

———. *Miracles*. 1st ed. London: Bless, 1947.

———. "Other-Worldly Faith." *Newsweek* 51 (1958) 64.

———. *Space Trilogy (Out of the Silent Planet, Perelandra, That Hideous Strength)*. London: Bodley Head, 1990.

Lewis, Nicola Denzey. *Cosmology and Fate in Gnosticism and Graeco-Roman Antiquity: Under Pitiless Skies*. Leiden: Brill, 2013.

Linzey, Andrew. *Animal Gospel: The Christian Defense of Animals*. London: Hodder & Stoughton, 1998.

Lohse, Edward. *Colossians and Philemon*. Philadelphia: Fortress, 1971.

Lonergan, Bernard. *Method in Theology*. Toronto: University of Toronto Press, 1990.

Long, James. "Aquinas and the Cosmic Christ." In *Medieval Masters: Essays in Memory of E. A. Synan*, edited by Rollen Houser, 233–48. Houston: Center for Thomistic Studies, 1999.

Lossky, Vladimir. "Orthodox Theology." In *Astrotheology: Science and Theology Meet Extraterrestrial Life*, edited by Ted Peters, 110–11. Eugene: Cascade, 2018.

Lovejoy, Arthur. *The Great Chain of Being: A Study of the History of an Idea*. Cambridge: Harvard University Press, 1936.

Lucretius. *De Rerum Natura*. Translated by W. H. D. Rouse. Revised by Martin F. Smith. LCL 181. Cambridge: Harvard University Press, 1924.

Luther, Martin. *Luther's Works, Volume 1: Lectures on Genesis Chapters 1–5*. Edited by Jaroslav Pelikan. St. Louis: Concordia, 1958.

Lynch, Daniel J. *Our Lady of Guadalupe and Her Missionary Image*. St. Albans: Missionary Image of Our Lady of Guadalupe, 1993.

Lynch, John. "Christians on Other Planets?" *Friar* 19 (1963) 26–29.

Lyons, Joseph. *The Cosmic Christ in Origen and Teilhard De Chardin*. Oxford: Oxford University Press, 1982.

Lytkin, Vladimir, et al. "Tsiolkovsky, Russian Cosmism, and Extraterrestrial Intelligence." *Quarterly Journal of the Royal Astronomical Society* 36 (1995) 369–76.

MacArthur, John. *The MacArthur Study Bible*. Wheaton: Crossway, 2010.

MacGowan, Roger, and Frederick I Ordway III. *Intelligence in the Universe*. Englewood Cliffs: Prentice Hall, 1966.

Machamer, Peter, ed. *The Cambridge Companion to Galileo*. Cambridge: Cambridge University Press, 1998.

Maimonides, Moses. *Guide for the Perplexed*. Translated by M. Friedlander. London: Routledge & Sons, 1919.

Martins, Zita, et al. "Extraterrestrial Nucleobases in the Murchison Meteorite." *Earth and Planetary Science Letters* 270 (2008) 130–36.

"Mars Global Surveyor: MOLA MEGDRs." NASA. August 25, 2017. https://pds-geosciences.wustl.edu/missions/mgs/megdr.html.

"Mars Pathfinder." NASA. https://mars.nasa.gov/mars-exploration/missions/pathfinder/.

Marsack, Leonard M. *The Achievement of Bernard le Bovier de Fontenelle*. New York: Johnson Reprint, 1970.

Martyr, Justin. *Dialogue with Trypho*. Pickerington: Beloved, 2015.

———, *Adv. Haer.* IV, 15, 2. "Justin Martyr on Monotheism and Heresy," *Dutch Review of Church History* NIEUWE SERIE 56.1 (1975) 210–234.

Maruyama, Magoroh, and Arthur Harkins. *Cultures Beyond the Earth*. New York: Random House, 1975.

Mascall, Eric. *Christian Theology and Natural Science: Some Questions in Their Relations*. North Haven: Archon, 1965.

Matson, John. "Meteorite That Fell in 1969 Still Revealing Secrets of the Early Solar System." February 15, 2010. http://www.scientificamerican.com/article/murchison-meteorite/?mobileFormat=true.

Matthews, Freya. *The Ecological Self*. London: Routledge, 1994.

Maunder, Edward. *Initia Doctrinae Physicae*, *Corpus Reformatorum* 13. Frankfurt: Minerva, 1963.

Mautner, Michael. "Planetary Bioresources and Astroecology: 1. Planetary Microcosm Bioessays of Martian and Meteorite Materials: Soluble Electrolytes, Nutrients, and Algal and Plant Responses." *Icarus* 158 (2002) 72–86.

Mayr, Ernst. *Animal Species and Evolution*. Cambridge: Harvard University Press, 1963.

———. "Probability of Extraterrestrial Intelligent Life." In *Extraterrestrials: Science and Alien Intelligence*, edited by Edward Regis Jr., 23–30. Cambridge: Cambridge University Press, 1985.

McColley, Grant. "Saint Bonaventure, Francis Mayron, William Vorilong, and the Doctrine of a Plurality of Worlds." *Speculum* 12 (1937) 386–89.

———. "The Seventeenth Century Doctrine of a Plurality of Worlds." *Annals of Science* 1 (1936) 393.

McGuckin, John Anthony. *The Westminster Handbook of Patristic Theology*. Louisville: Westminster John Knox, 2004.

McHugh, L. C. "Life in Outer Space?" *Sign: A National Catholic Monthly Magazine* 41 (1961) 28.

———. "Others out Yonder." *America: A Catholic Review of the Week* 104 (1960) 296–97.

———. "Other Worlds, Other Beings?" *Newsweek* 60 (1962) 112–15.
———. "Space Theology." *Time* 66 (1955) 81.
McMullin, Ernan. "Galileo on Science and Scripture." In *The Cambridge Companion to Galileo*, edited by Peter Machamer, 271–347. Cambridge: Cambridge University Press, 1998.
———. "Life and Intelligence Far From Earth: Formulating Theological Issues." In *Many Worlds: The New Universe, Extraterrestrial Life and the Theological Implications*, edited by Steven Dick, 151–75. West Conshohocken: Templeton, 2000.
———. "Persons in the Universe." *Zygon* 15 (1980) 69–89.
McNamara, Patrick. *The Neuroscience of Religious Experience*. Cambridge: Cambridge University Press, 2009.
McNulty, Robert. "Bruno at Oxford." *Renaissance News* 13 (1960) 300–5.
Meech, Karen, et al. "Commission 51: Bioastronomy – Search for Extraterrestrial Life." *IAU Transactions* 26 (2007) 171–74.
Menut, Albert, and Alexander Denomy, eds. *Le Livre du Ciel et du Monde*. Madison: University of Wisconsin Press, 1968.
Merati, Angelo. *"Il Sommario del Processo di Giordano Bruno, con Appendice de Documenti Sull' Eresia e L'inquisizione a Modena Nel Secolo XVI."* Città del Vaticano: Biblioteca apostolica vaticana, 1942.
Michael, Donald M. "Proposed Studies on the Implications of Peaceful Space Activities for Human Affairs." Washington, DC: Brookings Institution, 1960. https://ntrs.nasa.gov/citations/19640053196.
Michaud, Michael. *Contact with Alien Civilizations: Our Hopes and Fears about Encountering Extraterrestrials*. New York: Copernicus, 2007.
Migne, Jacques Paul, ed. *Augustine: Commentaries on St. John*. Patrologica Latina 35. Paris, 1924.
———, ed. "Epistola XI ad Bonifacium." Patrologiae Cursus Completus. Paris: Garnier, 1850.
———, ed. *Glossa Ordinaria*. Patrologia Latina 113. Montrouge: SEU Petit, 1831.
———, ed. *Patrologia Graeca*. Chrysostom, John. "*In Genesis*" 3, 8 (Homily L7, 1): 53, 134. Paris: Migne, 1860.
Miller, Hugh. *Geology Versus Astronomy: or, The Conditions and the Periods; Being a View of the Modifying Effects of Geologic Discovery on the Old Astronomic Inferences Respecting the Plurality of Inhabited Worlds*. Glasgow: Collins & Co., 1855.
Milne, Edward. *Modern Cosmology and the Christian Idea of God*. Oxford: Oxford University Press, 1952.
Mok, Alex. "Humanity, Extraterrestrial Life, and the Cosmic Christ in Evolutionary Perspective." *Australian eJournal of Theology* 4 (2005).
Moltmann, Jürgen. *God in Creation: An Ecological Doctrine of Creation*. San Francisco: Harper and Row, 1981.
———. *Jesus Christ for Today's World*. Minneapolis: Fortress, 1994.
Montgomery, John Warwick. *Christ at Center and Circumference: Essays Theological, Cultural, and Polemic*. Eugene: Wipf & Stock, 2012.
Moore, Patrick. "Life in the Universe." *Astronomy Encyclopedia*. London: Philips, 2002.
Moravec, Hans. *Mind Children: The Future of Robot and Human Intelligence*. Cambridge: Harvard University Press: 1988.
More, Henry. *Divine Dialogs*. Vol. 1. London: Flesher, 1668.

Morison, Ian. *Introduction to Astronomy and Cosmology*. Hoboken: Wiley-Blackwell, 2014.

Moritz, Joshua. "Redeeming Animals and ET: That Which Has Been Assumed Has Also Been Saved." In *Astrotheology: Science and Theology Meet Extraterrestrial Life*, edited by Ted Peters, 341–43. Eugene: Cascade, 2018.

Morris, Simon Conway. *Life's Solution: Inevitable Humans in a Lonely Universe*. Cambridge: Cambridge University Press, 2004.

Morris, Thomas. *The Logic of God Incarnate*. New York: Cornell University Press, 1987.

Morrison, Philip, et al. *The Search for Extraterrestrial Intelligence*. Washington, DC: NASA, 1977.

Motte, Andrew. *Sir Isaac Newton's Mathematical Principles of Natural Philosophy and His System of the World*. Oakland: University of California Press, 1966.

Moule, C. F. D. *The Epistles to the Colossians and to Philemon*. Cambridge: Cambridge University Press, 1962.

Murphy, George. "Cosmology and Christology." *Science and Christian Belief* 6 (1994) 101–11.

Murphy-O'Connor, Jerome. "Tradition and Redaction in Col 1:15–20." *Revue Biblique* 102 (1995) 237–41.

Murray, Robert. *The Cosmic Covenant*. London: Sheed and Ward, 1992.

Nares, Edward. *An Attempt to Shew How Far the Philosophical Notion of a Plurality of Worlds Is Consistent, or Not So, with the Language of the Holy Scriptures*. London: F&C, 1801.

"NASA Announces Mars 2020 Rover Payload to Explore the Red Planet as Never Before." NASA. July 31, 2014. http://www.nasa.gov/press/2014/july/nasa-announces-mars-2020-rover-payload-to-explore-the-red-planet-as-never-before/.

"NASA's Exoplanet Archive KOI Table." Exoplanetarchive.ipac.caltech.edu.

"NASA's Kepler Releases New Catalog—2,321 Planet Candidates." NASA. March 2, 2012. www.nasa.gov/mission_pages/kepler/news/kepler-newcatalog_prt.htm.

"NASA's Kepler Marks 1,000th Exoplanet Discovery, Uncovers More Small Worlds in Habitable Zones." NASA. January 6, 2015. http://www.nasa.gov/press/2015/january/nasa-s-kepler-marks-1000th-exoplanet-discovery-uncovers-more-small-worlds-in/.

Neuman, Matthias, and Thomas Valters. *Christology: True God, True Man*. Chicago: Loyola, 2001.

Newman, John Henry. *An Essay on the Development of Christian Doctrine*. Notre Dame Series in the Great Books 4. Notre Dame: University of Notre Dame Press, 1994.

Newman, Phil. "New Energy Source 'Wrings' Power from Black Hole Spin." http://www.gsfc.nasa.gov/topstory/20011015blackhole.html.

Newman, William, and Carl Sagan. "Galactic Civilizations: Population Dynamics and Interstellar Diffusion." *Icarus* 46 (1981) 293–327.

Newton, Isaac. *Le Traite 'De l'infini' de Jean Mair*. Edited by and translated by Hubert Elie. Paris: Vrin, 1938.

———. *Principia*. Texas: Snowball, 2010.

Noble, Samuel. *Astronomical Doctrine of a Plurality of Worlds Irreconcilable with the Popular Systems of Theology, but in Perfect Harmony with the True Christian Religion*. London, 1838.

Noguchi, Masafumi. "Early Evolution of Disk Galaxies: Formation of Bulges in Clumpy Young Galactic Disks." *Astrophsysical Journal* 514 (1999) 77–95.

Nordgren, Tyler. *Sun Mood Earth: The History of Solar Eclipses from Omens of Doom to Einstein and Exoplanets*. New York: Basic, 2016.
Norden, Eduard. *Die antike Kunstprosa vom VI. Jahrhundert v. Chr. bis in die Zeit der Renaissance*. Vol. II. Leipzig: Teubner, 1923.
Norris, Ray. "How Old is ET?" In *When SETI Succeeds: The Impact of High-Information Contact*, edited by Allen Tough, 103–5. Bellevue: Foundation for the Future, 2000.
Norris, Pippa, and Ronald Inglehart. *Sacred and Secular: Religion and Politics Worldwide*. Cambridge: Cambridge University Press, 2011.
Ockham, William. *Opera Plurima: Super 4 Libros Sententiarum, in Sententiarum I*. Vol. III. Belgium, Gregg, 1962.
O'Collins, Gerald, Fr. *Christology: A Biblical, Historical, and Systematic Study of Jesus*. Oxford: Oxford University Press, 1995.
———. "The Incarnation: The Critical Issues." In *The Incarnation*, edited by Stephen Davis et al., 1–30. Oxford: Oxford University Press, 2002.
O'Donovan, Leo. "Making Heaven and Earth: Catholic Theology's Search for a Unified View of Nature and History." In *Theology and Discovery: Essays in Honor of Karl Rahner, S.J.*, edited by William J. Kelly, 269–99. Milwaukee: Marquette University Press, 1980.
O'Leary, Bede. *Our Lady of Guadalupe, Hope of America*. Lafayette: Trappist Abbey of Our Lady of Guadalupe, 1949.
O'Meara, Thomas. "Christian Theology and Extraterrestrial Intelligent Life." *Theological Studies* 60 (1999) 3–30.
———. *Vast Universe: Extraterrestrials and Christian Revelation*. Collegeville: Liturgical, 2012.
O'Malley-James, Jack, et al. "Swansong Biospheres: Refuges for Life and Novel Microbial Biospheres on Terrestrial Planets Near the End of their Habitable Lifetimes." *International Journal of Astrobiology* 12.2 (2013) 99–112.
Orbe, Antonio. "Homo Nuper Factus: En Torno a S. Ireneo, *Adv. Haer*. IV, 38, 1." *Gregorianum* 46 (1965) 481–84.
Origen. *De Principiis*. https://www.newadvent.org/fathers/04122.htm.
———. *Homilies on Leviticus*. Translated by Gary Wayne Barkley. Fathers of the Church Series. Washington, DC: Catholic University of America Press, 1990.
Ortolan, Theophile. *Astronomie et Theologiei ou L'erreur Géocentrique. La Pluralité des Mondes Habités et le Dogme de L'incarnation*. Paris: Bloud, 1894.
Overbye, Dennis. "Far-Off Planets Like the Earth Dot the Galaxy." *New York Times*. November 4, 2011. https://www.nytimes.com/2013/11/05/science/cosmic-census-finds-billions-of-planets-that-could-be-like-earth.html.
Paine, Thomas. *The Age of Reason*. In *Thomas Paine: Collected Writings*, edited by Eric Foner. New York: Library of America, 1995.
Palmer, Craig, et al. "ET Phone Darwin: What Can an Evolutionary Understanding of Animal Communication and Art Contribute to Our Understanding of Methods for Interstellar Communication?" In *Civilizations Beyond Earth: Extraterrestrial Life and Society*, edited by Douglas Vakoch and Albert Harrison, 214–26. New York: Berghan, 2011.
Palmer, Jason. *Antimatter Caught Streaming from Thunderstorms on Earth*. BBC News. January 11, 2011. https://www.bbc.com/news/science-environment-12158718.
———. "Exoplanets Are Around Every Star, Study Suggests." *BBC News*. January 11, 2012. https://www.bbc.com/news/science-environment-16515944.

Pannenberg, Wolfhart. *Systematic Theology*. Grand Rapids: Eerdmans, 1994.
Pantin, Isabelle. "New Philosophy and Old Prejudices: Aspects of the Reception of Copernicanism in a Divided Europe." *Studies in History and Philosophy of Science* 30 (1999) 237–62.
Papagiannis, Michael. *The Search for Extraterrestrial Life: Recent Developments*. Dordrecht: Reidel, 1985.
Pawl, Timothy. "Thomistic Multiple Incarnations." *The Heythrop Journal* 57 (2016) 359–70.
Peacocke, Arthur. "The Challenge and Stimulus of the Epic of Evolution to Theology." In *Many Worlds, The New Universe, Extraterrestrial Life, and the Theological Implications*, edited by Steven Dick, 108–15. Philadelphia: Templeton, 2000.
———. *God and Science: A Quest for Christian Credibility*. London: SCM, 1996.
———. *Theology for a Scientific Age*. Minneapolis: Fortress, 1993.
Pennington, Johnathan T. *Heaven and Earth in the Gospel of Matthew*. Leiden: Brill, 2007.
Perego, Angelo. "Origine Degli Esseri Razionali Estraterreni." *Divus Thomas* 61 (1958) 3–24.
———. "Possibilitá di una Redenzione Cosmica." In *Origini l'Universo, la Vita, L'intelligenze*, edited by Francesco Bertola et al., 121–40. Padue: Il Poligrafo, 1994.
———. "Rational Life beyond the Earth?" *Theology Digest* 7 (1959) 178.
Peter, Carl J. "The Position of Karl Rahner regarding the Supernatural: A Comparative Study of Nature and Grace." *Proceedings of the Catholic Theological Society of America* 20 (1965) 81–94.
Peters, Ted. "Astrotheology and the ETI Myth." *Theology and Science* 7.11 (2009) 3–30.
———. *Astrotheology: Science and Theology Meet Extraterrestrial Life*. Eugene: Cascade, 2018.
———. *The Evolution of Terrestrial and Extraterrestrial Life: Where in the World is God?* Proceedings of the Seventh Annual Goshen Conference on Religion and Science. Edited by Carl S. Helrich. Kitchener: Pandora, 2008.
———. "Exo-Theology: Speculations on Extraterrestrial Life." In *The Gods Have Landed: New Religions from Other Worlds*, edited by James R. Lewis, 187–206. Albany: State University of New York Press, 1995.
———. "Implications of the Discovery of Extraterrestrial Life for Religion." *Philosophical Transactions of the Royal Society A* 369 (2011) 644–55.
———. *Peters ETI Religious Crisis Survey*. Berkeley: Pacific Lutheran Theological Seminary, 2008.
———. *Science, Theology, and Ethics*. Aldershot: Ashgate, 2003.
———. *UFOs: God's Chariots? Spirituality, Ancient Aliens, and Religious Yearnings in the Age of Extraterrestrials*. Pompton Plains: New Page, 2014.
———. "Would the Discovery of ETI Provoke a Religious Crisis?" In *Astrobiology, History, and Society: Life Beyond Earth and the Impact of Discovery*, edited by Douglas A. Vakoch. Heidelberg: Springer, 2013.
Petigura, Erik, et al. "Prevalence of Earth-Size Planets Orbiting Sun-Like Stars." *PNAS* 110.48 (2013) 19273–78. www.pnas.org/content/110/48/19273.abstract.
Petty, Michael W. *A Faith That Loves the Earth: The Ecological Theology of Karl Rahner*. Lanham: University Press of America, 1996.

Pew Research Center. "Global Christianity: A Report on the Size and Distribution of the World's Christian Population." December 19, 2011. http://www.pewforum.org/Christian/Global-Christianity-exec.aspx.

———. "Spirit and Power: A 10-Country Survey of Pentecostals." October 5, 2006. http://www.pewforum.org/surveys/pentecostal.

Pieris, Aloysius, SJ. *An Asian Theology of Liberation*, Edinburgh: T&T Clark, 1988.

Pinker, Steven. "The Evolutionary Psychology of Religion." Presented at the Freedom from Religion Foundation. Madison, WI: October 29, 2004. https://ffrf.org/about/getting-acquainted/item/13184-the-evolutionary-psychology-of-religion.

"The Pioneer Missions." NASA. Last updated December 14, 2017. https://www.nasa.gov/centers/ames/missions/archive/pioneer.html.

Pittenger, Norman. *The Word Incarnate: A Study of the Doctrine of the Person of Christ*. London: Nisbet, 1959.

Planck Collaboration. "Planck 2013 results. I. Overview of Products and Scientific Results." Revised June 5, 2014. http://arxiv.org/abs/1303.5062.

Plato. "Timaeus. 31b-34b: The Body of the Cosmos." In *Philo of Alexandria and the Timaeus of Plato*, edited by David Runia, 117–98. Philosophia Antiqua. Leiden: Brill, 1986.

Poe, Rahasya. *To Believe or Not to Believe: The Social and Neurological Consequences of Belief Systems*. Bloomington: Xlibris, 2009.

Pogge, Richard W. "Some Notes on the Theological Response to Copernicus." February 16, 2006. http://www.astronomy.ohio-state.edu/~pogge/Essays/Copernic.html.

Pohle, Joseph. *Christology: A Dogmatic Treatise on the Incarnation*. St. Louis: Herder, 1913.

———. *Die Sternenwelten und ihre Bewohner*. Cologne: Verlag Gmbh & Company, 1885.

Poidevin, Robin Le. "Multiple Incarnations and Distributed Persons." In *The Metaphysics of the Incarnation*, edited by Anna Mormodoro and Jonathan Hill. Oxford Scholarship Online, 2011. https://oxford.universitypressscholarship.com/view/10.1093/acprof:oso/9780199583164.001.0001/acprof-9780199583164-chapter-1.

Polkinghorne, John. *One World: The Interaction of Science and Theology*. Princeton: Princeton University Press, 1986.

———. "Reason and Reality." *Scottish Journal of Theology* 46.3 (1991) 403–4.

———. *Science and the Trinity: The Christian Encounter with Reality*. New Haven: Yale University Press, 2004.

Pollard, Thomas. "Colossians 1:12–20 A Reconsideration." *NTS* 27 (1981) 572–75.

Pollard, William. "The Prevalence of Earth-Like Planets." *American Scientist* 6 (1979) 653–59.

"Polycyclic Aromatic Hydrocarbons: An Interview with Dr. Farid Salama." *Astrobiology Magazine*. 2000. https://gitso-outage.oracle.com/thinkquest.

Pope John Paul II. *Cosmology and Fundamental Physics*. Pontifical Academy of Science, October 3, 1981.

———. *Fides et Ratio*. Boston: Pauline, 1998.

Pope John Paul XXIII. *Ad Gentes*, 5. Second Vatican Council. 1965. http://www.vatican.va/archive/hist_councils/ii_vatican_council/documents/vat-ii_decree_19651207_ad-gentes_en.html.

Pope Leo XIII. *Providentissimus Deus*. 1893. https://www.papalencyclicals.net/leo13/l13provi.htm

Pope Paul VI. *Apostolicam Actuositatem*. Second Vatican Council. 1965. http://www.vatican.va/archive/hist_councils/ii_vatican_council/documents/vat-ii_decree_19651118_apostolicam-actuositatem_en.html.

———. *Dei Verbum*. Second Vatican Council. New York: Pauline, 1965.

———. *Gaudium et Spes*. Second Vatican Council. 1965. http://inters.org/Gaudium-et-Spes.

Pope Pius IX. *De Filius*. Denzinger-Schönmetzer, *Enchiridion Symbolorum, Definitionum et Declarationum de Rebus Idei et Morum*. Freiburg im Breisgau: Herder,1965.

Pope Pius XII. *Divino Afflante Spiritu*. Boston: Daughters of St. Paul, 1943.

———. *Humani Generis*. August 12, 1950. http://www.vatican.va/content/pius-xii/en/encyclicals/documents/hf_p-xii_enc_12081950_humani-generis.html.

Porteus, Beilby. "On the Christian Doctrine of Redemption." In *Works*, vol. III. London, 1811.

Potvin, Thomas. "Congar's Thought on Salvation outside the Church: Missio ad Gentes." *Science et Espirit* 55 (2003) 139–63.

Powell, Baden. *The Unity of Worlds and of Nature: Three Essays on the Spirit of Inductive Philosophy; the Plurality of Worlds; and the Philosophy of Creation*. Whitefish: Kessinger, 2009.

Powell, Russell. "From Humanoids to Heptapods: The Evolution of Extraterrestrials in Science Fiction." May 1, 2017. http://www.extinctblog.org/extinct/2017/5/1/from-humanoids-to-heptapods-the-evolution-of-extraterrestrials-in-science-fiction.

Primack, Joel, and Nancy Abrams. *The View from the Center of the Universe*. New York: Riverhead, 2007.

Puccetti, Roland. *Persons: A Study of Possible Moral Agents in the Universe*. London: Macmillan, 1968.

Quintana, Elisa V., and Jack J. Lissauer. *Terrestrial Planet Formation in Binary Star Systems*. May 23, 2007. http://arxiv.org/abs/0705.3444.

Race, Margaret. "Societal and Ethical Concerns." In *Planets and Life: The Emerging Science of Astrobiology*, edited by Woodruff T. Sullivan III and John A Baross, 483–97. Cambridge: Cambridge University Press, 2007.

Rahner, Karl. "Christology Within an Evolutionary View of the World." In *Theological Investigations*, vol. V, translated by Karl Kruger, 183–84. Baltimore: Helicon, 1966.

———. "De Personali Unione Duarum Naturam in Christo." *The Heythrop Journal* 47.2 (2006) 275–90.

———. *Foundations of the Christian Faith: An Introduction to the Idea of Christianity*. New York: Crossroad, 1982.

———. "Landung auf dem Mond." In *Kritisches Wort: Aktuelle Probleme in Kirche und Welt*, 233–34. Freiburg: Herder, 1970.

———. "Naturwissenschaft und vernünftiger Glaube." In Bd. 15 of *Karl Rahner: Schriften zur Theologie*, 24–62. Zürich: Benziger, 1983.

———. "Natural Science and Reasonable Faith." *Theological Investigation* 21 (1983) 48–49.

———. "Revelation." In *Dictionary of Theology*, 2nd ed, edited by Karl Rahner and Herbert Vorgrimler, 444–49. New York: Crossroad, 1985.

———. "Sternenbewohner: Theologisch." In *Lexikon Für Theologie und Kirche, vol. 9*, 2nd ed, edited by Josef Hofer and Karl Rahner, 1061–62. Freiburg: Herder, 1964.

———. *Theological Investigations I: God, Christ, Mary, and Grace*. London: Darton, Longman, and Todd, 1961.

———. "Theology and Anthropology." In *The Word in History*, edited by Patrick Burke, 15. New York: Sheed & Ward, 1966.
Raible, Daniel. "Missionaries to Space." *Newsweek* 55 (1960) 90.
———. "Rational Life in Outer Space?" *America: A Catholic Review of the Week* 103 (1960) 352.
———. "Theology of Saucers." *Time* 18 (1952) 39.
Rainaud, Armand. *Le continent austral*. Charleston: Nabu, 2011.
Ramm, Bernard. *The Christian View of Science and Scripture*. Grand Rapids: Eerdmans, 1954.
Rana, N. C. and Wilkinson, D. A. "Molecular Hydrogen and the Chemical Evolution of the Galaxy." *Monthly Notices of the Royal Astronomical Society* 226 (1987) 395–422.
Randolph, Richard, et al. "Reconsidering the Theological and Ethical Implications of Extraterrestrial Life." *Center for Theology and Natural Sciences Bulletin* 17.3 (1997) 1–8.
Reisz, Robert, and Jörg Fröbisch. "The Oldest Caseid Synapsid from the Late Pennsylvanian of Kansas, and the Evolution of Herbivores in Terrestrial Vertebrates." *PLoS ONE* 9.4 (2014). doi:10.1371/journal.pone.0094518.
Rengers, Christopher. *Mary of the Americas, Our Lady of Guadalupe*. New York: Alba, 1990.
Reynaud, Jean. *Terre Et Ciel*. Charleston: Nabu, 2012.
Richerson, Peter, and Robert Boyd. "Build for Speed, not for Comfort: Darwinian Theory and Human Culture." *History and Philosophy of the Life Sciences (Special Issue on Darwinian Evolution Across the Disciplines)* 23 (2001) 423–63.
Robelo, Cecilio Agustín. "Biblioteca Porrúa: Imprenta del Museo Nacional de Arqueología, Historia y Etnología." In *Diccionario de Mitología Nahua*, edited by Eusebio Dávalos Hurtado. Mexico City: MéxiCompany, 1905.
Robinson, Daniel, et al., eds. *Human Nature in Its Wholeness: A Roman Catholic Perspective*. Washington, DC: Catholic University of America Press, 2006.
Robinson, Joan. "A Formal Analysis of Colossians 1: 15–20." *Journal of Biblical Literature* 76 (1957) 270–87.
Robitaille, Thomas, and Barbara A. Whitney. "The Present-Day Star Formation Rate of the Milky Way Determined from Spitzer-Detected Young Stellar Objects." *The Astrophysical Journal Letters* 710.1 (2010) 11–15.
Rochberg-Halton, Francesca. "Elements of the Babylonian Contribution to Hellenistic Astrology." *Journal of the American Oriental Society* 108 (1988) 51–62.
Rothery, David, and Iain Gilmour. *An Introduction to Astrobiology*. Cambridge: Cambridge University Press, 2011.
Rosbury, Robert, and Fei Yan. "High Resolution Transmission Spectrum of the Earth's Atmosphere-Seeing Earth as an Exoplanet using a Lunar Eclipse." *International Journal of Astrobiology* 14 (2015) 255–66.
Rosen, Edward. *Kepler's Conversation with Galileo's Sidereal Messenger*. New York: Johnson, 1965.
Royer, Franchón. *The Franciscans Came First*. Paterson: St. Anthony Guild, 1951.
Russell, Mary Doria. *Children of God*. Cambridge: Black Swan, 2009.
———. *The Sparrow*. New York: Villard, 1996.
Russell, Robert John. "Life in the Universe: Philosophical and Theological Issues." In *First Steps in the Origin of Life in the Universe*, edited by Chela-Flores et al., 365–74. Dordrecht: Kluwer Academic, 2001.

———. "What Are Extraterrestrials Really Like?" In *God for the 21st Century*, edited by Russell Stannard, 65–70. Philadelphia: Foundation, 2000.

———, et al., eds. *John Paul II on Science and Religion: Reflections on the New View from Rome*. Notre Dame: University of Notre Dame Press, 1990.

Sagan, Carl. *Contact*. London: Orbit, 1997.

———, and Frank Drake. "The Search for Extraterrestrial Intelligence." *Scientific American*. January 6, 1997. https://www.scientificamerican.com/article/the-search-for-extraterre/.

Sahney, Sarda, and Michael Benton. "Recovery from the Most Profound Mass Extinction of all Time." *Proceedings of the Royal Society* B275 (2008) 759–65.

Salaverri, Joaquín. "La Possibilidad de Seres Humanos Extraterrestres ante el Dogma Catolica." *Razon y Fe* 148 (1953) 23–43.

Salvestrini, Virgilio. *Bibliografia di Giordano Bruno*. Firenze: Sansoni Antiquariato, 1958.

Schiaparelli, G. V. *La vita sul pianeta Marte. Tre Scritti su Marte e i Marziani*. Edited by Pasquale Tucci et al. Milano: Mimesis, 1998.

Schmaus, Michael. *Dogma 3: God and His Christ*. 1st ed. New York: Sheed and Ward, 1971.

Schneider, Susan. "Alien Minds." In *The Impact of Discovering Life Beyond Earth*, edited by Steven J. Dick. Cambridge: Cambridge University Press, 2015.

———. *Science Fiction and Philosophy: From Time Travel to Superintelligence*. Hoboken: Wiley, 2009.

———. *Superintelligent AI and the Postbiological Cosmos Approach*. Cambridge: Cambridge University Press, 2017.

Schillebeeckx, Edward. *Christ: The Christian Experience in the Modern World*. Translated by John Bowden. London: SCM, 1988.

Schopf, J. William, and Anatolly Kudryavtsev. "Evidence of Archean Life: Stromatolites and Microfossils." *Precambrian Research* 158 (2007) 141–55.

Schutz, Bernard. *A First Course in General Relativity*. New York: Cambridge University Press, 1985.

Schweizer, Eduard. *The Letter to the Colossians*. London: SPCK, 1982.

Scott, Ian. *Paul's Way of Knowing: Story, Experience, and the Spirit*. Grand Rapids: Baker Academic, 2008.

Scotus, Duns. *Reportatio Parisiensis III*. http://individual.utoronto.ca/pking/editions/SCOTUS.RepPar.1A.d3.q4.ed.pdf.

Searle, John. "Biological Naturalism." In *The Blackwell Companion to Consciousness*, 2nd ed., edited by Max Velmans and Susan Schneider, 327–37. Hoboken: Wiley-Blackwell, 2017.

———. "Minds, Brains, and Programs." *Behavioral and Brain Sciences* 3.3 (1980) 417–57.

Second Lateran Council. "The Canons of the Second Lateran Council, 1123." Eighth session. https://sourcebooks.fordham.edu/basis/lateran2.asp.

Second Ancona Ufological Congress. "Alien Civilization: Between Doubt and Reason." April 17, 2000.

Second Vatican Council. *Lumen Gentium*. November 24, 1964. https://www.vatican.va/archive/hist_councils/ii_vatican_council/documents/vat-ii_const_19641121_lumen-gentium_en.html.

Seely, Paul. "The Firmament and the Water Above." *The Westminster Theological Journal* 53 (1991) 227–40.

Shults, LeRon. *Reforming Theological Anthropology: After the Philosophical Turn to Relationality*. Grand Rapids: Eerdmans, 2003.

Sindoni, Elio. *Esistono gli Extraterrestri?* Milano: Il Saggiatore, 1997.

Sklovskii, Losoff, and Carl Sagan. *Intelligent Life in the Universe*. Boca Raton: Emerson-Adams, 1999.

Shapire, Barbara. *John Wilkins, 1614–1672. An Intellectual Biography*. Berkeley: University of California Press, 1969.

Shen, Shu-zhong, et al. "Calibrating the End-Permian Mass Extinction." *Science* 334 (2011) 1367–72.

Shermer, Michael. *The Science of Good and Evil: Why People Cheat, Gossip, Care, Share, and Follow the Golden Rule*. New York: Times, 2004.

Shostak, Seth. *Progress in Search for Extraterrestrial Life*. San Francisco: Astronomical Society of the Pacific, 1995.

———. *Sharing the Universe: Perspectives on Extraterrestrial Life*. Berkeley: Berkeley Hill, 1998.

Singer, Dorothea. *Giordano Bruno: His Life and Thought, with Annotated Translation of His Work On the Infinite Universe and Worlds*. New York: Schuman, 1950.

Slotten, Ross. *The Heretic in Darwin's Court: The Life of Alfred Russel Wallace*. New York: Columbia University Press, 2004.

Smart, Ninian. *Dimensions of the Sacred: An Anatomy of the World's Beliefs*. London: Harper Collins, 1996.

Smolin, Lee. *The Life of the Cosmos*. Oxford: Oxford University Press, 1997.

Sobel, Dava. *Galileo's Daughter*. London: Fourth Estate, 2000.

Sorabji, Richard. *Simplicius: On Aristotle's Physics 1.5-9*. London: Bloomsbury, 2012.

Souers, Clark. *Hydrogen Properties for Fusion Energy*. Berkeley: University of California Press, 1986.

Stark, Rodney. "Secularization, R.I.P." *Sociology of Religion* 60 (1999) 249–73.

Steenberg, M. C. "Children in Paradise: Adam and Eve as 'infants' in Irenaeus of Lyons." *Journal of Early Christian Studies* 12.1 (2004), 1–22.

Steidl, Paul. *The Earth, the Stars, and the Bible*. Grand Rapids: Baker Book House, 1979.

Stenmark, Mikael. "Ways of Relating Science and Religion." In *Science and Religion*, edited by Peter Harrison, 278–95. Cambridge: Cambridge University Press, 2010.

Stevens, Clifford J. *Astrotheology: For the Cosmic Adventure*. Techny: Divine Word, 1969.

Strick James. *The Living Universe: NASA and the Development of Astrobiology*. New Brunswick: Rutgers University Press, 2004.

Strickberger, Monroe. *Genética*. Barcelona: Omega, 1978.

Stump, Eleanore. *Aquinas*. New York: Routledge, 2003.

Sturch, Richard. *The Word and the Christ: An Essay in Analytic Christology*. New York: Clarendon, 1991.

Sullivan, Walter. *We Are Not Alone: The Continuing Search for Extraterrestrial Intelligence*. New York: Plume, 1994.

Swedenborg, Emanuel. *Earths in Our Solar System Which Are Called Planets in the Starry Heavens; Their Inhabitants, and the Spirits and Angels There, from Things Heard and Seen*. New York: Swedenborg, 1951.

Sydney Morning Herald. "Life on Other Planets?" November 8, 1959.

Sylla, Edith Dudley. "Aristotelian Commentaries and Scientific Change: The Parisian Nominalists on the Cause of the Natural Motion of Inanimate Bodies." *Vivarium* 31.1 (1993) 37–83.

Tanner, Norman P. *Decrees of the Ecumenical Councils, Vol. 1.* Washington, DC: Georgetown University Press, 2017.

Tarter, Jill. "SETI and the Religions of the Universe." In *Many Worlds: The New Universe, Extraterrestrial Life, and the Theological Implications*, edited by Steven Dick, 143–49. Philadelphia: Templeton, 2000.

Teilhard de Chardin, Pierre. *Christianity and Evolution.* London: Collins, 1971.

———. *Heart of the Matter.* New York: Harcourt, 1978.

———. *Hymn of the Universe.* Modern Spiritual Masters Series. London: Fontana, 1971.

———. "Life and the Planets." In *The Future of Man.* New York: Harper and Row, 1964.

———. *Phenomenon of Man.* Translated by Bernard Wall. New York: Harper and Row, 1959.

Terrasson, Abbe Jean. *Traité de L'infini Créé, Avec Explication de la Possibilité de la Transubstantiation et un Petit Traité de la Confession et de la Communion.* Amsterdam: Marc-Michel Rey, 1769.

Tertullian. *Contra Marcion.* Book II. https://www.newadvent.org/fathers/03122.htm.

Than, Ker. "Antimatter Found Orbiting Earth - A First." *National Geographic.* August 10, 2011. http://news.nationalgeographic.com/news/2011/09/110810-antimattter-belt-Earth-trapped-pamela-space-science/.

Tillich, Paul. "A Multidisciplinary Approach to Original Sin." *Theology and Science* 7 (2009) 67–83.

———. *Systematic Theology.* London: Nisbe, 1953.

Tipler, Frank. "Extraterrestrial Intelligent Beings Do Not Exist." In *Extraterrestrials: Science and Alien Intelligence*, edited by Edward Regis, 133–50. Cambridge: Cambridge University Press, 1985.

Todhunter, Isaac, and William Whewell. *An Account of His Writings with Selections from His Literary and Scientific Correspondence.* 2 vols. New York, Johnson Reprint, 1970.

Tollefsten, Torstein. *The Christocentric Cosmology of St. Maximus the Confessor.* New York: Oxford University Press, 2008.

Tough, Allen. *Alien Worlds: Social and Religious Dimensions of Extraterrestrial Contact.* Syracuse: Syracuse University Press, 2007.

———. "How to Achieve Contact: Five Promising Strategies." In *When SETI Succeeds.* edited by Allen Tough, 115–25. Bellevue: Foundation for the Future, 2000.

———. "What Role will Extraterrestrials Play in Humanity's Future?" *Journal of British Interplanetary Society* 39 (1986) 491–98.

———. *When SETI Succeeds: The Impact of High-Information Contact.* Washington, DC: Foundation for the Future, 2000.

Traphagan, John. "Anthropology at a Distance: SETI and the Production of Knowledge in the Encounter with an Extraterrestrial Other." In *Archaeology, Anthropology, and Interstellar Communication*, edited by Douglas Vakoch, 131–43. San Francisco: Astronomical Society of the Pacific, 2011.

Tumminia, Diana G. *Alien Worlds: Social and Religious Dimensions of Extraterrestrial Contact.* Syracuse: Syracuse University Press, 2007.

Turnor, Edmund. *Collections for the History of the Town and Soke of Grantham Containing Authentic Memoirs of Sir Isaac Newton.* London: Bulmer and Company, 1806.

Twitchell, David E. *Global Implications of the UFO Reality.* Haverford: Infinity, 2003.

Tylor, Edward Burnett. *Primitive Cultures: Researches into the Development of Mythology, Philosophy, Religion Art, and Custom.* 2 vols. London: Murray, 1871.

UCS Satellite Database. *Union of Concerned Scientists.* Updated April 1, 2020. https://www.ucsusa.org.

Ulmschneider, Peter. *Intelligent Life in the Universe: Principles and Requirements Behind Its Emergence.* New York: Springer, 2006.

Vakoch, Douglas. *Civilizations Beyond Earth: Extraterrestrial Life and Society.* New York: Berghahn, 2011.

———. "Predicting Reactions to the Detection of Life beyond Earth." Paper presented at the Societal Implications of Astrobiology Workshop, NASA-Ames Research Center, 1999.

———. "Reactions to Receipt of a Message from Extraterrestrial Intelligence: A Cross-Cultural Empirical Study." *Acta Astronautica* 46 (2000) 737–44.

———. "Roman Catholic Views of Extraterrestrial Intelligence: Anticipating the Future by Examining the Past." In *If SETI Succeeds: The Impact of High Information Contact*, edited by Allen Tough, 165–74. Bellevue: Foundation For the Future, 2000.

———. *Social Implications of the Detection of an Extraterrestrial Civilization.* Mountain View: SETI, 1999.

Van Huyssteen, Jacobus Wentzel. *Alone in the World?: Human Uniqueness in Science and Theology.* The Gifford Lectures. Grand Rapids: Eerdmans, 2005.

Van Kooten, George H. *Cosmic Christology in Paul and the Pauline School.* Tübingen: Mohr Verlag, 2003.

Vatican Council I. *Dogmatic Canons and Decrees.* Rockford: Tan, 1977.

———. *On Faith and Reason.* http://www.vatican.va/content/john-paul-ii/en/encyclicals/documents/hf_jp-ii_enc_14091998_fides-et-ratio.html.

Vinge, Vernor. "The Coming Technological Singularity: How to Survive in the Post-Human Era." In *Vision-21: Interdisciplinary Science and Engineering in the Era of Cyberspace*, edited by Geoffrey Landis. NASA, 1993.

"Voyager Enters Interstellar Space - NASA Jet Propulsion Laboratory." NASA. https://www.jpl.nasa.gov/news/nasas-voyager-2-probe-enters-interstellar-space.

Walker, James. *Evolution of the Atmosphere.* New York, Hafner. 1977.

Wall, Mike. "Super-Earth Alien Planet May Be Habitable for Life." *Fox News.* January 11, 2012. http://www.foxnews.com/science/2012/11/08/super-Earth-alien-planet-may-be-habitable/.

Wallace, Anthony. *Religion: An Anthropological View.* New York: Random House, 1966.

Ward, Keith. *God, Faith, and the New Millennium: Christian Belief in an Age of Science.* London: Oneworld, 1998.

Washam, Erik. "Cosmic Errors: Martians Build Canals!" *Smithsonian Magazine.* December 2010.

Wawrykow, Joseph Peter. *The Westminster Handbook to Thomas Aquinas.* Louisville: Westminster John Knox, 2005.

Webb, Steven. *If the Universe Is Teeming with Aliens . . . Where is Everybody?: Seventy-Five Solutions to the Fermi Paradox and the Problem of Extraterrestrial Life.* New York: Springer International, 2015.

Weigel, Margaret, and Kathryn Coe. "Impact of Extraterrestrial Life Discovery for Third World Societies: Anthropological and Public Health Considerations." In *Astrobiology, History, and Society: Life Beyond Earth and the Impact of Discovery*, edited by Douglas Vakoch, 227–57. Springer, Heidelberg, 2013.

Weintraub, David. *Religions and Extraterrestrial Life: How Will We Deal with It?* London: Springer, 2014.

Weissmahr, Bela. "The Affinity of Evolutionary Worldview and Christian Belief Highlighted by Karl Rahner." In *The Philosophical Sources of the Theology of Karl Rahner*, edited by Harald Schöndorf, 175–80. Freiburg: Herder, 2005.

West, Barbara A., and Francis T. Murphy. *A Brief History of Australia*. New York: Facts on File, 2010.

Weston, Frank. *The Revelation of Eternal Love: Christianity Stated in Terms of Love*. London: Mowbray, 1920.

Wiegert, Paul, and Matt Holman. "The Stability of Planets in the Alpha Centauri System." *The Astronomical Journal* 113 (2007) 1445–50.

Wiker, Benjamin. "Alien Ideas, Christianity, and the Search for Extraterrestrial Life." *Crisis* 20.10 (2002) 26–31.

White, Ellen. *The Story of Patriarchs and Prophets*. Nampa: Pacific, 1890.

White, Leslie. *The Evolution of Culture: The Development of Civilization to the Fall of Rome*. Abingdon: Routledge, 2007.

White, Vernon. *Atonement and Incarnation: An Essay in Universalism and Particularity*. Cambridge: Cambridge University Press, 1991.

Whittet, Douglas. *Dust in the Galactic Environment*. Boca Raton: CRC, 2002.

Wilkins, John. *The Discovery of the World in the Moone, or a Discourse Tending to Prove, That 'Tis Probable There May Be Another Habitable World in That Planet*. London: E. G., 1638.

Wilhelm, Joseph, and Thomas B. Scannell. *A Manual of Catholic Theology*. London, Paul, Trench, Trubner & Company, 1906.

Wilkinson, David. *Alone in the Universe? The X-Files, Aliens, and God*. Crowborough: Monarch, 1997.

———. *Science, Religion, and the Search for Extraterrestrial Intelligence*. Oxford Scholarship Online, 2013. https://oxford.universitypressscholarship.com/view/10.1093/acprof:oso/9780199680207.001.0001/acprof-9780199680207.

———. "Searching for Another Earth: The Recent History of the Discovery of Exoplanets." *Zygon* 51.2 (2016) 414–30.

Witze, Alesandra. "Claims of Earth's Oldest Fossils Tantalize Researchers." *Nature*. August 31, 2016. https://www.nature.com/news/claims-of-earth-s-oldest-fossils-tantalize-researchers-1.20506.

Wöhlert, Hans. *Das Weltbild in Klopstocks's Messias*. Halle: Niemeyer, 1915.

Wolpert, Lewis. *Six Possible Things Before Breakfast: The Evolutionary Origins of Belief*. New York: Norton & Company, 2008.

Woods, Thomas. *How the Catholic Church Built Western Civilization*. Washington, DC: Regnery, 2005.

Wright, Edward J. *The Early History of Heaven*. Oxford: Oxford University Press, 2000.

Wurchterl, Guenther. "Planet Formation Towards Estimating Galactic Habitability." In *Astrobiology: Future Perspectives*. New York: Kluwer Academic, 2004.

Zoltán, Galántai. "Life, Intelligence, and the Multiverse." July 19, 2016. http://arxiv.org/pdf/1607.06114.

Zubek, Theodore. "Theological Questions on Space Creatures." *The American Ecclesiastic Review* 145 (1961) 393–99.

Index

Abraham, 191n9, 234, 236, 242n192, 244, 253n20, 254n222
acheiropoieta, 244, 244n196, 261
Adam, 57, 60, 70, 94, 116, 171, 173, 211n66, 233, 238, 239n186
Alexander Report, xix, 80, 120n225, 222
Ancient Hebraic Cosmology (Illus.), 54
animals, 19, 27, 53, 61, 62, 92, 109, 109n183, 130, 131, 132, 146, 155, 160, 213, 217, 233, 235n159, 234, 235n159, 236, 280
angels
    and cosmology, 202, 202n35, 203–6
    fallen, 100, 101, 173, 182, 225
    heavenly, xiv, xvii, 10, 44, 53, 54, 55, 75, 144, 144n72, 154, 165, 166, 170, 170n149, 173, 176–79, 191n5, 194n15, 195, 213, 213n77, 214, 218–21, 221n107, 226n126, 230, 244–45, 247, 247n204, 248, 253n220, 254, 256
Anselm, St. 102, 156n100, 212, 276
anthropocentrism, xiii, 14, 32, 48, 78, 83, 89n117, 92, 93, 102, 110, 124, 152, 190, 193, 204, 222, 262, 275, 277, 279, 280
anthropomorphism, 51, 73, 83, 89n117, 92–93, 124, 140, 154, 183n163, 189, 191, 202, 208
    and angels, 220
    historical theological, xviii

antipodes, people of, 56–57, 73, 74, 94, 113
Apollo missions, 49, 121
Aquinas, Thomas, xix, 1, 5, 38, 58, 60, 74, 87, 100, 102, 107, 108, 119, 123, 203, 209, 212
    and *Exclusive* view, 84–86
    and fittingness of Incarnation for salvation of the human race, 194, 249
    and multiple incarnations, 214–17
    and necessity of Incarnation for salvation of the human race, 213
    and purpose of incarnation, 217–18
    and scriptural interpretation, 199, 240
    and state of pure nature, 168–69
    and supernatural revelations, 252
    on angels, 218–20
    on Church's denouncement of teaching on denial God could create other worlds, 58
    on knowing God through his effects, 166
    teaching against plural worlds, 57
    theological relevance for exotheology, 8–12
Ark of the Covenant, 236, 244, 253, 253n221
Aricebo radio telescope, 32
Aristotle, 36–39, 58–60, 65–66, 69, 77, 78 n100, 87, 202 n31, 268 n264,

INDEX

Aristotelian cosmology, 38, 55, 57–59, 62–63, 65–67, 94, 123, 192, 201, 207, 274, 277, 280–81
    arguments against, 69, 77
artifacts, extraterrestrial, xviii, 28, 29, 221, 222n112
artificial intelligence, 127, 142–43, 143n70, 146
assumption of mediocrity, 132
astrobiology, xiv, xv, xviii, xix, 2, 13, 19–23, 27, 47, 79–80, 83, 102, 121, 272
    engagement with theology, 3, 6, 8, 12, 107, 196
    and extraterrestrial morphologies, 126, 129
atheism, 184, 224
atomism, Greek, 35–37, 43, 55, 62, 81
atonement, 97, 108, 111, 113, 176, 178n156, 192, 235–36, 245
Augustine, St., 7n23, 56n11, 56n12, 74, 87, 108, 245
    and scriptural interpretation, 199
    and the *Exclusive* view, 84
    and the *Varied* view, 119
    denial of plural worlds, 55
    denial of the existence of people in the Antipodes, 56
    on fall of the angels, 178
    on the economy of salvation, 246
Avatar (2009) film, 135, 167
Averroes, 268n264
Aztecs, 160, 261

bacteria, 21, 22, 90
Badham, Paul and Linda, 106
Bailey, Cyril, 35
Balducci, Monsignor Corrado, 93–94, 122, 170
Barbour, Ian, 117, 200n27
Barth, Karl, 110
Basil of Caesarea, 211
beatific vision, 166n139, 167, 171, 173, 175, 176, 213n77, 216, 219
Bentley, Richard, 44, 70, 72–73, 82
biblical criticism, xiv, 3, 12, 241
Bieri, Robert, 129, 133, 135

Big Bang, xiv, 14, 16, 18, 23, 26, 49, 127, 219, 278
bilateral symmetry of extraterrestrials, 130, 133
biosignatures, 13, 50
black holes, 78, 148, 148n88, 278
Blumenberg, Hans, 224
Bonaventure, St., 58, 87 n113, 102–3, 107–8, 111, 124, 210 n62, 218
Bonting, Sjoerd, 95–96, 209 n59
Bostrom, Nick, 143–44
Brahe, Tycho, 40
Braine, David 114–15
Breig, Joseph, 84–85, 189
Brewster, David, 44, 46, 72–73, 81–82
Broms, Allen, 131
Brookings Report, 28–30, 121, 189n1, 222
Bruno, Giordano, 43, 65–67, 69, 277
Bruns, J. Edgar, 109
Bultmann, Rudolf, 205
Buridan, Jean, 59, 60
Burns, Patout J., 245–46
Burroughs, Edgar Rice, 47

Campanella, 69 n65, 81 n103, 82, 106
carbon, 18, 19, 21, 23, 50
    based life-forms, 20, 76, 95, 130
Caro, Lucrezio, 177
Cartesian universe, 39, 41–42, 43, 276
Cartesian vortices, 42
Cassini-Huygens mission, 51
Chalmers, Thomas, 71–72, 81n103, 82, 104n107
Christ
    as universal redemptive figure, 60, 64, 73, 81, 84–97, 99, 102, 109–11, 122, 188, 190, 192
    preexistence of, xiv
Christie, W.H.M., 47
Christocentrism, xvii, xiii, 10, 85, 87, 89, 92–93, 94, 96, 99, 121–22, 123–24, 193, 231
christology, xx, xxi, 1, 5, 6, 7, 12, 74, 83, 86, 87, 90–3, 95, 97, 104n164, 105–6, 109–11, 115, 122–23, 126, 189, 191, 194, 203–4, 206–8, 212

Chrysostom, John, 203, 259, 260, 261n262, 268
Clarke, Arthur, 119
Clement of Alexandria, 225
climate change, 21–2
cloning, 136, 137, 142, 143, 164
Cockell, Charles, 131
colonialism, European, xix, 28, 223
colonization of extraterrestrials, 33, 142, 149, 150
communion
　with creatures, 112,
　with God, 90, 109, 111, 168, 177, 181, 194, 230, 231, 236, 249
Concilio, Januarius de, 75, 81n103
Conduitt, John, 44
cognitive universals, 139
collective consciousness, 136, 139, 143, 146n78, 155
Congar, Yves, 97, 101–2, 123, 124, 215, 232n134
Connell, Francis, 101–2
Consolmagno, Br. Guy, xvi, 2n9, 87, 90–92, 94, 122
Contact (1997 film), 145
contact with extraterrestrials, xiii, xvi, xxi, 16, 29, 29n86, 32, 33, 78, 80, 185, 119n223, 155, 159, 221–29, 263, 271
Copernican Principle, xiv
Copernicus, xiv, 39, 64–65, 78, 276, 280
Copernicanism, xv, xvii, 5, 14, 36, 40, 42–43, 45, 65, 67–69, 78–79, 113, 115, 192, 198, 276
coronagraph, 27
Cosmic Christ, 87, 90n119, 91, 94, 95, 122, 201, 204, 270
　and Colossians hymn, 90, 191, 193
cosmology, xiii, xiv, xix, 3, 8, 34, 36, 43, 69, 79, 94, 95, 98n143, 111, 15
　against heliocentric, 68
　and Giordano Bruno, 65–67
　and Nicolas of Cusa, 61–62
　aristotelian, 38, 123
　cartesian vortex, 42
　copernican, 39
　early Christian, 55, 201

einsteinian, xiv, xv, 79, 119, 193, 201, 278
egyptian, 274, 280
epicurean, 35, 65, 274, 279
greek atomistic, 35–37
hebraic, 53–54
judeo-christian, 53–55, 78
modern advances in, 13–19
ptolemaic, 38
scriptural interpretation in light of modern, 200–202, 205–8
stoic, 202–4
theological implications of modern, 198, 263
theological limitations of pre-scientific, 123, 190
Council of Trent, 168
　and theological timeline, 274–81
covenant, 109n183, 163, 182, 232, 240, 241, 252, 258, 267, 268
　abrahamic, 233–34
　adamic, 233
　cosmic, 232
　davidic, 236–37
　deuteronomic, 235
　everlasting, 210, 237–38, 260
　mosaic, 234–36
　noahic, 233
　priestly, 173
　sinaitic, 235
Coyne, George, 91
creation
　ex nihilo, 5, 39
　humanity's special, 5, 40, 55, 79, 85
Crisp, Oliver, 96–97, 103, 209n59
Crowe, Michael, 76
Curiosity Rover, 50
Cusa, Nicholas of, 43, 61–64, 69
Cyril of Alexandria, 212

Dales, Richard, 58
Darwinian Theory, xv, 5, 45, 127, 128–31, 138
Davies, Paul, xviii2, 48, 49n152, 120, 122, 209n59
Davis, John Jefferson, 3n11, 93–94
Davison, Andrew, 106–8
Dawkins, Richard, 133

Dei Verbum, 3, 240
development of doctrine, 3, 3n12, 4, 12, 267
   in accordance with scientific discoveries, 5–6, 120, 229, 267
Delio, Ilia, 2n9, 97, 102–6, 124, 194, 218n97, 232n134
Democritus, 35
demons (See angels, fallen)
deoxyribonucleic acid (DNA), 19, 23, 135, 141n63, 143
Descartes, René, 81n98, 276, 281
DeVito, Carl, 129
Dick, Steven, xviin2, 20, 32, 34, 40n119, 56n13, 59n25, 60n30, 64n44n, 66n57, 76n95, 119n223, 121n232, 126–27, 138, 141n62, 185, 201n28, 225n121, 226
Dick, Thomas, 71
dignity, 168, 176, 194n15, 207n58, 213, 214, 228, 243
Di Noia, Joseph, 94, 191
divine
   action, 2, 6, 10, 86. 105, 108, 114, 118, 120, 162, 181, 187, 194, 196–97, 208, 210, 213, 220, 229, 231, 231n132, 232, 240, 241, 242, 248, 249, 252, 256–59, 261–62, 263, 266, 271
   communication, 124, 180, 181n159, 240
   immanence, 197
   transcendence, xviii, 102, 158, 174, 197, 229, 262
Divine Prerogative, principle of, 108, 194, 240, 249, 266, 271
   definition of, 154
Dobzhansky, Theodosius, 128–29
Doppler effect, 278
Doppler shift method of detecting exoplanets, 15
Drake Equation, 30–31
Drake, Francis, xviin2, 30, 33, 135, 145
Duhem, Pierre, 58
Dwight, Timothy, 71
Dyson, Freeman, 127n5, 147n85, 149
Dyson's sphere, 147

Earth chauvinism, 113, 190n3,
eastern religions, 29, 53, 155, 185, 186, 225
economies of revelation, 252–58
economies of salvation,
   angelic, xvii, 177, 213n77, 218, 219, 221, 233, 248
   extraterrestrial, xvii, xviii, 10, 75, 94, 109, 113, 124, 125, 158, 192, 194, 218, 220, 232, 243, 250, 267
   human, xx, 93, 94, 113, 191, 235–42, 244, 245, 267, 268
ecumenism, xiv, 1n1, 5, 101, 122
Egyptian
   mythology, 201
   religion, 160n115, 161, 259
Einstein, Albert, xv, 278
   Einsteinian universe, xiv, xv, 79, 119, 193, 201, 278
Eiseley, Loren, 133
Eisenhower, President Dwight, 49
Elevated nature, state of, 165, 171–73
Empedocles, 36–37
Enceladus (Saturnian moon), 51
enlightenment, 5, 70, 79
Epicurus, 35
epicycles, 39, 276, 277, 280
epistemology, 249
eschatology, 84, 154, 205, 221, 255n234
ethics, xviii, 5, 38, 139, 150, 153, 154, 182, 186, 227
Europa (Jovian moon), 51
European Extremely Large Telescope (EELT), 27
European Mars Express mission, 50
European Space Agency, 31
Eusebius, 55
evangelization in space, 88, 95, 121, 131, 192
evolution, xxi, 20, 21, 22, 92, 126, 135, 137, 155, 177, 189, 197, 198, 227, 233
   and angels, 144n72, 178
   and convergence, 130, 131–34, 137
   and divergence, 129, 130, 138
   and extraterrestrial psychology, 138
   and religious consciousness, 156, 156n99, 157, 159, 162

artificially directed, 141, 149
cosmic, 16, 18, 21, 27, 88, 185
cultural, 127, 141, 153, 184, 225, 225n123, 228, 261
of intelligence, 5, 19, 21, 92, 126, 128, 139, 145, 147–48, 250
planetary, 5, 25
theistic, 5
Exclusive soteriological view, xx, xxi, 6, 81–82, 84–86, 92, 107, 121, 188–90, 200–201, 207–8, 270
ExoMars mission, 50
exoplanets, xvi, 23
    detection methods, 14–16
    detection of, xv, 6, 14–16, 26, 27, 31, 78, 80, 83, 121, 270, 272, 278
    direct imaging of atmospheres for extraterrestrial civilization biomarkers, 27
    earth analogs, 271
    number confirmed, 14
    number in galaxy, 14, 271
    number of candidates, 14
    tidally locked, 161, 162n124
exotheological metanarrative, 261–67
    diagram, 265
exotheology, xvii, xx, xxi, 3, 7, 10, 12, 121, 125, 154, 187, 193, 195, 208
    and the *Varied* hermeneutic, 200
    and the *Varied* view, 269
    definition of, 8, 196, 259, 263, 272
extraterrestrial artifacts, xviii, 28, 29, 221, 222n112
extraterrestrial intelligent life
    advanced, xix, xxii, 21, 28–9, 30, 31, 33, 77, 117, 136, 139, 140–41, 142, 145, 147, 149–55, 170, 175, 182, 184, 243n194, 250, 258, 262–63
    afterlife of, 157, 158, 169, 171, 173, 175, 176
    aggression of, 153, 184
    anthropology of, xix, 7, 20, 48, 125–44, 193
    civilizations of, xxi, xxii, 10, 27, 30–33, 81, 87, 88, 95, 108, 110, 113, 120, 122, 138, 140, 141, 144–53, 171, 178, 181, 190, 193, 220, 228, 267, 269, 271, 272
    consciousness, 140, 142, 142n67, 144, 149, 156, 156n101, 186, 225, 263
    divine communication with, 124, 180, 181n159, 240, 252, 258–59, 260n261, 261
    evil and, xvii, 75, 89, 102–3, 140, 154, 172n152, 173, 181–82, 192, 210, 218, 220, 228, 263
    biology of, 12, 20, 32, 109, 126–38, 141–44, 149, 151–52, 156, 193, 250
    deities of, 140, 152, 155, 157–58, 161, 163, 164n137, 165, 181, 182, 185, 197
    immortality of, 142, 145, 175
    lesser-advanced, xix, 140, 145, 150, 151, 153, 155, 159, 178, 181, 182, 223, 224, 225, 227, 228, 267
    malevolence of, 173, 182, 222
    mediators of, 179, 195, 218, 243
    morphology of, 2, 131, 128, 133, 134, 138, 141–44, 196, 243
    mystical religions and states of, 158, 165, 174–75, 180–81, 183, 244, 248, 249, 256
    non-interference of, 151, 153
    primitive religions of, 154, 158, 161–62, 167, 183, 224, 224n116
    predation of, 131, 133, 134, 173
    preternatural gifts of, 144, 165–67, 170–72, 174, 176, 217n90, 245, 251
    psychology of, 20, 126, 129, 131, 138–41, 151, 152, 153, 172, 196, 259
    religions of, 154–64, 183–87, 227, 231
    religious archetypes of, 164, 196, 198, 261–67
    self-annihilation of, 153
    sociological compositions of, 144–53, 184

extraterrestrial intelligent life (*continued*)
  technology of, 30, 31, 33, 127, 134, 136, 140, 141–53, 158n105, 159, 167, 173, 181, 182, 184, 222n108, 225, 228, 250
  theological anthropology of, xvii, 6, 11, 12, 125, 164–80, 217, 218, 243, 248–49

fallen nature, state of, 165, 173–75, 179, 192, 220, 221, 226
Feinberg, Gerald, 131
Ferguson, James, 45
Fermi, Enrico, 33
Flammarion Engraving (Illus.), 46
Flammarion, Camille, 45–48, 81n103
Fontenelle, Bernard le Bovier, 41–43, 70, 82
Ford, Lewis, 3n11, 117–18, 125
fossils, 21, 22n60
freedom, 90, 105, 140, 170, 171, 174, 181, 216, 228, 243
  of God, 7, 8, 58, 77, 107, 114, 116, 118, 154, 165, 194, 196, 210, 215, 220n105, 240, 251, 262, 269, 270
Freitas, Robert A. Jr., 132, 135n40
Frontispiece of Fontenelle's Entretiens (Illus.), 42
Funes, Fr. José Gabriel, xvi, 2n9, 117–18, 124

galactic communities, 145, 147, 148, 150, 151, 152, 159, 269
Galántai, Zoltán, 129
Galilei, Galileo, xv, 2n8, 34, 39, 40, 198–99, 268n264, 281
  heliocentrism controversy, 65, 67–69, 277
Galileo mission, 51
genetic engineering, 127, 136, 137, 138, 140n60, 142n63, 146, 151, 250
geocentrism, xiii, xiv, xv, xviii, 6, 14, 51, 64, 67, 73, 75, 78, 89, 92, 102, 109, 112–13, 124, 189, 190–91, 207–8, 222, 262, 268, 275, 277, 279, 281
  aristotelian, 39, 78n100
  ptolemaic, 68
George, Marie, 2n9, 81, 85–86, 121, 189, 209n59
giantism, 135
Giant Magellan Telescope, 27
Glashow, Sheldon, 135
Glossia Ordinaria, 260n260
Gordley, Matthew, 203
Greeks, ancient, xv, 34–35, 39, 55, 77, 78n100, 81, 160n115, 163, 202, 242, 275, 280
Great Chain of Being, 62, 63 (illus.), 69, 77
Gregersen, Neils Henrik, 109
Gregory of Nazianzus, 110, 211, 212

habitable zones, xvi, 14, 16, 18, 24–25, 31, 43, 76, 83, 130, 162n124
Hale telescope, 83
Hart, H.L.A., 136–37
Hartmann, Lars, 21n54, 22n55, 203, 204n44
Hayes, Zachary, 96, 103n162, 209–10, 218n95
heaven
  Hebrew conceptions of, 53, 53n5, 202, 202n31, 202n33
  Islamic conceptions of, 161n31
  location of, 208
Hebblethwaite, Brian, 85, 91, 106, 121, 189, 190, 209n59
heliocentrism, xiv, 39–40, 64–65, 67–69, 77, 79, 81, 116, 277
  acceptance of, xiv, 39, 40, 69, 77, 79, 81
  condemnation of, 65, 67–68
helium, 17, 18, 147
Herschel, William, 14n7, 44–45, 49, 70, 71, 278
Hess, Peter, 110
Hippolytus, 55
Horowitz, Norman, 129n13
Hubble, Edwin, 49
Hubble telescope, 13, 24, 27, 83, 121, 270
  Deep Field Image (HDF), 25
  Deep Field South Image, 25

Extreme Deep Field Image, (XDF), 25
Ultra Deep Field Image, (HUDF), 14, 26 (image)
humanity,
  argument for centrality of, 67, 82, 85, 92, 93, 95, 105, 109, 122, 189, 192
  argument against centrality of, 89, 116, 178, 193–94, 193, 198, 210, 213, 216, 228, 270
  fallen condition of, 103, 173, 178, 206
  place and role in cosmic context, xiii, xv, xxii, 6, 16, 51, 69, 73, 75, 77, 88, 101, 207, 208, 263, 264, 269, 271–72
  uniqueness of, xiv, 2, 8, 12, 80, 107, 269
Humani Generis, 241
human nature, 73, 75, 85n110, 97n141, 98, 99, 103, 113, 114, 167–69, 190, 194, 194n15, 210, 213–16, 249, 250, 25n233, 257, 259
Huygens, Christiaan, 42–43, 45, 46, 128
hybridization of extraterrestrials, 142, 250–51
hydrogen, 17–19, 31, 146–47

Imago Mundi (Illus.), 52–53
incarnation, 79
  arguments against multiple, 71, 73, 85, 91, 95, 114–25, 120, 194–95, 208–9, 213, 217, 221
  arguments unique to Earth, xviii, 64, 67, 73, 74, 81, 82, 85–86, 86–97, 188–89, 214
  as archetype of divine activity, 232, 243, 260, 268, 271
  as redemption of all biological life, 110, 192
  as self-communication and completion, 102–5, 107–8, 111, 218, 267
  cosmic significance of, 85, 86–97, 111
  earthly insufficient for salvation of extraterrestrials, 112, 189, 230
  fittingness of, 249
  likelihood Earth not single site for, 113
  multiple view of, xviii, 10, 75, 82, 85, 89, 90n119, 93, 97–114, 116, 123–4, 208–21, 223
  necessity of, 104–5, 194
  non-necessity of, 114–25, 194–95, 208–9, 213, 217, 221, 249
  revelational view of, 110–12
Inclusive soteriological view, 6, 10, 85, 86–97, 99, 102, 104–5, 106, 107, 109, 111, 112, 121–22, 193, 200, 201, 207, 270
  and cosmology, 207–8
  definition of, xxi, 81, 188, 190
Incommensurability problem, 138–39
Infrared Space Observatory, 23
Ingenuity Mars helicopter, 51
inquisition, Spanish, 66
Insight mission, 50–51
Integral nature, state of, 165, 170–71
Intelligence Principle, 127
interdisciplinary study of extraterrestrials, xix, 3, 9, 12, 196
International Academy of Astronautics, 32
International Astronomical Union, 20
Irenaeus, St., 110n184, 210–11, 212n72, 239, 239n186
Islam, 120, 161, 183n163, 186, 224
Israel, 160n114, 163, 191n9, 202, 232–42, 255n234, 267

James Webb Space Telescope, 13, 26
Jastrow, Robert, 145
Jerusalem, 55, 71, 74, 84, 163, 191n9, 237, 252
Jewish
  apocalyptic tradition, 202–3
  cosmology, 202, 207, 276, 279, 281
  theology, 201, 232
Johuan, R.M., 82
Judaism, 111, 182, 186, 197n17, 201, 205, 243, 251n213, 254n222, 257n251, 266, 269

INDEX

Kaku, Michio, 148, 148n90
Kardashev, Nikolai, 146
  scales, 146–48
Kecskes, Csaba, 148, 148n91
Keill, John, 82
Kepler, Johannes, 40, 69, 121, 270, 281
Kepler telescope, xv, xvi, 13–15, 24–26, 31, 31n93, 34, 80, 83
Kereszty, Roch, 90, 90n120, 209n59
Kracher, Alfred, 140, 140n61

Lamm, Norman, 158
Lee, Marco, 131
Lemaître, George, 49
Lemarchand, Guillermo, 138–39, 139n57, 146, 147n81
Leo the Great, 211, 212n69
Lérins, Vincent de, *Commonitorium*, 3–4
Leucippius, 35
Lewis, C.S., 97n141, 115, 115n206, 207, 124, 209
Logos, xxi, 5n20, 63, 82, 88, 91, 92, 95, 97–8, 104–5, 109, 111, 112, 123–24, 188, 193, 198, 207, 214, 231, 255
Lombard, Peter, 59
Lomberg, Jon, 138, 139n57
Lonergan, 10n35
Lossky, Vladimir, 108
Lovejoy, Arthur, 36, 62, 64, 77n96
Lucretius, 35–36
Luther, Martin, 64, 65n47

MacGowan, Roger, 132
magisterium, xiii, 1, 2n6
Magnus, Albertus, 57
Mariner missions, 47n148, 49
Mars
  artifacts on, 28
  atmosphere of, 50
  images of, 50
  life on, 47, 51
  water on, 50
Mars Atmosphere and Volatile Evolution (MAVEN) orbiter, 50
Mars Exploration Rover mission, 50
Mars missions, 28, 38, 45, 47–51, 121

Mars Odyssey mission, 50
Mars Reconnaissance Orbiter, 50
Mars Science Laboratory mission, 50
Malthusian Catastrophe, 149
martian canals, 47–48
Mascall, E.L., 84, 88, 98–99, 106, 109, 120, 123, 192n12, 209n59, 232n134
Maximus, St., 214
Mayron, Francis, 58
McHugh, 84, 100n150
McMullin, Ernan, 40n120, 119, 120
Melanchthon, Philip, 64
Mercati, Cardinal Giovanni, 67
Mesoamericans (See Aztecs)
messianic age, 198, 25, 257, 260
Milky Way Galaxy, xv, 14, 17, 24, 30, 120, 148, 159, 161n124, 271, 278, 281
microbial extraterrestrial life, 27, 51, 222n111
microlensing, 14–15, 31
microorganisms, 19n36, 22
Milky Way galaxy, xv, 14, 17, 24, 148, 159, 161n1241, 271, 278, 281
military action, 143, 184, 224
Millerite movement, 73
Milne, Edward, 87–88, 92, 95, 99, 109, 121–22, 192n10
mind
  angelic, 177
  human, 30n27, 164, 271
  extraterrestrial, 101, 138–39, 141–42, 145, 155, 170, 173, 175, 225, 243
mind reading, 175
miracles, 175, 177, 244, 248, 254n222, 255n234, 256, 257n252
Mok, Alex, 94–95, 109, 116, 122, 192
Montignuez, de Monseigneur, 74, 81
Moltmann, Jürgen, 3n11, 87, 92–95, 192
Moon, 27, 28, 32, 37, 40, 45, 49, 53, 57, 63, 69–70, 77, 121, 149n95, 159, 160n110, 161, 190, 201, 274, 276, 280
Moritz, Joshua, 109–10
mormonism, 74, 80

Morris, Simon Conway, 80, 128n10, 130n18, 131–33, 209n59
Moses, 160n114, 198, 235, 236, 244, 245, 253n220, 257n249
Moule, C.F.D., 203
Multiple soteriological view, 6, 10, 82n106, 85, 88, 95–96, 97–116, 121–24, 200–201, 209, 214, 216, 270
   arguments against, 72, 91, 114, 194, 232, 250n212
   description of, xxi, 81–82, 188, 193
Murphy, George, 9
mythology,
   egyptian, 201
   hellenic, 201, 207

nanotechnology, 127, 138, 141
National Aeronautics and Space Administration (NASA), 23, 27–28, 31, 32, 34, 49, 121, 271
natural religion, 9n33, 71, 155, 185, 189, 201, 225, 227, 247n203, 251, 266
natural selection, 48, 128–32, 156
natural theology, xxi, 6, 8, 10–11, 12, 39, 47, 64, 67, 72, 77, 80–81, 196, 272
Newman, John Henry, 4–6
Newton, Isaac, xv, 43–44, 61, 278
newtonian universe, 39, 79
Nicene Creed, 10
nitrogen, 18, 19, 27
Noah, 109n13, 233, 236, 244

Ockham, William of, 59–60, 79, 82
O'Collins, Gerald, 94
O'Connor, Jerome Murphy, 203
O'Meara, Thomas, 118–19, 124, 203
omni-properties of God, 7, 12, 85, 91, 125, 154, 190, 194, 201, 229, 243, 270
O'Neill colonies, 149
Opportunity Mars rover, 50, 121
Ordway, Frederick, 132n26
Oresme, Nicole, 59, 60
Origen, xx, 79, 211, 245, 252
   and plurality of worlds, 55
original justice, 171–72, 233

original sin, xvii, xviii, xx, xxi, 57, 60, 69, 73, 74, 86, 90, 94, 122, 178, 179, 192, 194n15, 208, 246, 259
   as universal, 86–87, 96, 110, 210
   incarnation necessary to overcome, 102–3
   non-necessity of, 70, 84, 94, 96, 100, 116, 117, 118, 124, 154, 172n152, 174, 189, 193, 216, 218
oxygen, 16, 18–19, 21, 51, 130

paganism, 162, 225, 233, 234, 259, 260n257
Paine, Thomas, xx, 9, 71–72, 79, 89, 120–22, 124, 185, 189, 201n28
Pannenberg, Wolfhart, 3, 92, 93, 95, 122
paradigm shift, xxi, 6, 198, 228, 245
particularity, scandal of, 112
Pathfinder missions, 50
Peacocke, Arthur, 117–18
Penrose Process, 148
Perego, Angelo, 115–16, 124, 224
Perseverance Mars rover, 51
Peter Apian's Cosmographia. Renaissance woodcut illustrating the Ptolemaic system, 1524 (Illus.), 38
Peters *ETI Crisis Survey*, xix, 29n86, 80, 120n225, 222
Peters, Ted, 3n11, 106, 108–9, 110–13, 117, 120, 189n1, 222
Phoenix mission, 50
photosynthesis, 135
Pieris, Aloysius, 224
Pioneer space missions, 20, 49, 51
Pittenger, Norman, 91, 97, 99, 122, 232n134
planet
   formation, 16–19
   number in galaxy, 24, 31
   number of Earth analogs in universe, 271
   number of extrasolar, xv, 14
planet-hopping savior, notion of, 87, 106
Plato, 39, 57, 62, 66, 77, 202n30, 202n31, 274

plurality of worlds concept, xv, xix, xx, 2, 34–42, 45–48, 55–56, 60–62, 64, 67, 69, 72, 76–81, 270, 275, 277, 279
Pohle, Joseph, 2n9, 74–75, 82
Pollard, W.G., 130, 205n47
polycyclic aromatic hydrocarbons (PAHs), 20n47, 23
polytheism, 155, 158, 163, 183, 183n163, 184, 233, 233n143, 234, 234n148, 236, 241
Pontifical Academy of Sciences, 2
Pope Gregory XIII, 65
Pope John XXI, 58
Pope John Paul II, xvi, 199n22, 200
Pope Leo XIII, 3, 268n234
Pope Paul III, 64
Pope Paul VI, 116n209
Pope Pius XI, 2n8
Pope Pius XII, 3, 242n191
Pope Urban VIII, 68–69
Pope Zachary, 56–57
post-biological intelligence, 139, 143
Power, Baden 82
predestination, 103
Primack, Joel, 136
prime directive, 151
principle of plenitude, 55, 62, 64, 70, 75, 77–78
    definition of, 36
protogalaxies, 16–17
psychological impact of extraterrestrial contact, 29
psychology of extraterrestrials, 126, 129, 138–41, 151, 153, 172, 196, 259
ptolemaic system, xv, 14, 38–39, 65, 68, 78, 192, 276, 280
Puccetti, Roland, 114n204, 120, 121, 134n39, 136, 137n51, 224n118
Pure nature, state of, 165–69
Pythagorean universe, 37, 274, 202, 280

radio frequency, contact via, xviii, 20, 28–32, 79, 83, 88, 95, 121–22, 192, 222
Rahner, Karl, 2n9, 5n20, 85n110, 104–6, 108, 123, 124, 167, 169, 193n14, 194, 209n59, 232n133

Raible, Daniel, 84, 99–100, 120, 176n154
Rare Earth argument, 48, 80, 107
Redeemed nature, state of 166, 174–75
redemption
    available for extraterrestrials outside domain of Earth, 98, 100, 101n154, 102, 112, 115, 118, 179, 188, 193, 249, 251
    Christ's central to all beings, 111, 188, 190, 192, 201, 208
    limited to humanity, 82, 84, 85, 88, 96, 117, 123, 189, 190, 193
    of all flesh included in the human, 109–10
    universality of, xviii, xxi, 55–56, 87, 89–91, 93, 95, 102, 188, 224
Reformation, Protestant, 64, 67–68, 79, 277
    counter, 65
renaissance, xvii, 83
reproduction
    sexual, 138, 145, 149, 250
    non-sexual, 138, 186n171
resurrection, 74, 90, 82, 94, 110, 174, 191n9, 224, 261
retrograde motion of planets, 39, 280n8
revealed religion, 9, 10, 11n38, 71, 98, 99, 117, 155, 163, 179, 180, 190, 198, 225–27, 230, 231, 263, 241, 243, 244, 266
Reynaud, Jean, 45–46
ribonucleic acid (RNA), 23
robots, 143
Roman Catholic Church, xiii, xix, 1, 3–12, 34, 199
    having no official teaching on extraterrestrial life, xiii
    intellectual tradition of relation with science in discerning truth, 2, 7–8, 12
    on condemnation of Giordano Bruno, 66–67
    on condemnation of heliocentrism, 65, 68–69, 78
    on interpretation of indigenous Mesoamerican religions, 225–26
    on plurality of worlds concept, 34

on scriptural interpretation, 199–200
notion of intelligent extraterrestrials not incompatible with teaching, xvi, 2, 83, 208
teaching and theology important in considering the extraterrestrial intelligence question, 1–12
Russell, Robert John, 110–13

sacred space, 162, 167
Sagan, Carl, xvin2, 145
Salaverri, Joaquin, 115
salvation
　angelic, 177
　as divine initiative, 108
　cosmic, 6, 10, 90, 91, 92, 96, 98, 105n170, 109n183, 118
　extraterrestrial, 75, 85, 86, 88, 94, 100, 103, 111, 113, 115, 117, 123, 124, 179, 180, 189, 190–95, 209, 217–21, 269
　human, 84, 97, 106, 111, 120, 183, 185, 188, 196, 200–208, 210, 217, 229, 231–51, 267–68
　of the antipodes, 74
salvation histories, 72, 88, 92, 102, 104, 113, 124, 177, 179–80, 217n93, 231, 241, 243, 248, 262–64, 268
scandal of particularity, 112
Schiaparelli, Giovanni, 47
science and religion, xiv, 2, 7–8, 12, 70, 102, 120, 196
science fiction, 128, 140, 146, 153
Search for Extraterrestrial Intelligence (SETI), 20, 28, 31–32, 80, 83, 129n12, 144, 145, 152, 185, 224, 228
Secchi, Angelo, 47, 74
Seventh-day Adventists, 80
Shapiro, Robert, 131, 132n27
Simulated large scale structure of the universe (Illus.), 273
sin, origins of, 192–93, 210
slavery, 170, 173, 234
Smart, Ninian, 186
Smith, Joseph, 74
sociobiology, 131

sociology of extraterrestrials, xix, 20, 126, 129, 144–54, 239
Sojourner Mars rover, 50
Sola scriptura, 3
solar system, xv, xvi, 16–18, 28, 31, 32, 40, 43, 44, 45n137, 48–49, 69, 75, 76, 120, 148, 159, 222n112, 226n126, 280, 281
soteriology, xvii, xxi, 1, 3, 7, 10, 12, 108, 111, 112, 114, 164–78, 187, 189, 191, 193, 195, 205, 224, 250
soul
　human, 5, 168
　powers of, 167, 169, 170, 171
space
　evangelization, 95, 121–22, 192
　exploration, xiv, xix, 28, 48, 79, 102, 120, 127, 145, 272
　missions, 13, 39–51
spectroscopy, 19, 26, 48, 278
Spirit Mars rover, 50, 121
spirituality of extraterrestrials, 156–57, 170, 181, 184, 196, 227, 228
Spitzer Space Telescope, 23
Sputnik launch, 121
star formation, xvi, 13, 17, 26, 31,
Star Trek series, 147, 151n96
Star Wars films, 148, 173
Steidl, Paul, 84–85, 189
stoicism, 204
Strong AI Argument, 127–28
Sullivan, Walter, 134n37
Summa Theologica (see St. Thomas Aquinas)
Sun, 15, 15, 38, 41, 483, 45, 53, 56, 57, 62, 63, 64, 65, 66, 68, 147, 149n95, 150, 159, 160–61, 162n124, 190, 198n19, 201, 263, 274, 280
super-Earths, 24, 134
superintelligence, 143
supernatural
　beings, 178, 179, 183n163, 217n90, 244
　destiny, 165, 168, 170, 174, 175, 220, 249
　economies of salvation, 243, 250
　elevation, 165, 168, 192, 217

supernatural *(continued)*
   gifts, 166–67, 172, 174, 176, 178, 219n101, 245
   grace, 11, 175, 197, 237, 247
   manifestations, 177, 229, 248, 252
   messages, 180, 181, 249
   order, 100, 170
   phenomenon, 10, 168, 183n163, 244, 257, 258, 266, 271
   religion, 6, 11, 197–98, 217n93, 227–28, 251, 266–68
   revelation, 10, 229, 232n133, 242, 252
   state, 175–77, 238
Supernature, state of, 159, 166, 175–77
supernova, 17, 18, 22, 127, 150, 161
survival
   human, 132,
   extraterrestrial, 139, 152–53, 169, 239
Swedenborg, Emanuel, 70–71, 73
Swedenborgians, 80
symbols, religious, 104, 116, 181, 181n160, 198
syncatabasis, 258–61, 266
synthetic life forms, 137, 143

Tarter, Jill, 185, 224
technology
   advances in, xiv,
   and cybernetics, 142–43
   and limitations detecting other civilizations, 76
   and observational methods, 24, 27, 78, 270
   extraterrestrial disinterest in human, 228
technology *(continued)*
   large-scale extraterrestrial, 144–53, 181, 184, 22n108
   radio, 79, 88
   worship of, 158n105
Teilhard de Chardin, Pierre, 2n9, 87–90, 96, 103, 210
teleology, 62, 77
Tempier, Etienne, 58
terraforming, 147, 150
Terrasson, Abbe Jean, 82

Tertullian, 211, 260n257
theocentrism, 112
Theodoret, 55
theophany, 114, 124, 166, 168, 180–83, 195, 226, 239, 242, 252, 259, 261, 269
theosis, 164–83
theotokos, 134
thermophiles, 22
Tillich, Paul, 106, 112, 116–18, 124
Titan (Saturnian moon), 51
Tough, Allen, 32, 149, 222
Transiting Exoplanet Survey Satellite (TESS), 26
Traphagan, John W., 144

UFO phenomenon, 93
universal divine plan, 230
universal evolutionary processes, 178
universal physical laws, 135, 158
universal religion, 224, 228, 247
universe,
   age of the, xv
   history of the early, 13–19
   size of the, 14

Vakoch, Douglas, 32, 222
van Huyssteen, Wentzel, 225
van Kooten, George, 202n35, 203n38, 204n46, 206, 206n57
Varied soteriological view, xxii, 7, 10, 12, 83, 124–25, 187, 209, 221
   and exotheological metanarrative, 265–67
   and syncatabasis, 259
   as development of doctrine, 6
   definition of, xxi, 81–82, 114–19, 124, 154, 188–89, 194–96, 229, 243, 249, 250n212, 259, 263, 267–70
   scriptural hermeneutic, 3, 200–208
vastness of universe, xiii, xv, xviii, 3, 6, 8, 32, 72, 78, 83, 85, 85n110, 86, 8, 93, 112, 116, 121n232, 123, 190, 193, 207, 208, 230, 231, 252, 262, 263, 268, 269, 270, 271
Vatican II, 3, 240
Vatican observatories, 2

Venus, 28, 38
Viking space missions, 49–50
violence, 131, 151
Virgil of Salzburg, 56–57
Vorilong, William, 58, 60, 64, 70, 81, 106
Voyager space probes, 20, 49, 51

Wallace, Alfred, 48
war, 150, 152, 153, 220
water
    in mythic cosmology, 53, 281
    in interstellar space, 19
    necessary for lifeforms, 18, 21, 43, 48, 127, 130, 160
    on Europa, 51
    on Mars, 50,
    worship of, 162
Weinberg, Steven, 135

Weston, Frank, 97–98
Whewell, William, 72–74, 82, 84
White, Ellen G., 73–74
Wide Field Infrared Survey Telescope (WFIRST), 27
Wilkins, John, 69, 70
Wilkinson, David, 3, 15n11–16, 16n17, 17n23, 19, 101n153, 111, 122
Wilkinson Microwave Anisotrophy Probe (WMAP), 126–27
Wilson telescope, 82–83
Worthing, Mark, 111

Xenobiology, 2, 126–38

Yerkes telescope, 82

Zubek, Theodore, 2n9, 97, 100–101, 120, 124, 176n154, 232n134

www.ingramcontent.com/pod-product-compliance
Lightning Source LLC
Chambersburg PA
CBHW052144300426
44115CB00011B/1507